GAME THEORY, DIPLOMATIC HISTORY AND SECURITY STUDIES

T0177573

Game Theory, Diplomatic History and Security Studies

FRANK C. ZAGARE

OXFORD
UNIVERSITY PRESS

OXFORD
UNIVERSITY PRESS

Great Clarendon Street, Oxford, OX2 6DP,
United Kingdom

Oxford University Press is a department of the University of Oxford.
It furthers the University's objective of excellence in research, scholarship,
and education by publishing worldwide. Oxford is a registered trade mark of
Oxford University Press in the UK and in certain other countries

First Edition published in 2019

Impression: 1

Published in the United States of America by Oxford University Press
198 Madison Avenue, New York, NY 10016, United States of America

British Library Cataloguing in Publication Data

Data available

Library of Congress Control Number: 2018956452

ISBN 978–0–19–883158–7 (hbk.)
ISBN 978–0–19–883159–4 (pbk.)

DOI: 10.1093/oso/9780198831587.001.0001

Printed and bound by
CPI Group (UK) Ltd, Croydon, CR0 4YY

To Tish
One more, for the road

ACKNOWLEDGMENTS

I have led a charmed academic life. Along the way, it has been my good fortune to interact with a large number of very talented and very generous individuals. Among them have been my teachers. First and foremost was my high school track and cross-country coach and mathematics instructor, Murtha P. Lawrence. Murt, as he was affectionately called behind his back, taught me to keep going, whatever the obstacles. For Mr. Lawrence, as we called him to his face, there was literally no reason, physical or otherwise, for missing a practice or a meet, or for underperforming on a geometry examination or in a race. At Fordham University, I was introduced to the subtleties of political decision making by Professor Stephen M. David. From Professor David, I learned that there was indeed a Republican and a Democratic way to pick up the garbage. Who knew? The late Oskar Morgenstern also played an important part in my career. Even as he himself was in his final days, my "benefactor," as he sometimes referred to himself, put an otherwise unemployed Ph.D. to work at his research firm, Mathematica, simply to keep me in the "game." Last, but most certainly not least, is Steven J. Brams. Space considerations preclude a listing of the innumerable ways that Steve has helped me, ways that Steve neither acknowledges nor even seems to understand. If I had not had lunch with Steve in June 2017, this book would not have been written.

As well, I have had the opportunity to interact with a number of truly professional colleagues. I learned a lot about game theory from D. Marc Kilgour. I owe Marc a great deal. My good friend, Jacek Kugler, has also been a constant source of new ideas, criticisms, some of which were even warranted, and encouragement. Jacek picked my file out of a large pile of applicants and hired me at Boston University in 1978 when no one else would. Since then, he has made it impossible for me to ease up. And, for that, I am most thankful. Steven Quackenbush, one of my former graduate students, has also made my academic life much richer than it would otherwise have been. Steven, who read and commented on the entire manuscript, has helped me understand the wider implications of my own work. His insights and friendship have been invaluable. Lisa J. Carlson also read the entire manuscript. Her thoughtful suggestions were particularly insightful. I also owe a huge intellectual debt to Glenn H. Snyder and my friend and one-time office neighbor, Paul Diesing. Their classic work, *Conflict Among Nations*, which was published by Princeton University Press in 1977, has been and continues to be the source of many of my ideas about interstate crisis behavior. When Paul retired from the Political Science Department of the University at Buffalo, he bequeathed to me a large trove of material, including a number of unpublished case studies, that he accumulated in the course of writing that book. It is my hope to organize this material and make it available to the peace science community in the near future.

In addition to my teachers, colleagues, and former students, a number of other individuals deserve special thanks or, as they might see it, blame. Unfortunately, several of them are no longer with us: Ken Organski, Paul Senese, Don Rosenthal, and Franco Mattei. The rest include, in no particular order, Bruce Bueno de Mesquita, Randy Siverson, John Vasquez, Vesna Danilovic, Bill Geasor, Stan Lechner, and Ray Dacey. No one can ask for a better group of friends. At Oxford University Press, Dan Taber's enthusiasm and editorial suggestions have been both helpful and much appreciated. I have also learned a great deal about strategic behavior from my three beautiful daughters, Catherine, Ann, and Elizabeth. I am so very proud of them. Finally, I want to thank my best friend, Patricia Zagare, to whom this book is dedicated. This one's for you, too.

CONTENTS

LIST OF FIGURES

LIST OF TABLES

Introduction

History, it is oftentimes said, is just one damned thing after another. Generally speaking, highly skilled diplomatic historians and security studies specialists have performed the task of describing each of these "things" with great acumen. Trachtenberg (1990/1991: 136), for example, convincingly and insightfully shows that the sudden change in German foreign policy on the eve of World War I was precipitated by Russia's partial mobilization and not, as is oftentimes argued, by a warning in Berlin by the German ambassador in London that Great Britain was unlikely to stand aside in any war that involved France.[1]

Trachtenberg's temporal analysis is a more than admirable *description* of the chain of events that led up to the Great War. But diplomatic historians, as a group, aspire to much more than mere description and would be taken aback if their enterprise were relegated to data gathering and preference amalgamation. Indeed, the best diplomatic histories aim to both describe *and* explain why one thing has, in fact, led to another. Christopher Clark's (2012) magisterial history of the outbreak of war in 1914 is a case in point. But though it comes close, even Clark's treatise falls short of the explanatory mark. Clark, like most diplomatic historians, is less than transparent in stating his assumptions. In consequence, the process by which he moves from assumptions to conclusions is unclear at best, and illogical at worst. All of which is to say that the causal mechanism that drives his analysis is difficult, if not impossible, to discern. The problem is even more acute when lesser historians attempt an explanation of a complex series of events.

To overcome the limitations of unmoored historical explanations, the analytic narrative project was developed at Harvard in the late 1990s (Bates et al., 1998, 2000a, b). Like most diplomatic histories, the analytic narratives approach rests on the assumption that states and their leaders act purposefully, that is, that they are instrumentally rational.[2] But, unlike most diplomatic histories, the explanations that are derived from an analytic narrative are driven by an explicit causal mechanism, that is, by a game-theoretic model, that provides a theoretical foundation that is used "to develop systematic explanations based on case studies" (Bates et al., 2000b: 696).[3]

---

[1] See, for example, Albertini (1952: vol. 2, 520–2) or Massie (1991: 871).
[2] Instrumental rationality should be distinguished from procedural rationality. For a discussion of the distinction, see Zagare (1990a). See also Gilboa (2010).
[3] See also Mongin (2018).

Game Theory, Diplomatic History and Security Studies. Frank C. Zagare. Oxford University Press (2019).
© Frank C. Zagare 2019. DOI: 10.1093/oso/9780198831587.001.0001

The advantages of using an explicit theoretical model to develop explanations of actual events or historical processes are many. But two stand out: theoretically grounded explanations are at once more transparent and less ad hoc than atheoretical or poorly specified explanatory frameworks. They are more transparent because formalization requires an explicit statement of assumptions and arguments. And they are less ad hoc because well-articulated theoretical frameworks severely constrain both the number and the cast of variables that can be called upon to provide a coherent explanation.

The main purpose of this book is to demonstrate, by way of example, the several advantages of using a formal game-theoretic framework to explain complex events and relationships. The two chapters in Part I set the stage. In Chapter 1, I lay out the broad parameters and major concepts of the mathematical theory of games and its applications in the security studies literature. The ability of game theory's formal structure to highlight the implications of initial assumptions, facilitate the identification of inconsistent conclusions, and increase the probability of logical argumentation is also established.

Chapter 2 explores a number of issues connected with the use of game-theoretic models to organize analytic narratives, both generally and specifically. First, a causal explanation of the Rhineland crisis of 1936 is developed within the confines of a game-theoretic model of asymmetric or unilateral deterrence. Then, some methodological obstacles that may arise in more complex cases are discussed, and suggestions for overcoming them offered, again in the context of a real world example: the decision of Germany's chancellor, Otto von Bismarck, to enter a defensive alliance with Austria in 1879, an "unlikely" alliance that seemed to offer Germany few tangible benefits.

The focus of Part II is on more detailed analytic narratives. Chapter 3 interprets the Moroccan crisis of 1905–6 in the context of an incomplete information game model, the Tripartite Crisis Game, and one of its proper subgames, the Defender–Protégé Subgame. British support of France during the conference that ended the crisis, the firm stand that France took at the conference, and the German decision to press for a conference are all explained in terms of the model's principal variables.

In Chapter 4, I survey and evaluate several prominent attempts to use game theory to explain the strategic dynamic of the Cuban missile crisis of 1962, including, but not limited to, explanations developed in the style of Thomas Schelling, Nigel Howard, and Steven J. Brams. All of the explanations are judged either incomplete or deficient in some way. Accordingly, in Chapter 5, I offer a general explanation that answers all of the foundational questions associated with the crisis within the confines of a single, integrated, game-theoretic model with incomplete information, the Asymmetric Escalation Game. This explanation addresses the shortcomings of both standard, idiosyncratic studies and those of the prominent game-theoretic examinations discussed in detail in Chapter 4.

Chapter 6 uses the same game form to develop a logically consistent and empirically plausible explanation of the outbreak of war in Europe in early August 1914. The utility of the Asymmetric Escalation Game model for answering a number of related questions about the origins of the Great War is established. I argue that while the war was most certainly unintended, it was in no sense accidental or inevitable. I also contend that most attributions of responsibility for the war are misguided.

Part III contains two chapters. Turning away from specific cases, I focus on general theories and some of the implications of using game theory to generate them. In Chapter 7, I introduce perfect deterrence theory and contrast it with the prevailing realist theory of interstate war prevention. The assumptions, empirical implications, and policy prescriptions of the two approaches to deterrence are compared and contrasted. I argue that the standard theory, which I call "classical deterrence theory," suffers from both logical and empirical problems. Perfect deterrence theory, by contrast, is not only logically consistent but also has impressive empirical support, support that includes each of the analytic narratives developed in Part II.

In Chapter 8, I address the charge made by some behavioral economists (and many strategic analysts) that game theory is of limited utility for understanding interstate conflict behavior. Using one of perfect deterrence theory's constituent models, I demonstrate that there is a logically consistent game-theoretic explanation for the absence of a superpower conflict during the Cold War era. I also argue that a predictively inaccurate or logically inconsistent game model in no way undermines the utility of game theory as a potentially powerful methodological tool. Along the way I examine (1) a prescription based on an incorrect prediction attributed to John von Neumann, one of the cofounders of game theory and (2) a logically inconsistent explanation of the long peace offered by Thomas Schelling, the game theorist many consider the most important strategic thinker in the field of security studies. I conclude with a few final thoughts.

PART I
Overview

1

. . . . .

Game Theory and Security Studies

1.1 Introduction

Game theory is the science of interactive decision making. It was created in one fell swoop with the publication of John von Neumann and Oskar Morgenstern's *Games and Economic Behavior* (1944) by Princeton University Press.[1] Widely hailed when it was published, the book became an instant classic. Its impact was enormous. Almost immediately, game theory began to penetrate economics—as one might well expect. But soon afterward, applications, extensions, and modifications of the framework presented by von Neumann and Morgenstern began to appear in other fields, including sociology, psychology, anthropology, and, through political science, international relations and security studies.[2]

In retrospect, the ready home that game theory found in the field of security studies is not very surprising. Much of the gestalt of game theory can easily be discerned in the corpus of diplomatic history and in the work of the most prominent theorists of international politics.[3] And its key concepts have obvious real world analogs in the international arena.

1.2 Primitive Concepts

The basic concept is that of a game itself. A *game* can be thought of as any situation in which the outcome depends on the choices of two or more decision makers. The term is somewhat unfortunate. Games are sometimes thought of as lighthearted diversions. But, in game theory, the term is not so restricted. For instance, most if not all interstate conflicts qualify as very serious games, including, but not limited to, trade negotiations, acute crises, and all-out wars.

---

[1] This chapter is based on Zagare (2008).
[2] For an autobiographical account of the critical role he played in introducing game theory to political scientists and international relations specialists, see Riker (1992).
[3] For the connections between realism and game theory, see Jervis (1988a).

Game Theory, Diplomatic History and Security Studies. Frank C. Zagare. Oxford University Press (2019).
© Frank C. Zagare 2019. DOI: 10.1093/oso/9780198831587.001.0001

In game theory, decision makers are called *players*. Players can be individuals or groups of individuals who, in some sense, operate as a coherent unit. Presidents, prime ministers, kings and queens, dictators, foreign secretaries, and so on can therefore sometimes be considered as players in a game. But so can the states in whose name they make foreign policy decisions. It is even possible to consider a coalition of two or more states as a player. For example, in their analysis of the July Crisis of 1914, Snyder and Diesing (1977) use elementary game theory to examine the interaction between "Russia–France" and "Austria–Germany."

The decisions that players make eventually lead to an *outcome*. In game theory, an outcome can be just about anything. Thus, the empirical content associated with an outcome will vary with the game being analyzed. Sometimes, generic terms such as "compromise" or "conflict" are used to portray outcomes. At other times, the descriptors are much more specific. Snyder and Diesing use the label "Control of Serbia" by Austria–Germany to partially describe one potential outcome of the July Crisis.

Reflecting perhaps the intensity of the Cold War period in the United States in the early 1950s, almost all of the early applications of game theory in the field of security studies analyzed interstate conflicts as *zero-sum games*. A zero-sum game is any game in which the interests of the players are diametrically opposed. In a zero-sum game, what one player wins, the other loses. Examples of this genre include an analysis of two World War II battles by O. G. Haywood (1954) and a study of military strategy by McDonald and Tukey (1949).

By contrast, a *nonzero-sum game* is an interactive situation in which the players have mixed motives, that is, in addition to conflicting interests, they may also have some interests in common. Two states locked in an economic conflict, for instance, obviously have an interest in securing the best possible terms of trade. At the same time, they both may also want to avoid the costs associated with a trade war. It is clear that, in such instances, the interests of the two states are not diametrically opposed.

The use of nonzero-sum games became the standard form of analysis in international politics toward the end of the 1950s, due in no small part to the scholarship of Thomas Schelling (1960, 1966) whose works are seminal. When Schelling's book *The Strategy of Conflict* was republished in 1980 by Harvard University Press, he remarked in a new preface that the idea that conflict and common interest were not mutually exclusive, so obvious to him, was among the book's most important contributions. In 2005, Schelling was awarded the Nobel Prize in economics for his work on game theory and interstate conflict. The award was well- deserved.[4]

Most studies also make use of the tools and concepts of noncooperative game theory. A *noncooperative game* is any game in which the players are unable to irrevocably commit themselves to a particular course of action, for whatever reason.[5] For example, the players may be unable to communicate with one another to jointly decide on an action plan. Or there may be some legal impediment to coordinated decision making. Since it is commonly

---

[4] To understand why, see Myerson (2009).
[5] By contrast, binding agreements are possible in a *cooperative game*.

understood that the international system lacks an overarching authority that can enforce commitments or agreements, it should come as no surprise that noncooperative game theory holds a particular attraction for theorists of interstate conflict.

1.3 Strategic Form Games and Nash Equilibria

Game theorists have developed a number of distinct ways to represent a game's structure. Initially, the *strategic form* (sometimes called the *normal* or the *matrix* form) was the device of choice. In the strategic form, players select *strategies* simultaneously, before the actual play of the game. A *strategy* is defined as a complete contingency plan that specifies a player's choice at every situation that might arise in a game. Figure 1.1 depicts a typical arms race game between two states, State A and State B, in strategic form.[6] Although the generic name for this game is Prisoners' Dilemma, it is referred to here as the *Arms Race Game*.[7]

In this representation, each state has two strategies: to *cooperate* (C) by not arming, and to *defect* from cooperation (D) by arming. If neither arms, the outcome is a compromise: a military balance is maintained, but at little cost. If both arm, then both lose, as an arms race takes place; the balance is maintained, but this time at considerable cost. Finally, if one state

		State B	
		Not Arm (C)	Arm (D)
State A	Not Arm (C)	*Tacit Arms Control* (3,3)	*B Gains Advantage* (1,4)
	Arm (D)	*A Gains Advantage* (4,1)	*Arms Race* (2,2)*

Key: (*x,y*) = payoff to State A, payoff to State B
  * = Nash equilibrium

Figure 1.1 Arms Race Game (Prisoners' Dilemma)

[6] For obvious reasons, such a game is called a two-person game. Games with three or more players are referred to as *n-person* games. The *Tripartite Crisis Game*, which is introduced in Chapter 2, is an example of an *n-person* game.

[7] "Prisoners' Dilemma" takes its name from a story that Albert W. Tucker, the chair of Princeton's psychology department, told his students to illustrate its structural dynamics. For the story and background, see Poundstone (1992: 116–21). The story itself is well known and can be found in most game-theory textbooks, including Zagare (1984).

arms and the other does not, the state that arms gains a strategic advantage, and the state that chooses not to arm is put at a military disadvantage.

Each cell of the matrix contains an ordered pair of numbers below the names of the outcomes. The numbers represent the payoff that the row player (State A) and the column player (State B), respectively, receives when that outcome obtains in a game. Payoffs are measured by a *utility* scale. Sometimes, as in this example, only *ordinal utilities* are, or need be, assumed. Ordinal utilities convey information about a player's relative ranking of the outcomes. In many studies of interstate conflict, however, *cardinal utilities* are assumed. A cardinal scale indicates both rank and intensity of preference.

In this example, the outcomes are ranked from best (i.e., "4") to worst (i.e., "1"). Thus, the ordered pair (4,1) beneath the outcome *A Gains Advantage* signifies that this outcome is best for State A and worst for State B. Similarly, the outcome *Tacit Arms Control* is next best for both players.

In game theory, the players are assumed to be instrumentally *rational*. Rational players are those who maximize their utility. Utility, though, is a subjective concept. It indicates the worth of an outcome *to a particular player*. Since different players may evaluate the same outcome differently, the rationality assumption is simply another way of saying that the players are purposeful, that is, that they are pursuing goals (or interests) that they themselves define.

Rationality, however, does not require that the players are necessarily intelligent in setting their goals. It may sometimes be the case that the players are woefully misinformed about the world and, as a consequence, have totally unreasonable objectives. Still, as long as they are purposeful and act to bring about their goals, they can be said to be instrumentally rational.[8]

Rationality also does not imply that the players will do well and obtain their stated objective, as is easily demonstrated by identifying the *solution* to the Arms Race Game. A solution to any strategic form game consists of the identification of (1) the best, or optimal, strategy for each player and (2) the likely outcome of the game. The Arms Race Game has a straightforward solution.

Notice first that each player (state) in the Arms Race Game has a *strictly dominant strategy*, that is, a strategy that is always best regardless of the strategy selected by the other player. For instance, if State B chooses not to arm, State A will bring about its next-best outcome (3) if it also chooses not to arm; however, it will receive its best outcome (4) if it chooses to arm. Thus, when State B chooses (C), State A does better by choosing (D). Similarly, if State B chooses to arm, State A will bring about its worst outcome (1) if it chooses not to arm; however, it will receive its next-worst outcome (2) if it chooses to arm. Again, when State B chooses (D), State A does better by choosing (D). Regardless of what strategy State B selects, therefore, State A should choose (D) and arm. By symmetry, State B should also choose to defect by arming. And when both players choose their unconditionally best strategy, the outcome is an *Arms Race*—which is next-worst for both players.

The strategy pair (D,D) associated with the outcome labeled *Arms Race* has a very important property that qualifies it to be part of the solution to the game of Figure 1.1. It is

---

[8] For an extended discussion of the rationality assumption, see Zagare (1990a).

called a *Nash equilibrium*—named after John Nash, the subject of the film *A Beautiful Mind* and a co-recipient of the Nobel Prize in economics in 1994, which, not coincidentally, was the fiftieth anniversary of the publication of von Neumann and Morgenstern's monumental opus. If a strategy pair is a Nash equilibrium, neither player has an incentive to switch to another strategy, provided that the other player does not also switch to another strategy.

To illustrate, observe that if both State A and State B choose to arm (D), State A's payoff will be its second-worst (2). But if it then decides to not arm (C), its payoff is its worst (1). In consequence, State A has no incentive to switch strategies if both states choose to arm. The same is true of State B. The strategy pair (D,D), therefore, is said to be stable, or *in equilibrium*.

There is no other strategy pair with this property in the Arms Race Game, as is easily demonstrated. For instance, consider the strategy pair (C,C) associated with the outcome *Tacit Arms Control*. This outcome is second-best for both players. Nonetheless, both players have an incentive to switch unilaterally to another strategy in order to bring about a better outcome. State B, for instance, can bring about its best outcome (4) by simply switching to its (D) strategy. Thus, the payoff pair (C,C) is not a Nash equilibrium. The same is true for the remaining two strategy pairs in this game, (C,D) and (D,C).

For reasons that will be more fully explained below, strategy pairs that form a Nash equilibrium provide a *minimum* definition of rational choice in a game. By contrast, strategy pairs that are not in equilibrium are simply inconsistent with rational choice and purposeful action. This is why only Nash equilibria can be part of a game's solution.

But notice that *both* players do worse when they are rational and select (D) than when *both* make an irrational choice and select (C). In other words, two rational players do *worse* in this game than two irrational players! Paradoxically, however, it is also true that *each* player always does best by choosing (D). All of which raises a very important question for the two states in our game: can they, if they are rational, avoid an arms race and, if so, under what conditions? More generally, can two or more states ruthlessly pursuing their own interests find a way to cooperate in an anarchic international system?

The definitive answer to this question is highly contentious. Suffice it to say that it is an issue that lies at the heart of the ongoing debate between realists and liberals about the very nature of international politics. That the (Prisoners' Dilemma) game shown in Figure 1.1 both highlights and neatly encapsulates such a core problem must be counted among game theory's many contributions to the field of security studies.[9]

Even though rational players do not fare well in this game, the game itself has a well-defined solution that helps to explain, inter alia, why great states sometimes engage in senseless and costly arms competitions that leave them no more secure than they would have been if they had chosen not to arm. The solution is well defined because there is only one outcome in the game that is consistent with rational contingent decision making by all of the players: the unique Nash equilibrium (D,D).

Not all games, however, have a solution that is so clear-cut. Consider, for example, the two-person game shown in Figure 1.2 that was originally analyzed by John Harsanyi

[9] A good place to start when exploring this and related issues is Oye (1986). Baldwin (1993) contains a useful collection of articles, many of which are seminal. Axelrod (1984), which provides one prominent game-theoretic perspective, should also be consulted.

	State B	
	Cooperate (C)	Defect (D)
State A Cooperate (C)	*Outcome CC* (1,3)*	*Outcome CD* (1,3)
Defect (D)	*Outcome DC* (0,0)	*Outcome DD* (2,2)*

Key: (x,y) = payoff to State A, payoff to State B
* = Nash equilibrium

Figure 1.2 Strategic form game with two Nash equilibria (Harsanyi's game)

(1977a), another 1994 Nobel Prize laureate in economics. As before, the two players, State A and State B, have two strategies: cooperate (C), or defect from cooperation (D). State A's strategies are listed as the rows of the matrix, while B's strategies are given by the columns. Since each player has two strategies, there are 2 × 2 = 4 possible strategy combinations and four possible outcomes. The payoffs to State A and State B, respectively, are again represented by an ordered pair in each cell of the matrix.

Of these four strategy combinations, two are Nash equilibria, as indicated by the asterisks (*). Strategy pair (D,D) is in equilibrium since either player would do worse by switching unilaterally to its other strategy. Specifically, were State A to switch from its (D) strategy to its (C) strategy, which would induce Outcome CD, State A's payoff would go from "2"— A's best—to "1"—its next-best outcome. And if State B were to switch to its (C) strategy, B's payoff would go from "2"—its next-best outcome—to "0"—its worst. Thus, neither player benefits by switching unilaterally to another strategy, so (D,D) is a Nash equilibrium. For similar reasons, strategy pair (C,C) is also a Nash equilibrium; neither player benefits by switching unilaterally to its (D) strategy. In contrast, neither of the remaining two strategy pairs is stable in the sense of Nash, because at least one player would gain by changing strategies.

The existence of two or more Nash equilibria in a strategic form game can confound analysis. When only one Nash equilibrium exists in a game, it is easy to specify a game's solution. But when two or more equilibria exist, it is clearly more difficult to identify the likely outcome of a game or the best strategy of the players—unless there are criteria that allow discrimination among equilibria and the elimination of some stable strategy pairs from the solution set.

Of course, the possible existence of multiple Nash equilibria in a strategic form game would not be problematic if all equilibria were *equivalent*—that is, if all extant equilibria have exactly the same consequences for the players—and *interchangeable*—in the sense that every possible combination of equilibrium strategies are also in equilibrium.

John Nash (1951) proved long ago that when multiple equilibria exist in a zero-sum game, all equilibrium pairs are both equivalent and interchangeable. But this is clearly not the case in the nonzero-sum game shown in Figure 1.2. The two equilibria are not equivalent simply because the player's payoffs are different under each equilibrium. For instance, State A's best outcome is associated with the strategy pair (D,D); however, its next-best outcome is associated with the strategy pair (C,C). The two equilibria are also not interchangeable. Although the strategy pairs (C,C) and (D,D) are in equilibrium, the pairs (C,D) and (D,C) are not. This means that the players cannot use the strategies associated with the two Nash equilibria interchangeably.

Although the two Nash equilibria in the game shown in Figure 1.2 are neither equivalent nor interchangeable, there is one way in which they are can be distinguished. Notice that State B's defection (D) strategy *weakly dominates* its cooperation (C) strategy, that is, it provides State B with a payoff that is at least as good, and sometimes better, than its other strategy, no matter what strategy State A selects.[10] Thus, there is a good reason to expect that State B will choose (D).

Notice also that if State B defects, State A does better by also defecting. Given that State B defects, State A will receive its highest payoff (2) by defecting, but only its second-highest payoff (1) by cooperating. Since the strategy pair (D,D) is associated with State B's unconditionally best (or *dominant*) strategy, and State A's best response to B's unconditionally best strategy, one might very well argue that it, and not strategy pair (C,C) is the equilibrium that best qualifies as the solution to Harsanyi's game.

But before this conclusion is accepted, there is one significant objection that must be considered: the fact that strategy pair (D,D) favors State A at the expense of State B. State B's payoff is clearly better under (C,C) than it is under (D,D), while it is the other way around for State A. Is there nothing that State B can do to induce the more preferred payoff associated with the equilibrium (C,C)?

One might argue that State B could do better in this game by threatening to choose (C) if State A selects (D), thereby inducing State A to choose (C) and bringing about State B's most-preferred outcome. But this line of argument is deficient. To understand why, we next explore an alternative representation of Harsanyi's game, the *extensive form*. As Morrow (1994: 58) notes, the extensive form "is the basic formalization of game theory."

1.4 Extensive Form Games, Backward Induction, and Subgame Perfect Equilibria

Figure 1.2 shows Harsanyi's game in strategic form; Figure 1.3 shows it in extensive form. There are a number of important differences between the two forms of representation. In the strategic form, the players select a strategy which, it will be recalled, is a complete plan of action that specifies what a player will do at every decision point in a game. As well, the

---

[10] By contrast, a *strictly dominant strategy* always provides a player with a higher payoff than any other strategy, no matter what strategies other players select. Both players in the Arms Race Game shown in Figure 1.1 possess strictly dominant strategies. For a further discussion of this and related concepts, see Zagare (1984).

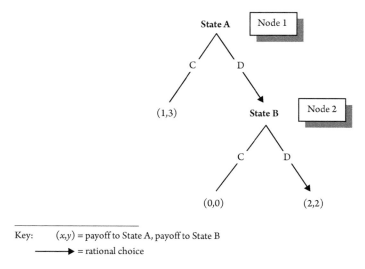

Key: (x,y) = payoff to State A, payoff to State B
        ——————▶ = rational choice

Figure 1.3 Extensive form representation of Harsanyi's game

players are assumed to make their choices simultaneously or, in what amounts to the same thing, without information about what strategy the other player has selected.

By contrast, in the extensive form, the players make *moves* sequentially, that is, they select from among the collection of *choices* available at any one point in time. In the extensive form, moves are represented by *nodes* on a game tree. The *branches* of the tree at any one node summarize the choices available to a player at a particular point in the game. The payoffs to the players are given by an ordered pair at each *terminal node*. In an extensive form game of *perfect information*, the players know where they are on the game tree whenever there is an opportunity to make a choice. Harsanyi's game is an example of a game of perfect information. In a game with *imperfect information*, the players may not always know what prior choices, if any, have been made.

Early applications of game theory to the fields of security studies and diplomatic history relied on strategic form representations. To some extent, this was an accident of history. But it was also because simple 2 × 2 strategic form games seemed to capture the dynamics of many issues that lay at the heart of interstate politics. Snyder (1971a), for example, considered the security dilemma and alliance politics to be, in essence, real world manifestations of a Prisoners' Dilemma game, while Jervis (1979: 292) argued that the game of Chicken (see Chapter 4, Figure 4.1) provided the basis for a "reasonable definition of deterrence theory."

Over time, however, the strategic form of representation gave way to the extensive form. To solve any extensive form game, a procedure known as *backward induction* must be used. As its name suggests, backward induction involves working backward up the game tree to determine, first, what a rational player would do at the last (or terminal) node of the tree, what the player with the previous move would do given that the player with the last move is assumed to be rational, and so on, until the first (or *initial*) node of the tree is reached. We will now use this procedure to analyze the extensive form representation of Harsanyi's game. More specifically, we now seek to establish why State B cannot rationally threaten to select (C) at Node 2 in order to induce State A's cooperation at Node 1, thereby bringing about State B's highest-ranked outcome.

To this end, we begin by considering the calculus of State A at the first node of the tree. At Node 1, State A can either select (C) and induce its second-best outcome, or select (D), which might result in either State A's best outcome or its worst outcome. Clearly, State A should (rationally) choose (C) if it expects State B to also select (C), since the choice of (D) would then result in State A's worst outcome. Conversely, State A should select (D) if it expects State B to select (D), since this induces State A's best outcome. The question is, what should State A expect State B to do? Before we can answer this question, we must first consider State B's choice at the last node of the tree.

If State A assumes that State B is rational, then State A should expect State B to select (D) if and when State B makes its choice at Node 2. The reason is straightforward: State B's worst outcome is associated with its choice of (C), while its next-best outcome is associated with its choice of (D). To expect State B to carry out the threat to choose (C) if A chooses (D), then, is to assume that State B is irrational. It follows that, for State A to expect State B to select (C), one must assume that State A harbors irrational expectations about State B. To put this in a slightly different way, State B's threat to choose (C) is not credible, that is, it is not rational to carry out. Since it is not credible, State A may safely ignore it.

Notice what the application of backward induction to Harsanyi's game reveals: State B's rational choice at Node 2 is (D). In consequence, State A should also choose (D) at Node 1. Significantly, the strategy pair (D,D) associated with these choices is in equilibrium, in the same sense that the two Nash equilibria are in the strategic form game shown in Figure 1.2: neither player has an incentive to switch to another strategy, provided the other player does not also switch. But, also significantly, the second Nash equilibrium (C,C) is nowhere to be found. Because it was based on an incredible threat, it was eliminated by the backward induction procedure.

The unique equilibrium pair (D,D) that emerges from an analysis of the extensive form game shown in Figure 1.3 is called a *subgame perfect equilibrium*.[11] The concept of subgame perfection was developed by Reinhard Selten (1975), the third and final recipient of the 1994 Nobel Prize in economics.[12] Selten's perfectness criterion constitutes an extremely useful and important refinement of Nash's equilibrium concept. It is a refinement because it eliminates less-than-perfect Nash equilibria from the set of candidates eligible for consideration as a game's solution. As well, Selten's idea of subgame perfection helps us to understand more deeply the meaning of rational choice as it applies to individuals, groups, or even great states involved in a conflictual relationship. Subgame perfect equilibria require that the players plan to choose rationally at every node of the game tree, whether they expect to reach a particular node or not.

It is important to know that all subgame perfect equilibria are also Nash equilibria, but not the other way around. As just demonstrated, Nash equilibria that are based on threats

---

[11] A *subgame* is that part of an extensive form game that can be considered a game unto itself. For a more detailed definition, with pertinent examples, see Morrow (1994: ch. 2).

[12] Recall that John Nash and John Harsanyi were the other two.

that lack credibility, such as the Nash equilibrium strategy pair (C,C) in the game shown in Figure 1.2, are simply not perfect. As Harsanyi (1977a: 332) puts it, these less-than-perfect equilibria should be considered deficient because they involve both "irrational behavior and irrational expectations by the players about each other's behavior."

1.5 Applications of Game Theory in Security Studies

Speaking more pragmatically, the refinement of Nash's equilibrium concept represented by the idea of a subgame perfect equilibrium and related solution concepts—such as *Bayesian Nash equilibria* and *perfect Bayesian equilibria*—permits analysts to develop more nuanced explanations and more potent predictions of interstate conflict behavior when applying game theory to the field of security studies.[13] It is to a brief enumeration of some of these applications, and a specific illustration of one particular application, that we turn next.

As noted earlier, applications, extensions, modifications, and illustrations of game-theoretic models began to appear in the security studies literature shortly after the publication of *Games and Economic Behavior* (1944). Since then, the literature has grown exponentially, and its influence on the field of security studies has been significant.[14] As Walt (1999: 5) observes,

> Rational choice models have been an accepted part of the academic study of politics since the 1950s, but their popularity has grown significantly in recent years. Elite academic departments are now expected to include game theorists and other formal modelers in order to be regarded as "up to date," graduate students increasingly view the use of formal rational choice models as a prerequisite for professional advancement, and research employing rational choice methods is becoming more widespread throughout the discipline.

Walt (1999: 7) went on to express the fear that game-theoretic and related rational-choice models have become so pervasive, and their influence so strong, that other approaches are on the cusp of marginalization. Although Martin (1999: 74) unquestionably demonstrates, empirically, that Walt's fear is "unfounded," there is little doubt that game-theoretic studies are now part and parcel of the security studies literature.

In security studies, subject areas that have been heavily influenced by game-theoretic reasoning include the onset (Bueno de Mesquita and Lalman, 1992) and escalation (Carlson, 1995) of interstate conflict and war, the consequences of alliances (Smith, 1995)

---

[13] Nash and subgame perfect equilibria are the accepted measures of rational behavior in games of *complete* information, in which each player is fully informed about the preferences of its opponent. In games of *incomplete* information, that is, games in which at least one player is uncertain about the other's preferences, rational choices are associated with *Bayesian Nash equilibria* (in strategic form games) and *perfect Bayesian equilibria* (in extensive form games). See Gibbons (1992) for a helpful discussion. Bayesian equilibria are discussed in Section 1.6; perfect Bayesian equilibria are discussed in Chapter 2. See also Table 1.2 in the addendum to this chapter.

[14] An insightful review of both the accomplishments and the limitations of the approach can be found in Bueno de Mesquita (2002). See also Brams (2002).

and alignment patterns (Zagare and Kilgour, 2003), the effectiveness of missile defense systems (Powell, 2003; Quackenbush, 2006), the impact of domestic politics on interstate conflict (Fearon, 1994), the dynamics of arms races and the functioning of arms control (Brams and Kilgour, 1988), the spread of terrorism (Bueno de Mesquita, 2005), the efficacy of economic sanctions for combating transnational terrorism (Bapat, De La Calle, Hinkkainen and McLean, 2016), the dangers of nuclear proliferation (Kraig, 1999), the implications of democratization for coercive diplomacy (Schultz, 2001), the characteristics of crisis bargaining (Banks, 1990), the operation of balance of power politics (Niou, Ordeshook and Rose, 1989), the role of reputation in diplomatic exchanges (Sartori, 2005), and the efficacy of military threats (Slantchev, 2011), to name just a few.[15] The large and influential research program that Reiter (2003: 27) refers to as the "bargaining model of war" has also relied heavily on formal (i.e., game-theoretic) reasoning "to expand and deepen the [program's] theoretical reach" (see also Powell, 2002, and Reed and Sawyer, 2013). And, as noted above, game-theoretic models have played a central role in the debate between realists and liberals about the relative importance of absolute and relative gains and about the possibility of significant great power cooperation (see footnote 9).

It is clear, however, that there has been no area of security studies in which game theory has been more influential than the study of deterrence. Accordingly, I now turn to a brief discussion of this subject and attempt to illustrate, with a simple example, how game theory can help not only to clarify core concepts but also to shed light on the conditions that lead to successful deterrence. A more systematic treatment of deterrence theory is given in Chapter 7.

Although it may be somewhat of a stretch to say that Thomas Schelling was the inventor of classical deterrence theory, as does Zakaria (2001), his work is a good place to start (for an overview, see Zagare 1996a). Like all classical deterrence theorists, Schelling's work is characterized by two core assumptions: (1) states (or their decision makers) are rational and (2) especially in the nuclear age, war or conflict is the worst possible outcome of any deterrence encounter. It is not difficult to demonstrate that these two assumptions are incompatible with the conclusion of most deterrence theorists that bilateral nuclear relationships, such as that between the United States and the Soviet Union during the Cold War, are inordinately stable.

To see this, consider now the *Rudimentary Asymmetric Deterrence Game*, which is shown in Figure 1.4. In this, perhaps the simplest deterrence game one can imagine, State A begins play at Node 1 by deciding whether to *concede* (C) and accept the status quo or to *demand* (D) its alteration. If State A chooses (C), the game ends, and the outcome is the *Status Quo*. But if State A defects, State B must decide at Node 2 whether to *concede* (C) the issue—in which case, the outcome is *A Wins*—or *deny* (D) the demand and precipitate *Conflict*. Notice that the endpoints of this simple deterrence game list outcomes rather than player payoffs. I list outcomes and not payoffs in this example in order to use the same game form to analyze the strategic implications of more than one payoff configuration.

[15] This listing is meant to be suggestive. It is by no means exhaustive. Useful reviews include O'Neill (1994a, b) and Snidal (2002). Zagare and Slantchev (2012) discusses the historical development of game-theoretic models in international relations and provides a more detailed overview of the current literature. See also O'Neill (2007).

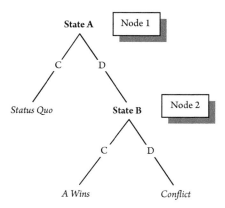

Figure 1.4 The Rudimentary Asymmetric Deterrence Game

Next, we determine what rational players would do in this game—given the assumption that *Conflict* is the worst outcome for both players—by applying backward induction to the game tree. Since the application of this procedure requires one to work backward up the game tree, we begin by considering State B's move at Node 2.

At Node 2, State B is faced with a choice between choosing (C), which brings about the outcome *A Wins*, and choosing (D), which brings about *Conflict*. But if *Conflict* is assumed to be the *worst* possible outcome, State B, if it is rational, can *only* choose to concede, since, by assumption, *A Wins* is the more preferred outcome.

Given that State B will rationally choose to concede at Node 2, what should State A do at Node 1? State A can concede, in which case the outcome will be the *Status Quo*, or it can defect, in which case the outcome will be *A Wins*—because a rational State B will choose to concede at Node 2. If State A has an incentive to upset the *Status Quo*, that is, if it needs to be deterred because it prefers *A Wins* to the *Status Quo*, it will rationally choose (D). Thus, given the core assumptions of classical deterrence theory, the *Status Quo* is unstable, and deterrence rationally fails.

To put this in a slightly different way, one can reasonably assume that states are rational, and one can also reasonably assume that war is the worst imaginable outcome for all the players, but one cannot make both these assumptions at the same time and logically conclude, as classical deterrence theorists do, that deterrence will succeed.[16]

Logically inconsistent theories are clearly problematic. Since *any* conclusion can be derived from them, inconsistent theories can explain *any* empirical observation. Inconsistent theories, therefore, are non-falsifiable and so of little practical use. When used properly, formal structures, like game theory, can help in the identification of flawed theory.

If the core assumptions of classical deterrence theory are inconsistent with the possibility of deterrence success, which assumptions are? It is easy to demonstrate that, in the

---

[16] One can, however, make both these assumptions and conclude that the status quo will survive rational play if one is willing to drop the standard realist assumption that all states are the same, that is, are undifferentiated (Waltz, 1979). For a further discussion, see Zagare and Kilgour (2000: 142). See also the discussion of the Unilateral Deterrence Game in Chapter 7.

Rudimentary Asymmetric Deterrence Game, the *Status Quo* may remain stable, and deterrence may succeed, but only if State B's threat is credible in the sense of Selten, that is, if it is rational to carry out.

To see this, assume now that State B prefers *Conflict* to *A Wins*. (Note that this assumption implies that *Conflict* is not the worst possible outcome for State B.) With this assumption, State B's rational choice at Node 2 changes. Given its preference, its rational choice at Node 2 is now to choose (D) and deny State A's demand for a change in the *Status Quo*.

But State B's rational choice is not the only rational choice that changes with this new assumption. The rational choice of State A is also different. Applying backward induction to State A's decision at Node 1 now reveals a choice between *Status Quo* and *Conflict*. This means that the *Status Quo* will persist, and deterrence will succeed, *as long as State A's preference is for peace over war*. On the other hand, it will fail whenever this latter preference is reversed, even when State B's Node 2 threat is credible.

At this juncture, two final observations can be made. The first is about the relationship between credible threats and deterrence success. Apparently, credibility is not, as Freedman (1989: 96) claims, the "magic ingredient" of deterrence. As just demonstrated, a credible threat is not sufficient to ensure deterrence success. Deterrence may rationally fail, even when all deterrent threats are rational to execute.

Still, in order to explain even the possibility of deterrence success in this simple example, a core assumption of classical deterrence theory had to be modified. But any analysis that proceeds from a different set of assumptions will constitute an entirely different theory. This is no small matter. As illustrated in the films *Sliding Doors* and *Run Lola Run*, and as demonstrated in Chapter 7, where the standard version of deterrence theory is contrasted with an alternative specification called perfect deterrence theory, ostensibly minor differences in initial assumptions can have important theoretical consequences and significant policy implications.

1.6 Strategic Form Games with Incomplete Information: Bayesian (Nash) Equilibria

Up to this point, we have considered only games of *complete information*, that is, games in which all the players know, for sure, the others' preferences. In the real world, however, it is not often the case that information is complete. In this section, we examine, briefly, strategic form games with *incomplete information*. When information is incomplete, at least one player does not know another's preference function. In a strategic form game with incomplete information, the accepted standard of rational play is called a *Bayesian (Nash) equilibrium*, which is defined as a strategy combination that maximizes each player's expected utility, given that player's (subjective) beliefs about the other players' types. In Chapter 2, I introduce the concept of a perfect Bayesian equilibrium, which is a natural extension of the concept of a Bayesian equilibrium to extensive form games with incomplete information.

Figure 1.5 depicts the Rudimentary Asymmetric Deterrence Game in strategic form. In this representation, the players' (ordinal) utilities are left unspecified. As before, if State

State B

	Cooperate (C)	Defect (D)
Cooperate (C)	*Status Quo* (a_{SQ}, b_{SQ})	*Status Quo* (a_{SQ}, b_{SQ})
Defect (D)	*A Wins* (a_{DC}, b_{DC})	*Conflict* (a_{DD}, b_{DD})

State A

Key: (a_{XX}, b_{XX}) = payoff to State A, payoff to State B at outcome XX

Figure 1.5 Strategic form of the Rudimentary Asymmetric Deterrence Game

A cooperates by choosing (C), the *Status Quo* (outcome SQ) obtains and the payoffs to the players, as expressed symbolically, are a_{SQ} and b_{SQ} to State A and State B, respectively. But if State A defects and chooses (D), the outcome of the game will depend on State B's choice. If State B chooses (C), the outcome is *A Wins* (outcome DC), with a payoff of a_{DC} to State A, and b_{DC} to State B. *Conflict* (outcome DD), of course, results when both players choose (D). In this case, State A's payoff is a_{DD}, and State B's is b_{DD}.

In the present analysis of the Rudimentary Asymmetric Deterrence Game, the assumption will be that State A, the *Challenger* in this game, most prefers *A Wins*, second-most-prefers the *Status Quo*, and least-prefers *Conflict*. State B, the *Defender*, is assumed to most prefer the *Status Quo*. The further assumption, however, is that State B may be one of two types: *Hard* and *Soft*. A Hard State B is one who prefers *Conflict* to *A Wins*. A Soft State B has the opposite preference. With respect to preferences, then, the assumptions will be as follows:

> **State A:** *A Wins* > *Status Quo* > *Conflict*
> **State B (Soft):** *Status Quo* > *A Wins* > *Conflict*
> **State B (Hard):** *Status Quo* > *Conflict* < *A Wins*

where ">" means "is preferred to."

Consider now Figure 1.6, which depicts the Rudimentary Asymmetric Deterrence Game when State B is known to be Soft, that is, when it prefers *A Wins* to *Conflict*. Although there are two Nash equilibria in this version of the game, the strategy pair (D,C), which is associated with the outcome *A Wins*, stands out: it is the product of State B's weakly dominant strategy (C) and State A's best response to it (D). And, unlike the other Nash equilibrium, strategy pair (C,D), which is associated with the outcome *Status Quo*, (D,C) is subgame perfect. All of which suggests that when *Conflict* is State B's least-preferred outcome, the *Status Quo* is unstable, deterrence fails, and *A Wins*.

By contrast, deterrence succeeds if and when State B is Hard. As Figure 1.7 indicates, the unique Nash equilibrium in the Rudimentary Deterrence Game when State B prefers *Conflict* to *A Wins* is the *Status Quo*. Under complete information, then, the players' choices

State B

	Cooperate (C)	Defect (D)
State A Cooperate (C)	*Status Quo* (2,3)	*Status Quo* (2,3)*
Defect (D)	*A Wins* (3,2)**	*Conflict* (1,1)

Key: (x,y) = payoff to State A, payoff to State B
    * = Nash equilibrium
    ** = subgame perfect Nash equilibrium

Figure 1.6 Strategic form of the Rudimentary Deterrence Game when State B is Soft

State B

	Cooperate (C)	Defect (D)
State A Cooperate (C)	*Status Quo* (2,3)	*Status Quo* (2,3)**
Defect (D)	*A Wins* (3,1)	*Conflict* (1,2)

Key: (x,y) = payoff to State A, payoff to State B at outcome XX
    ** = subgame perfect Nash equilibrium

Figure 1.7 Strategic form of the Rudimentary Deterrence Game when State B is Hard

are clear: when State B is Hard, it will rationally choose (D); when it is Soft, it will choose (C). State A's best strategy depends on State B's type. When State B is Hard, State A will rationally choose (C); however, when State B is Soft, State A's best strategy is (D).

But what if State A does not know which type of State B it is facing? To answer this question, assume now that the payoffs to the players are cardinal utilities, that is, they are utility measures that reflect both the rank and the intensity of a player's preferences. Also assume that State A believes, with probability p_B, that State B is Hard and, with probability $(1 - p_B)$, that State B is Soft. If State A defects, then it will receive payoff a_{DD} with probability p_B, and payoff a_{DC} with probability $(1 - p_B)$. Of course, if it cooperates, it will receive payoff a_{SQ} with certainty. Clearly, State A should choose (D) if and only if its expected utility from choosing (D) exceeds its expected utility of choosing (C), that is, if and only if

$$p_B(a_{DD}) + (1 - p_B)(a_{DC}) > a_{SQ}, \tag{1.1}$$

which can be rewritten as

$$p_B(a_{DD} - a_{DC}) > (a_{SQ} - a_{DC}). \tag{1.2}$$

By assumption, both $(a_{DD} - a_{DC})$ and $(a_{SQ} - a_{DC})$ are negative. Multiplying both sides by -1 and reversing signs renders them positive and yields

$$p_B(a_{DC} - a_{DD}) < (a_{DC} - a_{SQ}). \tag{1.3}$$

Rearranging terms by dividing both sides of the equation by the coefficient of p_B reveals that State A should choose (D) if and only if

$$p_B < \frac{(a_{DC} - a_{SQ})}{(a_{DC} - a_{DD})} = a_m. \tag{1.4}$$

Otherwise, State A should cooperate and choose (C).

The expression on the right side of equation 1.4 defines a threshold value (a_m), here called the *deterrence threshold*, which specifies the conditions under which each of exactly two Bayesian equilibria in the Rudimentary Asymmetric Deterrence Game exist:

1. The *Deterrence Equilibrium*, under which State A always cooperates and State B always defects if it is Hard and always cooperates if it is Soft, will exist if and only if State A believes that State B is Hard with probability $p_B > a_m$. The outcome under this equilibrium will always be the *Status Quo*.

2. The *Attack Equilibrium*, under which State A always chooses (D), will exist if and only if State A believes that State B is Hard with probability $p_B < a_m$. The outcome under this equilibrium depends on State B's type. When it is Soft, State B will cooperate, and the outcome will be *A Wins*; but when it is Hard, State B will defect, and a *Conflict* will occur.

Table 1.1 summarizes the technical details of these two equilibria. As Table 1.1 indicates, the Deterrence Equilibrium and the Attack Equilibrium cannot coexist, that is, they are unique.[17]

Since the two Bayesian equilibria in the Rudimentary Asymmetric Deterrence Game exist under unique parameter conditions, point predictions and after-the-fact explanations about likely behavior in this game are straightforward and uncomplicated. Put somewhat differently, these equilibria represent two distinct rational strategic possibilities, either of

---

[17] One *transitional equilibrium* is ignored. An equilibrium is transitional if it exists only when the parameters of a model satisfy a specific functional relationship (i.e., an equation). The justification for ignoring transitional equilibria is that, however the parameter values are obtained, they are very unlikely to satisfy any specific equation.

Table 1.1 Forms of Bayesian equilibria and existence conditions for the Rudimentary Asymmetric Deterrence Game with one-sided incomplete information.

Equilibrium	Strategic Variables			Existence Condition
	State A	State B		
	(x)	(y_H)	(y_S)	
Deterrence	0	1	0	$p_B > a_m$
Attack	1	1	0	$p_B < a_m$

Key: x = probability that State A chooses (D)
  y_H = probability that State B chooses (D), given that it is Hard
  y_S = probability that State B chooses (D), given that it is Soft
  p_B = probability that State B is Hard
  a_m = deterrence threshold

which might come into play. Which one actually does depends primarily on State A's beliefs about what type of State B it is facing, that is, on the credibility or believability of State B's threat. When State B's threat is credible enough, State A is deterred. But when the credibility of State B's threat is below the deterrence threshold, the *Status Quo* will not survive rational play.

Additional insight into the dynamics of deterrence is provided by the relationship of the component variables that define the deterrence threshold: State A's utility for *A Wins* (a_{DC}), the *Status Quo* (a_{SQ}), and *Conflict* (a_{DD}). Specifically, ceteris paribus, the greater State A's utility is for the *Status Quo*, relative to its utility for the other outcomes, or the greater the cost of *Conflict* is, the lower a_m is, and the easier it is for State B to deter State A. Conversely, the more State A values *A Wins*, the higher the deterrence threshold and the higher must be State B's credibility to deter State A.

Of course, none of this is particularly surprising. What is surprising, however, is how often some of the straightforward policy implications of this simple model are overlooked. Generally speaking, the focus of the mainstream deterrence literature has been on the relationship between the costs of conflict and deterrence success. To some extent, this is understandable. But as the present analysis reveals, an opponent's evaluation of the status quo is also an important strategic variable. Clearly, policies that make things worse for an opponent can cut both ways. All of which strongly suggests that great care should be exercised when implementing coercive policies, but especially those that are strictly motivated by a desire to avoid an all-out conflict. Policies designed to deter an opponent that focus on one strategic variable, to the exclusion of others, are liable to backfire.

Finally, it should be pointed out that the analysis of the Rudimentary Asymmetric Deterrence Game with incomplete information offers an explanation of how deterrence can fail, and fail disastrously, even when both players prefer the *Status Quo* to *Conflict*, both players know this, and both players know that the other player knows this.[18] In other words,

---

[18] Technically, *Conflict* is a non-Pareto-optimal outcome. See the Glossary of Basic Concepts for a definition.

even under ostensibly benign conditions, a complete breakdown of deterrence will always remain a distinct possibility, a possibility that the rationality assumption, by itself, cannot, and does not, eliminate. The existence conditions associated with the Attack Equilibrium specify precisely the circumstances wherein this unsettling possibility is actualized. By isolating and highlighting these conditions, the analysis of the Rudimentary Asymmetric Deterrence Game (together with the collection of closely related models described in later chapters) brings additional clarity to what is oftentimes an opaque debate among security studies experts.

1.7 Coda

This chapter provides a gentle introduction to the key concepts and assumptions of game theory as it applies to the field of security studies. The examples used to illustrate many of these terms are meant to be suggestive, not definitive. Additional details will be introduced as they become relevant in the chapters that follow. Nonetheless, it should be clear that the securities studies literature that draws on, or has been influenced by, game-theoretic reasoning is not only vast but influential as well. And it is not difficult to explain why: game theory's formal structure facilitates the identification of inconsistent assumptions, highlights the implications of initial assumptions, and increases the probability of logical argumentation.

In Chapter 2, I delve a bit deeper into the theory of games to consider extensive form games with incomplete information, and the concept of a perfect Bayesian equilibrium. At the same time, I explore a number of issues connected with the use of game-theoretic models to organize analytic narratives, both generally and specifically. I begin with an easy case and develop a causal explanation of the Rhineland crisis of 1936. Then, some methodological obstacles that sometimes arise in more complex cases are discussed, and suggestions for overcoming them offered. Finally, the advantages of using game models to more fully understand real world events are highlighted.

Addendum

Table 1.2 Accepted standards of rational play in static and dynamic games.

Game Form	Information	
	Complete	Incomplete
Strategic form (static)	Nash equilibrium	Bayesian equilibrium
Extensive form (dynamic)	Subgame perfect equilibrium	Perfect Bayesian equilibrium

2

. . **.** . .

Game Theory and Diplomatic History

2.1 Introduction

Just before the turn of the last century, in an article designed to provoke, Walt (1999: 33) charged that "empirical testing is not a *central* part of the formal theory enterprise—at least, not in the subfield of security studies—and probably constitutes its most serious limitation" (emphasis added).[1] To buttress this claim, Walt ignored a number of important empirical applications and denigrated those attempts that he did not (Zagare, 1999). Of course, he was carefully parsing his words. But that is not the only reason why Walt had a point. Until recently, systematic empirical investigations were the exceptions, and not the rule, in the mainstream formal (i.e., game-theoretic) literature of international relations, security studies, and diplomatic history. Today, however, it is no longer uncommon to find game-theoretic models investigated rigorously. Quackenbush's (2010a) exacting test of perfect deterrence theory (Zagare and Kilgour, 2000) and Signorino and Tarar's (2006) work on extended immediate deterrence are two good examples.[2]

While sophisticated statistical analysis is now part and parcel of the formal literature in security studies, it remains true that carefully constructed analytic narratives, organized around an explicit game-theoretic model, remain rarae aves. One reason why game-theoretic and other types of formal models are seldom used to guide a case study may be the importance placed on both theory development and generalization by peace scientists and security studies analysts alike. To be sure, case studies can be used to generate theory (Büthe, 2002). But the existence of a well-articulated game model implies the presence of at least the rudiments of a prior theory.

Case studies, though, also have an important role to play in the testing of theory (George and Bennett, 2005). Thus, this explanation does not fully suffice. A second and likely more fundamental reason why game models have not more often been used to structure case analyses is the inherent difficulty of bringing the theoretical concepts of an abstract formal model into an isomorphic relationship with the nuanced reality of an interpersonal, intergroup, or interstate relationship. These practical problems may also

---

[1] This chapter is based on Zagare (2011a).
[2] See also Bennett and Stam (2000), Palmer and Morgan (2006), Sartori (2005), and Schultz (2001).

Game Theory, Diplomatic History and Security Studies. Frank C. Zagare. Oxford University Press (2019).
© Frank C. Zagare 2019. DOI: 10.1093/oso/9780198831587.001.0001

help explain why historians have failed to exploit the potential of game-theoretic models to generate causal explanations of singular real world events and processes (Riker, 1990). Historical narratives, which typically lack the element of necessity that is essential for establishing causality (Fischer, 1970: 104), would be much improved were they built on a firm theoretical foundation (Hanson, 1958: 90; Trachtenberg, 2006: 28).

One purpose of this chapter, therefore, is to discuss some of the problems associated with using a game model as a template for organizing an analytic narrative. But another is to highlight some of the benefits of doing so, both generally and specifically. The contention here is that many of these advantages are less than apparent.

2.2 Game Models and Historical Narratives: An Easy Case

As noted in Chapter 1, a game can be thought of as any situation in which the outcome depends on the choices of two or more actors, that is, when the choices of these actors, called *players*, are interdependent. Players may be individuals or groups of individuals who act as coherent units. The players are assumed to be *rational*. Simply put, this means that they are purposeful, that is, that they have objectives and that they act to bring them about. It does not mean, however, that the players are necessarily intelligent in the sense that their objectives are wise, realistic, or even admirable. It also does not mean that a player will be successful. In game theory, as in life, players are oftentimes misguided, shortsighted, imprudent, and unsuccessful (Riker, 1995).

The players in a game are also assumed to make choices that lead to outcomes that the players can evaluate on either an ordinal or a cardinal utility scale. To fully specify a game, these choices, the associated outcomes, and the players' utility functions must also be determined. More nuanced game models will also detail the private information each player possesses, information that is common to all the players, and each player's belief about what is likely to take place as the game is played out. Each of these elements can be illustrated with a straightforward example that will serve as an explanatory baseline (i.e., a reality test) for assessing some of the benefits and pitfalls of constructing a case study around an explicit model, game-theoretic or otherwise: the Rhineland crisis of 1936.

Under Article 42 of the Treaty of Versailles of 1919, which concluded World War I, Germany was "forbidden to maintain or construct any fortifications either on the left bank of the Rhine or on the right bank to the west of a line drawn 50 kilometers to the East of the Rhine." Like Article 231, the so-called war guilt clause, Article 42 was almost universally detested within Germany. For the French, however, it was strategically vital. Because control of the Rhineland opened Germany to a French attack, a demilitarized Rhineland made it difficult for Germany to invade any country to its south or east (i.e., Austria, Czechoslovakia, or Poland). Thus, after 1919, French policy makers were motivated to see that the stipulations of the Versailles Treaty were enforced (i.e., to maintain the status quo) in the Rhineland, while German leaders sought to undermine them. Until 1936, the status quo held.

According to Kagan (1995: 355), Adolf Hitler, the German Chancellor, first spoke about the possibility of remilitarization in the summer of 1935. But he did not act until March 7,

1936, when a small contingent of German soldiers crossed the Rhine. The French government protested (halfheartedly), but did not respond militarily.

Since the Rhineland crisis is straightforward and relatively transparent, it is easy to model. The two core players, the decision makers who constituted the German and French governments, are manifest. Their choices are also apparent. Rather than contesting the military status quo in the Rhineland, Hitler could have decided to accept it—at least for the time being. In the contingency that he decided to unilaterally upset the status quo, however, French leaders would have to decide whether or not to resist. Finally, the historical record reveals that Hitler also considered what he would do if the French responded militarily. He had to consider this contingency since his initial choice would be made without certain knowledge of how the French would react. Thus, as is oftentimes the case in international affairs, the Rhineland crisis of 1936 was a game of *incomplete information*. Neither player knew for sure the other's preferences.

These choices and the general strategic environment that existed in March 1936 (and previously) are captured by the extensive form game given in Figure 2.1, which also summarizes, verbally and symbolically, the likely consequences (i.e., the outcomes) of the various choices, contingent or otherwise, available to the players.

The easy and natural correspondence between the simple game form depicted in Figure 2.1 and the strategic situation that existed in 1936 suggests a straightforward analysis

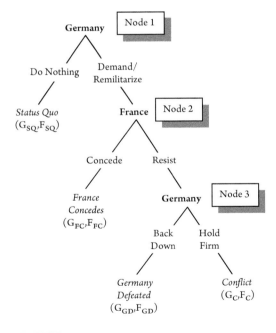

Key: (G_K, F_K) = payoff to Germany, France at outcome K

Figure 2.1 The Rhineland Crisis of 1936 (Unilateral Deterrence Game)

and explanation of the crisis. To be sure, the analysis is facilitated by the fact that the game form of Figure 2.1 is a generalized "off-the-shelf" game model now called the *Unilateral Deterrence Game* by Zagare and Kilgour (1993, 2000), who have analyzed it under both complete and incomplete information.[3] But before fully accepting this correspondence, the remaining elements of the game, that is, the players' preferences and their beliefs about each other's preferences, must first be spelled out. In the case of the Rhineland crisis, this critical modeling step is not difficult.

Consider first German preferences. Preference specification is oftentimes the most difficult step in modeling a real world interaction, for it is generally the case that the historical record is ambiguous, incomplete, or both. For example, Levy (1990/1991: 162) offers four different interpretations of German preferences during the July Crisis of 1914 (his own, and three that he associates with other historical schools), while Zagare (2011b) develops yet another. Additionally, it is easy to make false inferences about a player's preferences after observing their actual choices. The choices that players make do not necessarily reveal their true preferences (they may be acting strategically) or their complete preferences (the action choice may have limited informational content).

Nonetheless, in retrospect, German preferences in 1936 are relatively easy to specify. At the time of the crisis, Germany's armed forces were patently outgunned, which is why Hitler's generals initially opposed the remilitarization. So to gain their support, he promised that he would withdraw should the French resist (Snyder and Diesing, 1977: 230), which would only be the case if Hitler (i.e., Germany) preferred the outcome labeled *Germany Defeated* (with payoff G_{GD}) to the outcome labeled *Conflict* (with payoff G_C). Assuming that Germany was a revisionist power in 1936 (a safe assumption), it follows that, for Germany, *France Concedes* $>_G$ *Status Quo* $>_G$ *Germany Defeated* $>_G$ *Conflict* where "$>_G$" means "is preferred by Germany to."

On the basis of the testimony of the French ambassador to Germany, we know that, well before the crisis, the French government had decided that it would not risk a conflict by reacting militarily should German troops cross the Rhine (Cairns, 1965; Shirer, 1962: 402), that is, that it preferred to concede rather than risk a conflict with Germany.[4] The remaining French preferences are apparent, so, rather than belabor the obvious, I will simply stipulate that *Status Quo* $>_F$ *Germany Defeated* $>_F$ *France Concedes* $>_F$ *Conflict* where "$>_F$" means "is preferred by France to."

Notice that, in March 1936, neither Germany nor France preferred to act on their (deterrent) threats. If the French resisted, German decision makers preferred to back down, while French leaders were similarly intent on avoiding the costs of another all-out war. In other words, given their preferences, neither player's threat to resist was credible to execute (or rational to carry out). In their analysis of this game, Zagare and Kilgour (2000) call a

---

[3] The nomenclature of the game in Figure 2.1 has been modified in this chapter to reflect some of the particulars of the 1936 crisis. For example, in the generalized model, the two players are called "Challenger" and "Defender."

[4] It should be noted that, in the actual play of the game, the French revealed their preference for *France Concedes* over *Conflict* by their action choice. Thus, even in the absence of pre-crisis evidence, a key element of the French preference order can be established ex post. As mentioned previously and as will be discussed below, however, action choices do not necessarily establish preferences.

player with an irrational (or incredible) threat *Soft*. By contrast, a player with a credible threat is called *Hard*. Clearly, in 1936, both Germany and France were Soft.

Of course, both players knew their own type, that is, whether they were Hard or Soft. But neither knew the other's type (preferences). Had they had this information, the status quo would have held, deterrence would have succeeded, and there would have been no crisis. Given complete information about German preferences, the French would have rationally stood firm (at Node 2), forcing the Germans to (rationally) back down (at Node 3). Knowing this, Hitler would have (again, rationally) postponed the remilitarization in order to avoid humiliation and, perhaps, as he feared, removal from office.

All of which is to say that the crisis of 1936 was a game of incomplete information. As discussed in Chapter 1, in a game of incomplete information, the accepted standard of rational play in a static (or strategic form) game is a Bayesian equilibrium.[5] The natural extension of this concept to a dynamic (or extensive form) game is called a perfect Bayesian equilibrium.[6] A perfect Bayesian equilibrium specifies an action choice for every type (in this example, Hard and Soft) of every player at every decision node (or information set)[7] belonging to the player; it must also indicate how each player updates its beliefs about other players' types in the light of new information obtained as the game is played out.[8]

In any analytic narrative moored to a game-theoretic model, causal explanations must be developed in the context of the game's equilibrium structure. Riker (1990: 175) explains why: "equilibria are ... identified consequences of decisions that are necessary and sufficient to bring them about. An explanation is ... the assurance that an outcome must be the way it is because of antecedent conditions. This is precisely what an equilibrium provides."[9]

As it turns out there are five perfect Bayesian equilibria in the Unilateral Deterrence Game, but only one that is both plausible and consistent with the preferences and the beliefs of German and French decision makers in 1936.[10] Hence, an explanation is (almost) at

---

[5] Recall that, in a static (strategic form) game with complete information, the standard equilibrium form is a Nash equilibrium, while the accepted measure of rational strategic behavior in a dynamic (extensive form) game with complete information is a subgame perfect (Nash) equilibrium.

[6] As Gibbons (1992: xii) points out, "one could [even] say that the equilibrium concept of interest is always perfect Bayesian equilibrium, ... but that it is equivalent to Nash equilibrium in static games of complete information, equivalent to subgame-perfection in dynamic games of complete (and perfect) information, and equivalent to Bayesian Nash equilibrium in static games of incomplete information."

[7] An *information set* is a graphical device that is used to indicate a player's knowledge of his or her place on a game tree. In a game of perfect information, where every player knows what prior choices, if any, have been made, all information sets are *singletons*, that is, information sets that contains only one node of a game tree. In a game of imperfect information, some information sets may contain two or more nodes of a game tree, indicating that the player whose turn it is to make a move is unaware of some or all prior choices made by at least one other player.

[8] In an extensive form game of incomplete information, the initial (or a priori) beliefs of the players are taken as givens. The assumption is that the players update their beliefs rationally (i.e., according to Bayes's rule) given the actions it observes during the play of the game (see Morrow (1994: chs. 6–8) for the technical details and instructive examples). The definition of a perfect Bayesian equilibrium, however, places no restriction on the players' updated (or a posteriori) beliefs "off the equilibrium path," that is, on beliefs at nodes that are never reached under rational play. It is sometimes the case that a perfect Bayesian equilibrium is supported by a posteriori beliefs that are inconsistent with a player's a priori beliefs. Perfect Bayesian equilibria that are based on internally inconsistent beliefs are implausible. In consequence, they will not be considered as rational strategic possibilities in this and in subsequent chapters.

[9] See also Bates et al. (2000b: 700).

[10] Hitler was a risk taker who (correctly) anticipated that the French would not march. For their part, French policy makers, who were risk averse, believed that over 35,000 German troops (about three divisions) had crossed

hand. For reasons that will shortly become obvious, Zagare and Kilgour call this equilibrium *Attack*.[11] Under the Attack Equilibrium, a Challenger (i.e., Germany) demands an alteration of the *Status Quo* (at Node 1), regardless of its type, but a Soft Challenger (which Germany was in 1936) plans to back down (at Node 3) in the event that the Defender (i.e., France) resists at Node 2.[12] For their part, Hard Defenders always resist at Node 2, and Soft Defenders (like France in 1936) always concede.

The explanation that is derived from the association of a plausible perfect Bayesian equilibrium of the Unilateral Deterrence Game with both the beliefs and the action choices of the players during the Rhineland crisis is unexceptional and conforms with standard explanations of the event: in 1936, Hitler was a risk taker who was bluffing;[13] his gamble that the French would accept a rollback of the status quo paid off handsomely. But this unexceptional explanation should not obscure the point of the exercise: to illustrate in the simplest possible way how game models provide causal explanations. Game-theoretic models map out the behavioral implications of various combinations of player preferences and, in the case of a game of incomplete information, beliefs. These implications specify the action choices that define an equilibrium and which, given the rationality assumption, should be observed when the game is actually played out. An explanation is achieved whenever predicted behavior and observed behavior are the same.

2.3   Some Factors that May Complicate Explanation in Less than Easy Cases

In Section 2.2, an "off-the-shelf" game model was used to develop what turned out to be a standard explanation of the 1936 Rhineland crisis. The model itself, which should be thought of as an explicit causal mechanism, was key to the explanation since it provides the element of necessity missing in most historical narratives or atheoretical case studies.[14] The case analysis was straightforward, not only because there was a close and natural fit between the events that took place in 1936 and a preexisting model but also because the players' preferences were transparent, their beliefs about each other's preferences were clear and well documented, and the strategic implications of the model were both intuitive and easy to explicate. But the fact that the explanation was unexceptional may obscure the potential value that game models can bring to historical studies.[15] I discuss some these benefits in

---

the Rhine when, in fact, the Germans had sent only a "token force." In consequence, they saw no reason for Germany to give way (Shirer, 1962: 401–3).

[11]   In addition to the Attack Equilibrium, there are two distinct *Deterrence Equilibria*, a *Separating Equilibrium* and a *Bluff Equilibrium* (see Zagare and Kilgour (2000: ch. 5) for further details.)

[12]   In the Unilateral Deterrence Game, Challenger's Node 3 decision is strictly determined by its type. Hard Challengers always hold firm. Soft Challengers never do.

[13]   See also Slantchev (2011: 18).

[14]   Hindmoor (2006: 211) argues that rational choice theory's main purpose is to show "how the mechanism of rational action generates stable equilibrium in various political settings."

[15]   Nonetheless, it is reassuring that a game-theoretic analysis of the Rhineland crisis so economically conforms with standard interpretations. Were this not the case, the power of these models to generate compelling causal explanations of more complex cases would seriously be cast into doubt.

Section 2.4. Before doing so, however, I highlight a number of problems that may render a case less than easy.

The first, and perhaps most daunting, problem that must be confronted before using a game-theoretic model to organize a case study is locating a suitable model, that is, a model that not only captures the essential features of the case but is also rich enough to bring added value to the exercise. When no such model exists, there is no choice other than building one de novo. For example, to study the "deterrence-versus-restraint" dilemma that sometimes arises in extended deterrence relationships, Snyder (1997) proposed two related but theoretically isolated models that he called the *Alliance Game* and the *Adversary Game*. In the Alliance Game, a Defender either *Supports* (C) or *Withholds Support* (D) from its Protégé. In the Adversary Game, a Defender either *Stands Firm* (D) or *Conciliates* (C) the Challenger. In general, the choice of (C) in one game implies a choice of (D) in the other, and vice versa. The linkage between the choices in the two games makes it difficult for a Defender to choose optimally in both, that is, to simultaneously deter the Challenger and restrain its Protégé.

Clearly, Snyder views the deterrence-versus-restraint dilemma as a difficult cross-game maximization problem. But as Crawford (2003: 18) observes, his analysis of the dilemma is informal and his separation of two games unduly artificial. In consequence, his conceptual synthesis, while provocative and insightful, remains intuitive and needlessly imprecise.

To overcome these limitations, Zagare and Kilgour (2003) developed a unified model called the *Tripartite Crisis Game* (see Figure 2.2), which they then used to explain the "unlikely" alliance (Massie, 1991: 79) between Germany and Austria in 1879.[16] When they did, the temptation that they had to resist was not one of "data fitting" but rather one of "theory fitting," that is, constructing a model that too closely conformed to the particulars of the political conundrum that the German Chancellor, Otto von Bismarck, faced in 1879. As will be seen in Chapter 4, models that are constructed to reflect a particular set of facts are unlikely to reveal much about those facts. Obviously, the generality of such models will also be severely curtailed.

The Tripartite Crisis Game model, however, is a generic model of crisis bargaining and extended deterrence that can be used to generate theoretical knowledge about an important class of real world events. But while the model proved useful in constructing an explanation of both the timing and the terms of the Austro-German alliance in 1879, its application to another member of that class, the Austro-Serbian stage of the July Crisis of 1914, was less than straightforward (Zagare, 2009a), despite the fact that the two crises presented German decision makers with a strikingly similar strategic problem.

In 1879 Bismarck feared that the rivalry between Russia and Austria in the Balkans might draw Germany into a war it wished to avoid. So, to help stabilize the status quo, Bismarck offered Austria a defensive alliance that helped dampen the competition between Russia and Austria. Of course, in 1914, the outcome was different, and the Germans not so fortunate. The blank check that German decision makers presented to their Austrian counterparts after the assassination of Archduke Franz Ferdinand was an important step on the road to the war with Russia that Bismarck so assiduously sought to avoid.

---

[16] The strategic structure of this game and the preference and information assumptions that are used to analyze it will be discussed in detail in Chapter 3.

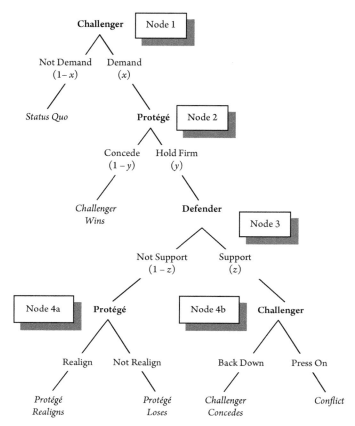

Figure 2.2 Tripartite Crisis Game

Given its ability to explain Bismarck's out-of-character behavior and the structural simi-larity of the strategic problem facing German decision makers in 1879 and 1914, one might well expect that an explanation of the divergent resolutions of the two crises could be gleaned from the Tripartite Crisis Game model. And indeed it can, but only after a few case-specific problems were solved.

The first obstacle involved the players. In 1879, the distinct roles and divergent strategic interests of Russia, Germany, and Austria corresponded closely to those of the players in the Tripartite Crisis Game—Challenger, Defender, and Protégé, respectively. But who made the first move in 1914—who was the Challenger? There is no compelling evidence to suggest that the Archduke's assassination was either initiated or sanctioned by either Serbia or Russia, the two usual suspects, nor are there any other empirically plausible candidate Challengers. Thus, a prima facie correspondence between the model and the case did not exist.

An inexact mapping between a model and a case is neither uncommon nor surprising. Whenever the problem arises, theoretical adjustment is the proper analytic response. As Niou and Ordeshook (1999: 93–94) point out, "to suppose that a formal model can wholly encompass a complex process…without resorting to some *ad hoc* discussion is ludicrous." In this instance, the only way to deal with the discrepant fit without taking undue license was to·treat the assassination as a random event (an act of nature) that simply precipitated the *proper subgame*[17] that commences with the Protégé's Node 2 choice, and to look to the equilibrium structure of the (Protégé–Defender) subgame for an explanation of the blank check. While this adjustment was clearly ad hoc, it is also entirely defensible within the empirical and theoretical confines of the case.

One problem that should be anticipated before analyzing any game's strategic dynamic is the coexistence of multiple nonequivalent and/or noninterchangeable equilibria, a problem that, incidentally, may arise in any theoretical framework, formal or informal, that assumes rational choice (Niou and Ordeshook, 1999: 89). When more than one rational strategic possibility exists, description is made more difficult; worse still, explanation and prediction may be frustrated.

Fortunately, none of the four plausible perfect Bayesian equilibria of the Protégé–Defender Subgame exist at the same time (see Table 2.1).[18] Better still, two of the equilibria, *Settlement* and *Bluff*, could be eliminated on empirical grounds,[19] which leaves only two perfect Bayesian equilibria, *Separating* and *Hold Firm*, as potential descriptors of the German and Austrian behavior during the first week of July in 1914.

Recall that a perfect Bayesian equilibrium must specify a rational choice for every type of every player at every decision point in a game. In the Tripartite Crisis Game model, there are two types of Defenders, Staunch and Perfidious, and two types of Protégés, Loyal and Disloyal. Notice from Table 2.1 that, under either the Separating or the Hold Firm perfect

Table 2.1 Action choices of perfect Bayesian equilibria for the Protégé–Defender Subgame with incomplete information.

Equilibrium	Strategic Variables			
	Protégé		Defender	
	y_D	y_L	z_S	z_P
Settlement	0	0	1	0
Separating	1	0	1	0
Hold Firm	1	1	1	0
Bluff	1	•	—	0

Key: y_D = probability that a Disloyal Protégé will choose to hold firm at Node 2
   y_L = probability that a Loyal Protégé will choose to hold firm at Node 2
   z_S = probability that a Staunch Defender will choose to support its Protégé at Node 3
   z_P = probability that a Perfidious Defender will choose to support its Protégé at Node 3
   "•" = fixed value between 0 and 1; "—" = value not fixed, although some restrictions apply
Source: Zagare and Kilgour, 2003

[17] A proper subgame is any part of an extensive form game that can be considered a game unto itself.
[18] The elements in this table will be explicated more fully in Chapter 3. For further details, see Zagare and Kilgour (2003).
[19] For a detailed explanation, see Zagare (2009a).

Bayesian equilibrium, Staunch Defenders always support Protégés at Node 2 with certainty (i.e., with probability $z_S = 1$), Perfidious Defenders never support Protégés (i.e., $z_P = 0$), and Disloyal Protégés (which German leaders presumed Austria to be) always hold firm at Node 2 (i.e., $y_D = 1$) and always realign at Node 4a should the Defender withhold support at Node 3. Thus, the two theoretically plausible, empirically viable, perfect Bayesian equilibria of the Protégé–Defender Subgame are distinguished only by the action choice of a Loyal Protégé, which Austria was not in 1914.

To put all this in a slightly different way, there are two rational strategic possibilities that are consistent with the behavior of Austria and Germany during the first stage of the July Crisis of 1914. And since they are distinguished only by behavior specified for a counterfactual contingency, there is no certain way to eliminate either one as a potential descriptor of a real world event. Whenever this occurs, one must accept the fact that an unambiguous conclusion is not possible. Nonetheless, this commonly encountered theoretical limitation is tempered by the fact that even when more than one plausible descriptor remains, others can generally be ruled out—logically, empirically, or both. In the end, additional causal insights into real world events or processes are obtained.

The final methodological obstacle to be discussed concerns preferences. As seen previously, German and French preferences were straightforward and relatively easy to ascertain in 1936. However, preference determination is potentially the most intractable and contentious element of any analytic narrative. To wit, historical assessments of German preferences during the run up to World War I run the gamut from the extremely sinister to the relatively benign. Copeland's (2000: 79) contention that, in 1914, German leaders preferred a world war (that included Great Britain) to all other possible resolutions of the July Crisis is the most extreme. Most other students of the Great War would reject Copeland's assertion, including Fischer (1967, 1975), whose controversial but now largely discredited argument that German policy makers deliberately provoked a continental war that they believed would not include the British is somewhat less baleful.[20] Finally, those who conclude that World War I was, in some sense, accidental or inadvertent rest their argument on the claim that decision makers in Berlin sought to avoid not only a world war, but a continental war as well.

That historians and political scientists should disagree about a player's preferences is hardly surprising. More often than not, the historical record is incomplete and, hence, frustratingly ambiguous. One way to attack the problem, however, is to examine secondary assumptions required to sustain an assessment. A case in point is Fischer's claim about German goals in 1914, as this rests on the further contention that German leaders considered British intervention very unlikely—a strong assertion that also lacks definitive empirical support. One should always be suspicious of strong claims that depend on a necessary yet unsubstantiated, and possibly convenient, subsidiary proposition. Interestingly, Copeland's even more extreme interpretation of German preferences, which, unlike Fischer's, does not depend on a specific German belief about British neutrality, is empirically more plausible. Copeland's (2000: 79) view is that "German leaders...saw the chance of Britain remaining neutral as very low."

---

[20] The issue of German preferences will be discussed more fully in Chapter 7.

It is easy to go astray when positing preferences. Drawing false inferences from observations of action choices is another common mistake. For example, Levy (1990/1991) has argued that, in 1914, Austrian leaders preferred a continental war to even a negotiated settlement of the crisis that satisfied almost all of their demands. Levy's conclusions are plausible. Throughout the crisis that proceeded the Great War, Austrian policy makers resisted every opportunity to settle the crisis diplomatically, even when they were strongly encouraged by the Germans to compromise. But did Vienna do everything it could to avoid mediation because it preferred a continental war, or was it because it simply did not believe that Russia would intervene and that it would have its way with Serbia (Jannen, 1996)?

None of this is to say that Levy's inferences are necessarily wrong; rather, it is to say that they cannot be established definitively. Whenever this occurs, the proper analytic response is to maintain some sense of theoretical contingency. Needless to say, conclusions that rely on contingent assumptions should always be evaluated differently than those that flow from preference assumptions that are less controvertible. Still, one of the advantages of using a formal model to organize an analytic narrative is that it permits contingent argument based on alternative preference assumptions or different model specifications. For example, in Chapter 6, I show that the argument that World War I was preventable depends critically on the plausible yet debatable premise that both German and Austrian leaders preferred the status quo to a negotiated settlement. Fischer, Levy, Copeland, and some proponents of the "inadvertent war" thesis argue otherwise. Since the premise remains unsettled, the conclusion that flows from it can only be considered provisional.

2.4 Why Use Game-Theoretic Models to Construct an Analytic Narrative?

There are many advantages to using a game-theoretic model to develop an analytic narrative. We have already seen that such models can reveal causal mechanisms that provide the element of necessity that most historical narratives lack, even if, as in the example of the Rhineland crisis, there may be no apparent need for one. Game-theoretic models also offer a rich environment for assessing both the logical consistency of an argument based on explicit assumptions and the possible reliance of these assumptions on hidden or secondary propositions. They encourage counterfactual or "off-the-equilibrium-path" reasoning and allow for contingent theorizing. And they highlight and reinforce an awareness of the interactive element of most conflict relationships. As King, Keohane, and Verba (1994: 45–6) remind us, any explanation of a complex event that "assumes the absence of strategic interaction and anticipated reactions" is likely to be deficient. For example, Fischer's (1967, 1975) explanation of the outbreak of the Great War has been criticized precisely because it focuses exclusively on German objectives. To be sure, Fischer's strong assumptions about Germany's preferences eliminate the necessity to take account of the policies of other states. But it is highly likely that it was Fischer's failure to take account of strategic interaction that led him to his more than questionable conclusion about the motivation of Germany's leadership group.

Another very good reason to examine specific events or processes within the confines of a well-articulated formal model is the additional organizing power that is gained by doing so. From the infinite variety of observations about an event or a process that might be made, a theory will single some out for special consideration. In the case of a game-theoretic model, these categories include, but are not limited to, the identification of the players, the choices they face, the set of possible outcomes, the players' preferences over the outcome set, the private information each player possesses, information that is common to all the players, and each player's beliefs about what is likely to take place as the game is played out. At the same time, the explicit use of a theoretical framework makes analysis more tractable by suggesting what information can or should safely be ignored.

Explorations of real world political phenomena are also rendered less ad hoc when they are theoretically informed. Theoretical frameworks work to severely limit not only the number but also the cast of variables that can be called upon to provide a coherent explanation. While this is true of all theories, it is especially true of formal theories, since formalization requires an explicit statement of assumptions and arguments. Formal theories, in other words, are more transparent. In consequence, they are at once subject to more intense scrutiny and less amenable to even unintended manipulation (Snidal, 2002: 80).

Explanations derived from deductive methodologies like game theory also have the added benefit of clarity. It is the relationship between the premises and the conclusions of game-theoretic models that provide an explanation of why something must be the case. As Kaplan (1964: 339) pointed out long ago: "the explanation shows that, on the basis of what we know, the something cannot be otherwise. Whatever provides this element of necessity serves as an explanation. The great power of the deductive model is the clear and simple way in which necessity is accounted for."

Theoretically based explanations of real world events are also more compelling than atheoretical accounts. When a seemingly unique event can be identified as an instance of a more general category that is part of some theory's empirical domain (Rosenau and Durfee, 2000: 3)—or, to use Hempel's (1965: 345) term, when its dynamic is placed under a covering law—a deeper understanding of an apparently singular event is achieved (Riker, 1990: 168).

2.5 Coda

It has been my purpose in this chapter to offer peace researchers and diplomatic historians an evenhanded assessment of both the benefits and the pitfalls of using game-theoretic models to organize an analytic narrative. Practical problems of application were highlighted, and some suggestions for overcoming some of them offered. To be sure, game-theoretic models are not panaceas. They may sometimes fail to provide compelling causal explanations of real world interactions, even when they are not misused. Intractable data problems or impenetrable strategic situations will always exist. Nonetheless, carefully calibrated game-theoretic models constitute powerful analytic devices that can be useful for uncovering the causal dynamic of complex, interactive strategic relationships.[21]

---

[21] For a similar argument, see Goemans and Spaniel (2016).

I hope to demonstrate this more convincingly in Part II, in Chapters 3 through 6, as I analyze the Moroccan crisis of 1905–6, the Cuban missile crisis of 1962, and the July Crisis of 1914. After that, in Part III, Chapter 7, I once again demonstrate the ability of a game-theoretic framework to both ensure logical consistency and uncover its absence, but this time in the context of a more pointed comparison of classical deterrence theory, the conventional wisdom of the field, and its theoretical competitor, perfect deterrence theory. Finally, in Chapter 8, I consider an argument that some behavioral economists have made that game-theoretic models are of little use for understanding interstate conflict behavior, and show that an explanation of the so-called long peace of the Cold War period can, in fact, be successfully explained within the confines of a rational choice framework.

PART II
Diplomatic History

3

The Moroccan Crisis of 1905–6

3.1 Introduction

The Moroccan crisis of 1905–6 was the first in a series of early twentieth-century great power confrontations that are generally considered the major causal incidents leading to World War I (e.g., see Taylor, 1954: 441). While it and two other intense interstate disputes, the Bosnian crisis of 1908 and the Agadir (or second Moroccan) crisis of 1911, were resolved short of war, the July Crisis of 1914 was not. In consequence, the first three crises are often-times thought of as dress rehearsals for the real thing, the clear implication being that the Great War was simply inevitable (Schroeder, 1972; Otte, 2014a: 87).[1]

Was World War I the result of a set of forces that, once set in motion, were destined to bring about the catastrophe of 1914, or were there fundamental structural conditions that set the July Crisis apart from the other three? Without an analytic framework in which to place these and similar crises and, therefore, to make meaningful cross-case comparisons, this question is almost impossible to answer (George and Bennett, 2005).

Historical examinations of the first Moroccan crisis abound, but they are generally long on description and short on theoretical context. And when they are not, their theoretical component is either opaque or entirely ad hoc. In this chapter, I hope to alleviate this problem by describing an incomplete information game model that potentially applies to all four crises and to show how it can be used to develop a logically consistent explanation, derived from an explicit set of preference and information assumptions, of the first Moroccan crisis. While the analytic narrative I construct is not necessarily at odds with the conclusions of some diplomatic historians, it is more powerful simply because it is explicit about the causal mechanism at work.

3.2 Background

While the first Moroccan crisis was ostensibly about economic and political control of a strategically important part of North Africa, it was actually about the evolving relationship

---

[1] This chapter is based on Zagare (2015b).

Game Theory, Diplomatic History and Security Studies. Frank C. Zagare. Oxford University Press (2019).
© Frank C. Zagare 2019. DOI: 10.1093/oso/9780198831587.001.0001

between Great Britain, France and, by extension, Russia that Germany saw as a long-term geopolitical threat. In 1890 the British and the French almost went to war over control of the upper Nile. But, shortly after the Fashoda crisis, they began to reconcile. In 1904 they signed a series of colonial agreements, called the Entente Cordiale, which brought their foreign policies into closer alignment. Their most significant understanding concerned Morocco. In exchange for recognizing British interests in Egypt, France was given a free hand in Morocco.

Initially, the Germans saw this arrangement as a positive. The German chancellor, Bernard von Bülow, believed that France's involvement in Morocco would both drain French resources and distract it from more volatile European issues. But when, in apparent violation of the Treaty of Madrid (1880), the French moved to assert political and economic dominance in Morocco without consulting Berlin, the German government took notice. A consensus quickly emerged at the Wilhelmstrasse that Germany was at risk of being encircled by hostile forces. Thereafter, German policy was directed at breaking the entente by putting it to the test (Schmitt, 1934: 68). In the end, however, it was Germany that had to stand down. At an international conference convened in Algeciras, Spain, at the insistence of the German government, the British backed the French to the hilt. Isolated diplomatically, Germany accepted minor concessions to end the matter. Worse still, the entente did not crumble, and the principal objective of the German government went unrealized.

Several key questions emerge from an examination of the historical record. First, why did Germany instigate a crisis in the first place? Second, why did the French eventually capitulate to the initial German demand but, when push came to shove, refused to back down on what were relatively minor issues? Third, why did the British risk war with Germany by unflinchingly supporting the French at Algeciras? And, finally, why did German policy fail so spectacularly— why did the Germans eventually capitulate? To answer these and related questions, I first attempt to place the crisis in a broader theoretical context.

3.3 Modeling the Crisis

Like the three interstate crises that followed it, the first Moroccan crisis was a case of a general extended deterrence failure.[2] Since almost all major power wars have arisen in the context of a breakdown of general extended deterrence (Danilovic, 2002), this fact alone justifies a re-examination of the Moroccan crisis via an appropriate theoretical framework that is prima facie applicable across crises. If Glenn Snyder was not the first to recognize this, he was at least the first to develop a transparent theoretical model to explore what he initially called the "deterrence vs. restraint dilemma" (Snyder and Diesing, 1977: 438) but which later he referred to as the "composite security dilemma" (Snyder, 1984, 1997). For Snyder, the dilemma at the heart of any extended deterrence relationship, general or immediate, is due to the conflicting pressures a defender faces when it simultaneously attempts to

---

[2] Both Huth (1988: 24) and Danilovic (2002: 66) also code it, correctly in my opinion, as instances of immediate extended deterrence success.

deter a challenger from attacking its ally (or protégé) and to deter its ally from acting irresponsibly and provoking a challenge it wishes to avoid.

As noted in Chapter 2, Snyder developed two simple game models to analyze the dilemma but treated them in theoretical isolation, leading to an impressionistic, albeit an insightful and provocative, understanding of their interactive dynamic. As well, his less than precise informal analysis necessarily failed to provide a systematic specification of the causally determinative contingencies. All of which is to say that Snyder's attempt to capture the strategic dynamic of the dilemma he so astutely describes remains somewhat muddled (Crawford, 2003: 18).

A different game model, called the *Tripartite Crisis Game*, which was briefly described in Chapter 2, offers a way to overcome some of the limitations of Snyder's analytic framework. This model, which is reproduced here for convenience as Figure 3.1, was designed specifically to integrate Snyder's two games into a single analytic framework and, therefore, to capture the mixed motives and contradictory impulses of extended deterrence relationships (Zagare and Kilgour, 2003). If Snyder is correct that the "deterrence vs. restraint dilemma"

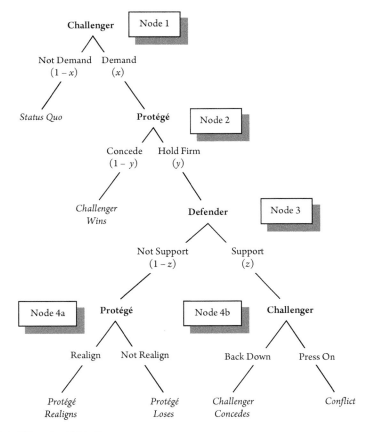

Figure 3.1 Tripartite Crisis Game

is central to most extended deterrence encounters, then the Tripartite Crisis Game model is an appropriate context in which not only to analyze the first Moroccan crisis but also to develop a general theory of interstate crisis initiation and resolution.

There are three players in the Tripartite Crisis Game: the Challenger, the Protégé, and the Defender. The roles they play and the decisions they face correspond closely to Germany's, France's, and Great Britain's, respectively, in 1905 and 1906. As Figure 3.1 shows, in this game form, the Challenger (i.e., Germany) begins play (at Node 1) by deciding whether to initiate a crisis. If it makes no demand of the Protégé, the game ends, and the outcome *Status Quo* is obtained. But if it initiates a crisis by demanding an alteration of the *Status Quo*, the Protégé (i.e., France) must choose (at Node 2) whether to Concede—in which case, the outcome is *Challenger Wins*—or Hold Firm, which brings about a difficult Node 3 decision for the Defender: whether to Support the Protégé or not.[3] Should the Defender (i.e., Great Britain) Support the Protégé, two outcomes, contingent on the Challenger's reaction at Node 4b, are possible: *Challenger Concedes* results when the Challenger decides to Back Down, but *Conflict* breaks out if and when the Challenger chooses to Press On. Two outcomes are also possible if the Defender does Not Support the Protégé. In either case, the Protégé is forced to capitulate to the Challenger's demand.[4] But, at Node 4a, the Protégé also reconsiders its relationship with a clearly unreliable Defender. If the Protégé severs its relationship with the Defender, the outcome is *Protégé Realigns*. But if it has few other options or if for some other reason it decides that its relationship with the Defender is worth maintaining, then the outcome is *Protégé Loses*.

3.4 Preference and Information Assumptions

A game is defined by both its rules and the preference and information assumptions that determine play. The game form of Figure 3.1 summarizes most of the elements that are generally considered to constitute the rules of the game. What remains to be specified, then, before a game is completely defined, is the set of assumptions about the players' preferences and what the players know about each other's preferences. Since different preference and information assumptions imply distinctly different games and, therefore, different strategic dynamics, great care must be taken in addressing this issue.

In general, when constructing an analytic narrative, there are two approaches that can be taken. Most historians rely on a technique that economists call *revealed preferences*. In short, this straightforward and intuitively appealing technique involves making an inference about an actor's preferences by observing its choice. For instance, if one is given a choice between

---

[3] Support, of course, can be signaled in many ways. For example, the Defender could mobilize its army or put its navy on alert in such a way that the Challenger would be forced to choose between war and peace; it could also make a visible and very public commitment to defend the Protégé; or it could offer diplomatic support at an international conference, as Britain (eventually) did at the Algeciras. For an innovative listing of these and related "commitment tactics," see Snyder (1972).

[4] The implicit assumption is that the Challenger is stronger than the Protégé so that a confrontation would be unfavorable to an unprotected Protégé, and that both the Protégé and the Challenger know this.

two alternatives, *a* and *b*, and *a* is chosen, then a reasonable conclusion would be that the actor prefers alternative *a* to alternative *b*.

There are many situations when such an inference is not inordinately problematic. One in particular is when an actor is faced with an either–or choice leading to the terminal outcome in a game. Nodes 4a and 4b in the game shown in Figure 3.1 are pertinent examples. At either of these nodes, a player makes a decision that effectively ends the game. Assuming rational play, the Protégé's choice at Node 4a, and the Challenger's choice at Node 4b, would not only depend solely on the actor's preferences, but once made, would most likely reveal them. It would only *most likely* reveal them, however, because even if a choice is observed, it remains possible that an actor who chooses one alternative over another is indifferent between them (Hausman, 2012: ch. 3).

At the other nodes of the tree, however, such inferences about preferences are even more difficult to justify. In addition to the possibility that a specific choice is a consequence of indifference and not strict preference, the observed choice may be part of a mixed strategy and, if so, does not necessarily reveal a preference relationship.[5] Additionally, and this consideration is especially pertinent when a choice is made mid game, a choice for one option over another may be the consequence of a strategic calculation. It is oftentimes the case that, strategically speaking, a rational actor will make a choice that runs counter to the one that is implied by a strict reading of a decision maker's preference function.

For all these reasons, I rely on a different procedure to specify preferences, at least initially: *posited preferences*. By positing or assuming preferences derived from a well-articulated theoretical point of view, one can avoid the potential inference problems associated with revealed preferences and, at the same time, avoid the temptation to fashion assumptions to fit a particular case (Morrow, 1997: 29).

Following Zagare and Kilgour (2003), then, I assume the specific preferences for the Challenger, the Protégé, and the Defender, as listed from most preferred to least preferred in the first, second, and third columns of Table 3.1, respectively. For example, the assumption is that the Challenger most prefers the outcome labeled *Challenger Wins*, then the outcome called *Protégé Realigns*, and so on.

Table 3.1 Preference assumptions for the Tripartite Crisis Game.

Challenger	Protégé	Defender
Challenger Wins	Status Quo	Status Quo
Protégé Realigns	Challenger Concedes	Challenger Concedes
Protégé Loses	Conflict	Challenger Wins
Status Quo	Challenger Wins	Protégé Loses
Conflict or Challenger Concedes	Protégé Loses or Protégé Realigns	Conflict or Protégé Realigns

[5] A *mixed strategy* should be distinguished from a *pure strategy*. A pure strategy involves the certain selection of a particular course of action. By contrast, a mixed strategy is composed of a probability distribution over the set of a player's pure strategies. As will be seen in Section 3.5, one perfect Bayesian equilibrium in the Protégé–Defender Subgame, the *Bluff Equilibrium*, involves a mixed strategy.

For the most part, the assumptions reflected in Table 3.1 are straightforward, easy to justify theoretically, and, as will be seen, easy to defend empirically.[6] In general, these assumptions take as a given the fact that the players prefer winning to losing and, because conflict is costly, that they also prefer to win, or if it comes to it, to lose, at the lowest level of conflict. For example, the Protégé is assumed to prefer *Challenger Wins*, which results when the Protégé Concedes immediately at Node 2, to either *Protégé Realigns* or *Protégé Loses*, each of which result when the Protégé capitulates under extreme duress at Node 4a.

Notice that the last cell of each column contains two outcomes. This reflects the fact that no specific assumption is made about the relative preference of these outcomes. Those preference assumptions that have been left open represent threats that the players may, or may not, prefer to execute. They also establish each player's type. For example, the Challenger could prefer *Conflict* to *Challenger Concedes*. Such a Challenger is called *Determined*. But the Challenger could also prefer *Challenger Concedes* to *Conflict*. A Challenger with these preferences is called *Hesitant*. Likewise, a Defender who prefers *Conflict* to *Protégé Realigns* is called *Staunch*. A *Perfidious* Defender has the opposite preference. And, finally, *Loyal* Protégés prefer *Protégé Loses* to *Protégé Realigns*. *Disloyal Protégés* do not. Table 3.2 summarizes the type designations of the players.

Notice also that the assumption is that the Defender prefers *Protégé Loses* to *Conflict*. This assumption reflects the presumption that the Defender is not heavily invested in the issues at stake. In other words, ceteris paribus, it would prefer that the Protégé conciliate the Challenger. For instance, in 1906, the British had no particular interest in how and by whom Morocco was controlled. And they surely would not have gone to war simply to ensure that the French controlled the country. For the British, then, what was at stake in the crisis was the entente, and, for that, they may well have taken on the Germans (Anderson, 1966: 371).[7] Significantly, the Germans did not see it this way.

Table 3.2 Player type designations.

Type/player	Preference
Determined Challenger	*Conflict* $>_{Ch}$ *Challenger Concedes*
Hesitant Challenger	*Challenger Concedes* $>_{Ch}$ *Conflict*
Loyal Protégé	*Protégé Loses* $>_{Pro}$ *Protégé Realigns*
Disloyal Protégé	*Protégé Realigns* $>_{Pro}$ *Protégé Loses*
Staunch Defender	*Conflict* $>_{Def}$ *Protégé Realigns*
Perfidious Defender	*Protégé Realigns* $>_{Def}$ *Conflict*

Key: "$>_{Ch}$" means "is preferred to" by the Challenger, "$>_{Pro}$" means "is preferred to" by the Protégé, and "$>_{Def}$" means "is preferred to" by the Defender.

---

[6] For a detailed discussion, see Zagare and Kilgour (2003: 591–4).

[7] On January 31, the British foreign minister, Sir Edward Grey, told the French ambassador, Paul Cambon, that Britain would not go to war over Morocco but would probably give assistance if Germany moved to disrupt the entente (Williamson, 1969: 81).

To be sure, there are situations where the Defender's interests are also at risk so that it, in fact, prefers *Conflict* to both *Protégé Loses* and *Protégé Realigns*. In 1914, for example, Russia clearly preferred to fight rather than allow Austria to dismember Serbia. In this instance and others like it, the Defender is not conflicted, and no dilemma can be said to exist. But, after 1905, it is clear that British foreign policy was conditioned by the conflicting objectives that constitute the deterrence vs. restraint dilemma (Kagan, 1995: 149). All of which suggests that the present assumption that the Defender prefers *Protégé Loses* to *Conflict* is justified not only theoretically but empirically as well.

In the informal analysis that follows, all relevant information is assumed to be *common knowledge*, except that the players may be uncertain about each other's type. Specifically, the players are assumed to be fully informed about the game defined by the rules of play, as reflected in the game tree of Figure 3.1, and the preference orderings, insofar as they are given in Table 3.1.

The Defender and the Protégé are also assumed to possess *private information* about their type, that is, they know their own type but have only probabilistic knowledge about each other's type. Specifically, the Defender is believed to be Staunch by both the Challenger and the Protégé with probability p_{Def}, and Perfidious with probability $(1 - p_{Def})$; and the Protégé is believed to be Loyal by both the Challenger and the Defender with probability p_{Pro}, and Disloyal with probability $(1 - p_{Pro})$.

The belief variables p_{Def} and p_{Pro} may be interpreted as measures of the Defender's and the Protégé's credibility, respectively. The greater p_{Def}, the more likely the Defender is to execute its deterrent threat, directed at the Challenger, to back the Protégé; the greater p_{Pro}, the more likely the Protégé is to execute its threat/promise, directed at the Defender/ Challenger to realign. Alternatively, the Defender's belief variable can be taken as measures of its reliability as an ally: the greater p_{Def}, the more reliable is the Defender, and conversely (Miller, 2012).

In using the Tripartite Crisis Game to analyze the Moroccan crisis, I initially assume that the Challenger is believed to be Determined, that is, that both the Defender and the Protégé believe that the Challenger prefers *Conflict* to *Challenger Concedes*. There are a number of good reasons for making this simplifying assumption. First, to assume that the Challenger's type is unknown to the other players would unduly complicate the analysis of the Tripartite Crisis Game. And to assume that the Challenger is known to be Hesitant would render the Tripartite Crisis Game theoretically trivial. To wit, regardless of the information structure of the game, the *Status Quo* is the only outcome that can be supported at any form of strategic equilibrium when the Challenger's preference for *Challenger Concedes* over *Conflict* is common knowledge.

Moreover, the simplifying assumption about the Challenger's type is consistent not only with Mercer's (1996) contention that adversaries are generally seen as resolute but also with the facts on the ground in 1905. Both the British and the French formulated policy during the crisis under the supposition that German decision makers favored war (Albertini, 1952: 2:168; Lebow, 1981: 310–11).[8] While this supposition eventually turned out to be incorrect,

[8] The French foreign minister, Théophile Delcassé, believed that the Germans were bluffing. But his view did not hold, and he was forced to resign midway through the crisis.

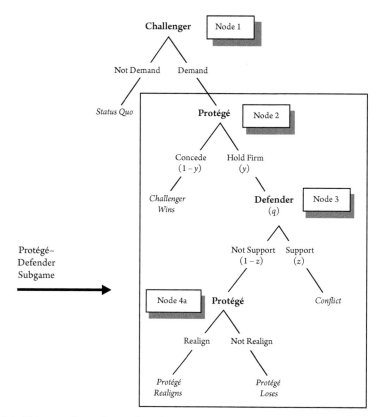

Figure 3.2 Tripartite Crisis Game when the Challenger is Determined

there was strong sentiment within the German government, especially among the military, for a war with France. The leading advocate for war was General Alfred von Schlieffen, the chief of the German general staff. But there were many others.[9]

---

[9] For reasons that are most likely obvious, the assumption that the Challenger is Determined will be relaxed below in order to explain actual German action choices. When push came to shove, the Germans capitulated. And when they did, they revealed their type. Clearly, Germany was Hesitant. On the other hand, the assumption that Germany was Determined conforms to the beliefs of both the French and the British governments when their policies were being formulated, and it is the behavior of the French, but especially of the British, that is problematic. Thus, as already noted, there are compelling empirical reasons for making this simplifying assumption right now. But there are pragmatic reasons as well. A formal analysis of the Tripartite Crisis Game given uncertainty about both the Challenger's and the Defender's preferences has yet to be developed. It has, however, been analyzed for two special cases. The first, which is summarized presently, assumes a Determined Challenger (Zagare and Kilgour, 2003). From the point of view of the British and French governments, this is the worst assumption that can be made about Germany's type. By assuming the worst, an all-too-easy explanation about British and French action choices is avoided. The second special case analysis of the Tripartite Crisis Game assumes that Defender is Staunch (Zagare and Kilgour, 2006). From the point of view of the German government, this is the worst that can be assumed about Britain's type. Nonetheless, German behavior is not difficult to explain even under these conditions. As will be discussed below, Hesitant Challengers may sometimes contest the status quo even when the Defender is known to be Staunch. The reason is straightforward: because they are caught between a rock and a

The assumption that the Challenger strictly prefers *Conflict* to *Challenger Concedes* implies that the outcome of the Tripartite Crisis Game will always be *Conflict*, should the play of the game reach Node 4b. Whenever this is the case, the game reduces to the game form given in Figure 3.2. Highlighted in this figure is one of the Tripartite Crisis Game's proper subgames. Zagare and Kilgour (2003) call this subgame the *Protégé–Defender Subgame*.[10] As shall be seen, solving the Protégé–Defender Subgame is key to understanding French and British behavior at Algeciras.

3.5 Behavioral Patterns

As described in Section 3.4, the Protégé–Defender Subgame is a game with incomplete information. In such a game, at least one of the players does not know for certain another player's type. The standard measure of rational play in a dynamic (i.e., extensive form) game of incomplete information is a perfect Bayesian equilibrium. Since a perfect Bayesian equilibrium specifies an action choice for every type of every player at every decision node or information set belonging to the player, it must specify the action choice of both a Loyal and Disloyal Protégé at Nodes 2 and 4a, and for both types of Defender at Node 3.

A perfect Bayesian equilibrium must also indicate how each player updates its beliefs rationally (i.e., according to Bayes's rule) about the other players' types in the light of new information obtained as the game is played out. For instance, should the Protégé Hold Firm at Node 2, the Defender will have an opportunity to reevaluate its initial beliefs about the Protégé's type before it makes a choice at Node 3. The assumption is that the Defender will rationally reassess its beliefs about the Protégé's type and, therefore, the Protégé's likely response at Node 4a, based on that observation. The information that the Defender obtains as a result of its observation of the Protégé's Node 2 choice will be useful. Much the same could be said of the Protégé's Node 4a choice, except that, in this case, any information that the Protégé can infer from the Defender's choice at Node 3 would be beside the point. Because it will end the game, the Protégé's Node 4a choice is strictly determined by its type (preferences).

Given these considerations, it follows that a perfect Bayesian equilibrium of the Protégé–Defender Subgame will consist of a five-tuple of probabilities, (y_D, y_L, z_S, z_P, q), where

- y_D is the probability that a *Disloyal* Protégé will choose to Hold Firm at Node 2
- y_L is the probability that a *Loyal* Protégé will choose to Hold Firm at Node 2
- z_S is the probability that a *Staunch* Defender will choose to Support its Protégé at Node 3

hard place, Staunch Defenders generally provide only halfhearted support for their Protégés. British foreign policy in 1914 is an instructive case in point (Steiner, 1977: 238). The tendency of a Hesitant Challenger to foment a crisis can only become more pronounced as this assumption about Defender's type is relaxed and the probability of a Staunch Defender withholding support increases. In consequence, an explanation of German action choices will be both easy and uncomplicated. And, significantly, it will also be consistent with the a priori belief at the Wilhelmstrasse that in the end the British would fold.

[10] Recall that a *subgame* is that part of an extensive form game that can be considered a game unto itself.

- z_P is the probability that a *Perfidious Defender* will choose to Support its Protégé at Node 3
- q is the Defender's updated probability that the Protégé is Disloyal, given that the Protégé Holds Firm at Node 2.

The first four probabilities are strategic variables describing the Protégé's and the Defender's choices, contingent on their type. The fifth probability is the a posteriori probability, updated by the Defender once the Protégé's choice to Hold Firm at Node 2 has been observed, that the Protégé will realign at Node 4a.

As Table 3.3 shows, there are four non-transitional perfect Bayesian equilibria in the Protégé–Defender Subgame.[11] Under the *Settlement* perfect Bayesian equilibrium, the Protégé always Concedes at Node 2 (i.e., y_D and $y_L = 0$), and the outcome is always *Challenger Wins*. A *Separating* perfect Bayesian equilibrium separates the players by type: Disloyal Protégés always Hold Firm (i.e., $y_D = 1$), and Loyal Protégés always Concede (i.e., $y_L = 0$); similarly, Staunch Defenders always Support Protégés (i.e., $z_S = 1$) while Perfidious Defenders never do (i.e., $z_P = 0$). Under the *Hold Firm* perfect Bayesian equilibrium, both types of Protégés Hold Firm at Node 2 (i.e., y_D and $y_L = 1$), but only Staunch Defenders Support the Protégé at Node 3 (i.e., $z_S = 1$ and $z_P = 0$). Finally, under the *Bluff* perfect Bayesian equilibrium, Disloyal Protégés always Hold Firm at Node 2 (i.e., $y_D = 1$), while Loyal Protégés sometimes do the same (i.e., $0 < y_L < 1$); similarly, only Staunch Defenders Support Protégés at Node 3, and only sometimes (i.e., $0 < z_S < 1$; $z_P = 0$).[12]

Table 3.3 Plausible perfect Bayesian equilibria and existence conditions of the Protégé–Defender Subgame with incomplete information.

Equilibrium	Strategic and Belief Variables					Existence Conditions*
	Protégé		Defender			
	y_D	y_L	z_S	z_P	q	
Settlement	0	0	1	0	$> d_1$	$p_{Def} < e_2$
Separating	1	0	1	0	1	$e_2 < p_{Def} < e_1$
Hold Firm	1	1	1	0	p_{Pro}	$p_{Def} > e_1, p_{Pro} > d_1$
Bluff	1	●	—	0	d_1	$p_{Def} > e_1, p_{Pro} < d_1$

* See Figure 3.3 for a graphical interpretation of these conditions.
Key: y_D = probability that a Disloyal Protégé will choose to Hold Firm at Node 2
  y_L = probability that a Loyal Protégé will choose to Hold Firm at Node 2
  z_S = probability that a Staunch Defender will choose to Support its Protégé at Node 3
  z_P = probability that a Perfidious Defender will choose to Support its Protégé at Node 3
  q = Defender's updated belief that its Protégé is Disloyal, given that the Protégé Holds Firm at Node 2
  "●" = fixed value between 0 and 1; "—" = value not fixed, although some restrictions apply

[11] One *transitional equilibrium* is ignored. As explained in Chapter 1, an equilibrium is transitional if it exists only when the parameters of a model satisfy a specific functional relationship. It is very unlikely that the parameter values of any model will satisfy any specific equation.
[12] Notice that only Loyal Protégés and Staunch Defenders have a mixed strategy in equilibrium. All other player types employ a pure strategy, that is, a strategy choice that involves the certain choice of a particular course of action.

Notice that, under the Separating Equilibrium, the Defender's updated belief that the Protégé is Disloyal, given that the Protégé Holds Firm at Node 2 (q), equals 1. The reason is straightforward. Under this equilibrium form, only Disloyal Protégés Hold Firm. The Protégé's Node 2 choice, then, reveals its type. By contrast, under the Hold Firm Equilibrium, the Defender's updated belief that the Protégé is Disloyal, given that the Protégé Holds Firm at Node 2 equals its initial belief (p_{Pro}). Since both Loyal and Disloyal Protégés Hold Firm at Node 2 when this equilibrium is in play, Protégé's actual choice has no information content. A perfect Bayesian equilibrium in which all types of the same player play the same strategy is called a *pooling* equilibrium.‾

A Defender's equilibrium behavior under a Bluff Equilibrium provides insight into an empirical puzzle raised by Fearon (1997). In Fearon's game model of costly foreign policy signaling, a player has two strategies to communicate its interests: it can signal that its "hands are tied" or that its "costs are sunk." Players never rationally bluff with either signal, leading Fearon (1997: 71) to wonder "why we sometimes observe halfhearted signals when convincing ones are possible?"

But this behavior arises naturally in the Tripartite Crisis Game. Under the Bluff Equilibrium in the Protégé–Defender Subgame, a Staunch Defender's strategy corresponds to a signal that is strong enough to deter all but the most Determined Challengers, yet not so strong that Loyal Protégés becomes intransigent and provoke crises. The Defender's rational objective is balance: too strong a commitment enflames the Protégé, whereas too weak a commitment incites the Challenger.

The signal is fuzzy, then, because it has two different audiences. By deterring the Challenger, the signal minimizes the risk of conflict and helps to stabilize the status quo; by restraining the Protégé, it reduces the risk of chain ganging (Christensen and Snyder, 1990) and protects the Defender's alignment relationship with the Protégé. What is surprising about this mixed message is that it is delivered by a Staunch Defender, one that would prefer to fight to save its alliance.

The "intentionally vague commitment" made by the United States in the Taiwan Relations Act of 1979 is a good example (Erlanger, 1996). To restrain China, the United States signaled its intention to back Taiwan. But to restrain Taiwan, the United States signaled that its support was not unconditional. In sum, the signal was halfhearted.

Each of the four perfect Bayesian equilibria exists under unique parameter conditions, that is, they do not coexist. As Figure 3.3 shows, the Settlement perfect Bayesian equilibrium uniquely exists when the Defender's credibility (p_{Def}) is low; the Bluff Equilibrium determines play when the Protégé's credibility (p_{Pro}) is low and the Defender's relatively high; the Hold Firm perfect Bayesian equilibrium exists only when the credibility of both the Defender's and the Protégé's threats are relatively high; and the Separating perfect Bayesian equilibrium occurs at intermediate levels of the Defender's credibility.[13]

The set of perfect Bayesian equilibria concisely summarize the range of rational strategic possibilities of the Protégé–Defender Subgame. In other words, they constitute the empirical

---

[13] Along the horizontal and vertical axes of this figure are graphed, respectively, the belief variables p_{Def} and p_{Pro}. The constants d_1, e_1, and e_2 are also indicated along the axes of Figure 3.3. These constants, whose technical characteristics will be ignored here, are convenient thresholds for categorizing the perfect Bayesian equilibria of the Protégé–Defender Subgame. For a detailed discussion, see Zagare and Kilgour (2003).

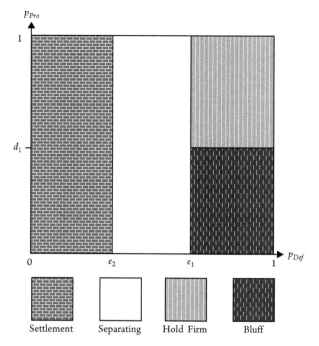

Figure 3.3 Location of perfect Bayesian equilibria in the Protégé–Defender Subgame

expectations (i.e., the predictions) of the model in the subgame.[14] When the behavior of the real world counterparts of the players in the model is found to be consistent not only with the action choices associated with one of the equilibria but also with the set of conditions that give rise to it, an explanation is derived. In Section 3.6, I use the theoretical characteristics of the perfect Bayesian equilibria of the Protégé–Defender Subgame to explain the underlying strategic dynamic of the first Moroccan crisis.

3.6 The Crisis

The first Moroccan crisis is usually understood to have begun on March 31, 1905. It was on this date that the German Emperor, Kaiser Wilhelm II, landed in Tangier and gave an address that made it clear that Germany was not ready to accept a dominant French role in Morocco. Shortly thereafter, the German government demanded the resignation of the French foreign minister, Théophile Delcassé, whose clumsy diplomacy brought about the crisis in the first place. Although Delcassé had reached an understanding with Great Britain, Spain, and Italy, he failed to consult Germany before he moved to consolidate French

---

[14] It should be pointed out that these are not always point predictions. Under the Bluff Equilibrium, both Loyal Protégés and Staunch Defenders employ mixed strategies. Thus, when this perfect Bayesian equilibrium is in play, the model's predictions are probabilistic (Bueno de Mesquita, 2002: 73; McGrayne, 2011: 250).

control of Morocco in January 1905. In consequence, Berlin threatened war unless German interests were recognized and due compensation received.

Since Delcassé believed that the Germans were bluffing—that their threat lacked credibility—he refused to budge. However, the French prime minister, Maurice Rouvier, who took the German threat at face value, wanted to avoid a land war in which France would be at a distinct disadvantage. For that reason, he recommended a policy of accommodation. When forced to choose, the French cabinet unanimously supported the prime minister. Rouvier took over the foreign ministry when Delcassé left office on June 5.

Delcassé's resignation was widely seen as a stunning German foreign policy success. Indeed, when Wilhelm heard the news, he immediately elevated Bülow to the rank of Prince of the German Empire. But Bülow, and others at the Wilhelmstrasse, were not satisfied. After all, German decision makers had little intrinsic interest either in Morocco or in a simple personnel change at the Quai d'Orsay. In the short term, their goal was to break the entente and thereby isolate France (Röhl, 2014: 347). Their longer-term objective was to induce both the Russians and the French into a continental alliance that would dominate Europe and stand in opposition to Great Britain. For this reason, the Germans pressed for an international conference, not only to gain additional concessions in Morocco, but also to expose what German leaders believed to be Great Britain's unreliability as an ally.[15] Without strong British support at a multilateral conference, the German expectation was that the entente would crumble and the French would have little choice but to realign, tilting toward Germany.

Delcassé's resignation marked the end of the first phase of the crisis (Snyder, 1971b: 26; Snyder and Diesing, 1977). In the second phase, it was Rouvier who refused to budge—much to the surprise of German decision makers. Rouvier's insistence on bilateral negotiations placed the Germans in an awkward position. If they accepted Rouvier's offer to negotiate, they might gain minor concessions in Morocco, at best. But their long-term policy goal would remain unfulfilled. As well, backing off from a public demand would result in a loss of face and a clear diplomatic defeat.

Forced to choose, Bülow decided to split the difference. To induce the French to negotiate multilaterally, he made a major concession. His ploy was successful. The French agreed to a conference once their "special interest" in Morocco was recognized by the Wilhelmstrasse (Anderson, 1966: 255–6).

The third and final phase of the crisis began on January 16, 1906, when the conference convened in Algeciras. Attending were representatives of all the major European powers, Spain, Italy, the United States, and a handful of minor powers. In the interim, a new government was formed in Great Britain.

Having already given way on the issue of French predominance in Morocco, there was little left to be gained by the Germans on that score. Nonetheless, the French government, backed by the British, offered only minor concessions on control of the police force.[16] Isolated diplomatically, the Germans accepted (and capitulated). The result was a stunning German

---

[15] To explain this German action choice, one must also assume that the French offer was so low that *Challenger Wins* was not Germany's most preferred outcome. The assumption, however, does not materially change the analysis that follows.

[16] Rouvier's government fell on March 7, and a new government was formed a week later. There was, however, no discernable policy shift or change in negotiating positions when the government changed hands (Anderson, 1966: 385).

foreign policy defeat. The British–French entente was not weakened, as the Wilhelmstrasse had hoped. By putting it to the test, Germany's demand for a conference had, in fact, strengthened it.

How can the dramatic reversal between the first and the final phase of the crisis be explained? Why was Germany able to successfully bully the French and force Delcassé's resignation, only to be upended at Algeciras? Why was British support of France lukewarm initially but steadfast when push came to shove? To answer these questions, I next refer to the changing political environment that characterized the first and last phase of the crisis, the evolving policy goals and beliefs that are associated with them, and the equilibrium structure of the Protégé–Defender Subgame that explains the action choices of the players and their consequences.

3.6.1 Diplomatic Foreplay: From the Visit to Tangier to the Fall of Delcassé

Although the underlying French and German foreign policy initiatives that would eventually lead to a confrontation were in the works for some time, the first Moroccan crisis became manifest only after Kaiser Wilhelm's landing in Tangier. In remarks that he made that day, Wilhelm proclaimed Germany's support for an independent Morocco and an open door policy. Subsequently, the German government called for an international conference to settle the status of Morocco and strongly suggested that Delcassé be replaced. These demands, accompanied by the explicit threat of war should they not be satisfied, presented a clear challenge to the French government of Maurice Rouvier.

The reaction of key officials in the French government is best explained in terms of their beliefs about the extent to which they expected British support. At the moment the crisis erupted, the Unionists controlled the British government. Arthur Balfour was the prime minister, but British policy abroad was formulated in the foreign office, where the foreign secretary, Lord Henry Lansdowne, operated with the full support of Balfour and the rest of the cabinet. Indeed, it was Lansdowne who engineered the entente agreement with France in 1904.

It is clear that Lansdowne valued the entente and that he was more than willing to cede control of Morocco to the French in return for a free hand in Egypt. But it is also clear that neither was his highest priority. In fact, the 1904 agreement was highly contingent. Among the contingencies, the most pertinent was Article VII, which stipulated that "the two governments agree not to permit the erection of any fortifications or strategic works on...the coast of Morocco."[17] Since areas of the coast that were controlled by Spain were specifically excluded by Article VII, and the French had already reached an understanding with Italy on Morocco, the only power that the stipulation applied to was Germany.

All of which is to say that Lansdowne placed a higher value on maintaining control of the Straits of Gibraltar than he did on the entente itself (Williamson, 1969: 35). Thus, his policy shifted as he saw the stakes shifting. For example, on April 25, he told Delcassé that the British government would stand with France should the Germans demand a port on the

---

[17] The full text of the agreement, formally titled the "Declaration between the United Kingdom and France Respecting Egypt and Morocco," is available at the World War I Document Archive (http://www.lib.byu.edu/~rdh/wwi/). It is also reprinted in Miller (2012: Appendix C).

Moroccan coast. A few weeks later, he suggested that the two governments meet to discuss other contingencies, but he would not make an explicit commitment to fight on France's behalf. Nonetheless, it is easy to understand why Delcassé might have "believed he was on the verge of an Anglo-French military alliance" (Massie, 1991: 361). Lansdowne, however, would later deny that any such alliance was in the works.

Lansdowne's commitment, in other words, was less than certain, and the message he conveyed (and the strategy he adopted) was mixed. Thus, it is also easy to understand why others in the French government and the German foreign office saw things differently. As discussed in Section 3.6.2, Rouvier discounted the British commitment (i.e., for him p_{Def} was low) and even feared that Lansdowne was maneuvering to instigate a war between France and Germany. The British foreign secretary, however, was trying to do just the opposite. His mixed strategy was itself a reaction to the "deterrence vs. restraint dilemma." By making a probabilistic commitment, he hoped to simultaneously restrain the French and deter the Germans, his supposition being that the game would be played out under the conditions that support the existence of the Bluff Equilibrium. Only the most highly motivated Challengers are not deterred whenever a Bluff Equilibrium is in play (Zagare and Kilgour, 2003: 603).

Just as British interests and beliefs determined their policy, the French interpretation of the extent to which the British were committed to their position would determine policy at the Quai d'Orsay. Seeing the glass as almost, if not in fact, full, Delcassé held firm and refused to compromise. His firm belief was that Germany was bluffing; but even if it was not, he was also convinced that the British had his back (i.e., that p_{Def} was very high). In contrast to Lansdowne, Delcassé saw the game playing out in either the uppermost area of the Bluff Equilibrium, where a Staunch Defender's support is very likely, or in the region of the Hold Firm Equilibrium, where a Staunch Defender's support is, in fact, certain. This was the reason for his policy.

As noted, Rouvier saw the glass as almost empty (Anderson, 1966: 229). Not only did he believe that the Germans were prepared to wage war but, as mentioned, he also discounted the sincerity of the British commitment (Williamson, 1969: 33, 36–7). Bülow had reminded the French ambassador in Berlin that even if the British entered a land war in support of France—a big if for Rouvier—France would still suffer enormously. Rouvier obviously agreed. "The British Navy does not run on wheels," he is said to have remarked (Massie, 1991: 362).

Fearing war, Rouvier forced the issue at a cabinet meeting on June 6. Without a single supporter at the meeting, Delcassé was compelled to resign. Rouvier believed that Delcassé's resignation would defuse the crisis, the game would end, and the Germans would be satisfied with the outcome *Challenger Wins*.

Of course, this was not the case. But why was Rouvier mistaken—why did the Germans press the issue in the first place and then why didn't they take "yes" for an answer and accept what was a clear diplomatic victory? Of all the questions that that might be addressed about the crisis, this is the easiest to answer. Given their strong belief that the British were Disloyal and that the German threat was highly credible, at least to Rouvier, Bülow believed that the British would find it difficult not to support Germany's call for an open door policy in Morocco and, therefore, would fail to fully back the French. The lack of British support, he believed, would mean the end of the entente and what he perceived to be Germany's grow-ing encirclement. Not only does this conjunction of beliefs easily explain the initial thrust of German foreign policy, but the endurance of this belief set explains why Bülow and others at the Wilhelmstrasse were not mollified when Delcassé resigned—why the German gov-ernment continued to demand an international conference to settle the status of Morocco.

3.6.2 Endgame: The Conference at Algeciras

The conference convened in Algeciras on January 16, 1906. There were two areas of conten-tion. The most salient was control of the police force and, by implication, of Morocco itself. The other involved the capitalization and control of a state bank.

Before the conference convened, the British government fell. Henry Campbell-Bannerman replaced Arthur Balfour as prime minister, and Sir Edward Grey took over for Lansdowne at the foreign office. Grey was a Liberal imperialist who, like Lansdowne, took a hard line against Germany. But, unlike Lansdowne, he was also more fully committed to preserving the entente (Williamson, 1969: 81; Mercer, 1996: 99). As will be seen, Grey's strong commitment to France would largely determine the play of the endgame once the obvious disconnect between preferences and beliefs among the players (see Section 3.6.1) was clarified by their action choices and negotiating positions at the conference.

After Delcassé's resignation, the British government came to doubt France's reliability. Afraid that the French might realign with the Germans and the Russians (Anderson, 1966: 371; Williamson, 1969: 37; Mercer, 1996: 102–5), Lansdowne moved to shore up the alliance by sending the British Atlantic fleet to Brest. Nonetheless, by both custom and long-standing policy, no firm commitment was made.

When Grey first assumed office, he did not change a thing (Anderson, 1966: 326–47). To simultaneously deter Germany and restrain France, he pursued a policy that has been variously characterized as "ambiguous," "unclear," "halfhearted," "semi-detached," and "tor-tuous." For example, when on January 10 he was asked by the French ambassador, Paul Cambon, what the British government would do should Germany declare war on France, the most that Grey would commit to was "a benevolent neutrality" (Grey, 1925: vol. 1, 71). But once the conference was underway, the British backed the French to the hilt on both the composition of the police force and the control of the bank and organized a coalition that blocked every German attempt to water down the specifics of the French position. In other words, Grey made good on his promise to Cambon at the end of January to "unreservedly" provide diplomatic support at Algeciras (Grey, 1925: vol. 1, 77).

For their part, the French refused to compromise at all. In the end, German preferences were almost certainly revealed when they accepted a negligible offer that gave the Swiss a minor role in the police force, rather than terminate the conference and further inflame the fabric of European affairs. Although the fig leaf they accepted allowed the Germans to claim victory, it was apparent to all that the game's outcome was *Challenger Concedes*.

So why did the British back the French at Algeciras, why did the French hold firm, and why did the Germans push for a conference only to be humiliated at it when they were even-tually forced to give way? British support might be explained by pointing to Article IX of the entente declaration, which stipulated that the two powers provide diplomatic support to one another on their agreement about the status of Morocco and Egypt. Of course, great powers do not consistently honor their treaty obligations, so this explanation, which is not totally without merit, is less than fully satisfying.[18]

A somewhat more compelling reason is that the British were forced to reveal their type. When votes are taken, hard choices have to be made. As Grey (1925: vol. 1, 100) himself put it, "the performance of our obligation to give diplomatic support to France was not hypothetical but actual." But the fact that the British hand was forced also does not fully explain their choice. For that, the model has to be consulted.

Recall that the British feared that the entente would crumble if their support was lukewarm at best. This was a view that was widely shared in the British foreign office, and a view that could only be reinforced by the hard line that Rouvier took shortly after Delcassé resigned. Under all but the Settlement Equilibrium, Disloyal Protégés Hold Firm. Thus, the belief of Grey and others that the French were Disloyal, that p_{Pro} was large, was entirely plausible. Nonetheless, in the Protégé–Defender Subgame, the credibility of the Protégé's threat/promise is not fully determinative of the Defender's choice. The constant (d_1) that separates the area of the Bluff Equilibrium, where the Defender's support is less than certain, from the area of the Hold Firm Equilibrium, where Staunch Defenders do just that, also depends on a Staunch Defender's utility for the outcomes *Protégé Loses*, *Protégé Realigns*, and *Conflict*. Ceteris paribus, the area occupied by the Bluff Equilibrium recedes relative to the area where the Hold Firm Equilibrium exists as Defender's utility for the outcome *Protégé Loses* increases, as its utility for the outcome *Protégé Realigns* decreases, and as its utility for *Conflict* increases (Zagare and Kilgour, 2003). As the crisis unfolded, all three of these utility shifts were trending in in a way that made the Defender's support more likely.

Take, for instance, the value of the outcome *Protégé Loses*. When Lansdowne headed the foreign office, strategic considerations in the Mediterranean held sway. Lansdowne was determined to prevent the Germans from obtaining a Moroccan port and since he feared that the Germans might demand one as compensation for acknowledging French predominance in Morocco, he placed a relatively high value on the loss associated with this outcome. Grey, by contrast, saw things differently. In fact, Grey (1925: vol. 1, 115) even played with the idea of offering the Germans a port or coaling station to settle the matter (Hamilton, 1977: 116). On the other hand, since Grey placed a much higher value on the entente itself, he saw the utility loss associated with the outcome *Protégé Realigns* to be larger than did Lansdowne. Finally, at the conference, the real world event associated with the outcome *Conflict* was the *risk* of war and not the certainty of war. This is a risk that Grey (1925: vol. 1, 109) was willing to take (Anderson, 1966: 370), especially since "it was diplomatic support only that was in question" at the conference. In other words, the immediate risk of war was tolerable.

Understanding the motivation underlying British behavior makes it easy to explain French action choices. Under the Hold Firm Equilibrium, the Protégé Holds Firm with certainty—regardless of type. In other words, even if the French were prepared to walk away from the entente, as the British at times feared they were, the certainty of British support dictated,

---

[18] Sabrosky (1980) finds that alliance partners honor a commitment to fight about 25 percent of the time. But Leeds, Long, and McLaughlin Mitchell (2000: 866) observe that alliance reliability rises to almost 75 percent when "the specific obligations included in alliance obligations" are taken into account.

rationally, that they too risk a more intense conflict. In fact, when, in late March, rumors surfaced in both Paris and London that the British government was less than resolute, Grey (1925: vol. 1, 102) wrote what he termed "an indignant telegram" to the British ambassador to France, Sir Francis Bertie, reaffirming British support in no uncertain terms. Once reassured, the French even refused to accept what amounted to a face-saving compromise offered by Austria on control of the police in Casablanca. When the issue was put to a vote, only Austria and Germany backed the Austrian plan. Isolated diplomatically, Germany capitulated.

It is important to note that British and French action choices at the conference were dynamically interactive. Not only was the French decision to resist the Austrian compromise entirely dependent on the extent to which it expected British support, but British support was also dependent on its expectations of French behavior. As Bertie told the French minister of war, Eugène Entienne, on March 14, "*if* the French Government were resolved not to accept the Austrian proposal about Casablanca His Majesty's Government would continue to support French views in the Conference as heretofore" (Grey, 1925: vol. 1, 104, emphasis added).

Similarly, German behavior at the conference was a consequence of its expectations about both British and French policy choices. As already mentioned, Bülow believed that the British would find it impossible not to support its call for an open door policy in Morocco and, therefore, would offer less than firm support of the French at Algeciras (Gooch, 1936: 247). In other words, his perception was that p_{Def} was low, that the British were likely Perfidious and that their failure to back the French at the conference would break the entente. But when his beliefs were put to the test, they proved to be incorrect. German preferences were most likely revealed when they were finally forced to make a hard choice at the conference, that is, Germany was almost certainly Hesitant.[19]

3.7 Coda

The alignment pattern that would define the European state system until 1914 came into sharp focus in the aftermath of the conference at Algeciras (Gooch, 1936). It is for this reason that, of the three prewar crises, the first Moroccan crisis is probably the most determinative. This chapter interprets the crisis of 1905–6 in the context of an incomplete information game model, the Tripartite Crisis Game, and one of its proper subgames, the Defender–Protégé Subgame. In the early stages of the crisis, the action choices of the players were shown to be consistent with the players' beliefs, but their beliefs were not tested. In the final phase, beliefs and action choices were brought into harmony. Throughout the crisis, the Germans contested the status quo because they believed that the British would fold. In the end, they gave way when it was more than apparent that their belief was incorrect.

For most of the crisis, British support of the French position was mixed. Indeed, even as the conference was about to convene, Whitehall refused to fully commit to aid the French should a war take place. Still, when push came to shove, British diplomatic support

---

[19] Bülow said as much after the conference (Massie, 1991: 356). Zagare and Kilgour (2006) show that a Hesitant Challenger might rationally contest the status quo even when the Defender is known to be determined, i.e., when $p_{Def} = 1$. Uncertainty about the Defender's type can only increase the propensity of a Hesitant Challenger to foment a crisis.

materialized because of their fear that a lack of support would destroy the entente, a relationship that took on even greater value once Edward Grey took over the foreign office.

Finally, French action choices fluctuated as underlying conditions changed. As long as Delcassé was in charge of foreign policy in Paris, France refused to conciliate Germany. Delcassé's firm belief was that Berlin was bluffing. But the prime minister, who initially also mistrusted the British, thought otherwise. Thus, Rouvier forced the foreign minister's resignation in the clearly mistaken belief that Germany would take yes for an answer.

Over time, Rouvier updated his beliefs about British resolve (Williamson, 1969: 43). And while he could never be certain, he placed a very high probability on their support at Algeciras (Gooch, 1936: 261). French action choices at the conference were fully consistent with his beliefs, as they had been previously.

The explanation derived from the Tripartite Crisis Game model is not necessarily at odds with consensus historical interpretations of the Moroccan crisis. Indeed, it draws heavily on the work of both diplomatic historians and strategic studies experts for both factual information and historical context. Nonetheless, the present examination of the crisis offers several advantages over standard, largely atheoretical or ad hoc descriptions. One clear advantage is the convenient framework the model provides for organizing information about the crisis around a common set of assumptions and concepts and for the clear way the most salient causal variables are highlighted.

But the advantages of the model are not limited to its greater organizing power. The model's ability to point to a logically consistent set of expectations about the connections between certain action choices and the beliefs that drive them means that its explanatory and predictive capability is also considerable. Finally, the model's clear applicability to an important and complicated watershed event is suggestive of its potential generality. To date, it has also been used successfully to explain Bismarck's decision to enter into what diplomatic historians considered an "unlikely" alliance (Massie, 1991: 79) with Austria–Hungary in 1879 (Zagare and Kilgour, 2003), British foreign policy in 1914 (Zagare and Kilgour, 2006), and Germany's decision to issue Austria a blank check just prior to the Great War (Zagare, 2009a, 2011b). It would also appear to be prima facie applicable to the Bosnian crisis of 1908–9, the second Moroccan crisis of 1911, and potentially the ongoing extended deterrence relationship among China, North Korea, and the United States. If, on further examination, this conjecture is confirmed, the model's potential as the basis for a generalized theory of extended deterrence crisis interaction will be realized, especially since its underlying structure is easily calibrated to take account of a wide variety of real world conditions that might distinguish one crisis from another.[20] In other words, the claim here is that the Tripartite Crisis Game model is a highly malleable theoretical construct and, therefore, a potent explanatory and predictive tool for both diplomatic historians and international relations theorists.

---

[20] Benson (2012: 8–9), who claims otherwise, overlooks the fact that the underlying model does not require the specific preference assumptions given by Table 3.1.

4

A Game-Theoretic History of the Cuban Missile Crisis

4.1 Introduction

"The history of science," writes Abraham Kaplan (1964: 354), "is a history of the successive replacement of one explanation by another." There is perhaps no clearer manifestation of this observation in the field of security studies than the attempts by game and decision theorists to explain the Cuban missile crisis, an event whose significance in international affairs almost defies hyperbole.[1]

More than half a century has passed since the crisis was settled; in the interim, researchers have gained access to a growing collection of primary sources. Key documents from the Soviet archives have been released, and secret recordings of the deliberations of the Kennedy administration have been made public.[2] In light of these developments, one might well expect that explanations of the crisis have been adjusted and refined. Indeed, this has been the case, both in the general literature of the crisis (e.g., Fursenko and Naftali, 1997; Allison and Zelikow, 1999; Fursenko and Naftali, 2006; Stern, 2012) and in the explanations constructed by game theorists. But the parallel game-theoretic literature also reflects controversies and refinements within game theory itself. As game theory has evolved, so have the explanations fashioned by its practitioners. The purpose of this chapter is to trace these explanatory refinements, using the Cuban crisis as a mooring. In Chapter 5, I construct a new interpretation of the crisis that exploits both the advances in game theory over time and the expanded evidentiary base.

4.2 Some Preliminaries

Before beginning, however, two preliminary questions must be addressed. The first, of course, is what is to be explained? There appears to be wide consensus in the literature

---

[1] This chapter is based on Zagare (2014).
[2] Many of the released Soviet documents are available online at the Cold War International History Project: http://www.wilsoncenter.org/program/cold-war-international-history-project. For the Kennedy tapes, see May and Zelikow (1997).

Game Theory, Diplomatic History and Security Studies. Frank C. Zagare. Oxford University Press (2019).
© Frank C. Zagare 2019. DOI: 10.1093/oso/9780198831587.001.0001

on this issue. Three questions are key. First, why did the crisis take place in the first place (i.e., why did the Soviets install medium and intermediate missiles in Cuba?)? Second, why was the US response measured (i.e., why did the United States respond with a blockade and not an air strike or an invasion?)? And, third, why was the crisis resolved short of war (i.e., why did the Soviets remove the missiles?)? The surveyed explanations will be evaluated by the extent to which all three of these questions are answered. As will be seen, partial or incomplete explanations are the norm.

There is also a consensus in both the wider and the game-theoretic literature that the bargain that resolved the crisis was a compromise (Gaddis, 1997: 261). In return for a public US pledge not to invade Cuba and a private assurance that US-controlled missiles in Turkey would eventually be dismantled, the Soviets agreed to withdraw the missiles. Although there are some who argue otherwise (e.g., Sorensen, 1965), this analysis will take as its starting point the fact that there was no clear winner of the crisis and that the key event to be explained was a political bargain in which both sides gave way. As will be seen, coding the outcome a "compromise" confounds explanation. One-sided victories are much easier to explain game theoretically.[3]

Second, what constitutes an explanation? An explanation, according to Kaplan (1964: 339), "shows that, on the basis of what we know, the something cannot be otherwise." As has been pointed out several times previously, within game theory, this task is delegated to the game's equilibria. And it is easy to understand why. Game theory takes as axiomatic the (instrumental) rationality of the players. Of all the outcomes in a game, only the equilibria are consistent with rational choices by all of the players. The assumption that the players in a game are rational, therefore, leads naturally to the expectation that they will make choices that are associated with some equilibrium outcome. Game-theoretic explanations and predictions derive from this expectation. When players in a real world game make choices that can plausibly be associated with an equilibrium outcome, as was the case in the Moroccan crisis of 1905–6, a game-theoretic explanation has been uncovered. Similarly, game-theoretic predictions about future play presume rational choice—that is, the assumption is that an equilibrium choice will be made by each of the players. In the discussion that follows, a variety of equilibrium definitions will be encountered. To keep things as simple as possible, the technical distinctions between different types of equilibria will be suppressed whenever possible. The interested reader should consult Morrow (1994) or another standard source on game theory for any omitted particulars.

4.3 Thomas Schelling and the "Threat That Leaves Something to Chance"

The benchmark against which all other explanations of the Cuban missile crisis should be measured is Thomas Schelling's. Schelling was the first game theorist to explore the strategic dynamic of the crisis and, if one takes the derivative literature seriously, his initial

---

[3] For an example, see Snyder and Diesing's (1977: 114–16) analysis of the crisis.

characterization was, and still is, the standard interpretation of the denouement of the crisis (Hesse, 2010; Dodge, 2012).

Strictly speaking, however, Schelling never actually offered a fully formed explanation of the missile crisis. In fact, in the preface to his widely read and very influential 1966 book *Arms and Influence*, he claimed that all of the real world examples he discussed were meant merely "to illustrate some point or tactic…[and that]…mention does not mean approval, even when a policy was successful." Nonetheless, an explanation can be pieced together from several lengthy passages about the crisis in the book. Trachtenberg (1985: 162) refers to this composite view as an explanation "à la Thomas Schelling".

Schelling (1966: 96, 176) saw the Cuban crisis, indeed, all crises, as a "competition in risk-taking." Lurking beneath this view of intense interstate confrontations are the structural dynamics of the 2 × 2 normal (or strategic) form game of Chicken (see Figure 4.1).[4] In Chicken, the two players, be they teen drivers or generic States A and B, are on a "collision course." A *Win* in this game occurs when one of the players cooperates ("Swerve," in the case of the drivers; "Wait," in the case of the states) by choosing (C) when the other defects from cooperation ("Not Swerve," in the case of the drivers; "Attack," in the case of the states) by choosing (D). A *Compromise* is reached if both cooperate. And finally, a disaster (i.e., *Conflict*) results if and when neither player cooperates.

In Chicken, the assumption is that each player most prefers to win and, failing that, to compromise. Chicken's defining characteristic, however, is the further assumption that *Conflict* is a mutually worst outcome. In other words, each player's preference is to back off and allow the other player to win rather than crash head on.

		State B	
		Cooperate (C) (Wait)	Defect (D) (Attack)
	Cooperate (C) (Wait)	*Compromise* (3,3)	*B Wins* (2,4)*
State A	Defect (D) (Attack)	*A Wins* (4,2)*	*Conflict* (1,1)

Key: (x,y) = payoff to State A, payoff to State B
4 = best; 3 = next best; 2 = next worst; 1 = worst
* = Nash equilibrium

Figure 4.1 Chicken (shown for states)

---

[4] As will be seen in Chapter 8, Schelling also used the Chicken analogy to model the strategic relationship of the superpowers during the Cold War era.

From a game-theoretic perspective, Chicken presents a number of perplexing analytic problems. There are two (pure strategy) Nash equilibria in Chicken (as indicated by the asterisks in Figure 4.1). But the two Nash equilibria are neither *equivalent*—that is, they are associated with different payoffs to the players—nor *interchangeable*, since the strategies associated with them do not have identical consequences. Needless to say, the existence of two or more nonequivalent and/or noninterchangeable equilibria, Nash or otherwise, confounds game-theoretic explanations and predictions (Harsanyi, 1977b: 3–4). This is especially true in Chicken, where the players have symmetric roles. On what basis might one make a prediction, and how, after the fact, might one explain why one player rather than the other has prevailed when more than one outcome consistent with rational choice exists?

Schelling's answer to both questions is that the player who is the first to commit to driving straight on will force the other player to (rationally) swerve and will thereby gain the advantage.[5] In a chapter appropriately entitled "The Art of Commitment," he offered several examples of commitment artists at work. Prime among them was President John F. Kennedy's televised speech to the nation on October 22. According to Schelling, Kennedy's promise of an automatic response against the Soviet Union should any nuclear missile be launched from Cuba was "effective."

Schelling went on to note, however, that a firm commitment was probably not necessary. Kennedy's threat might still have been effective if he had merely raised "the possibility that a single Cuban missile, if it contained a nuclear warhead and exploded on the North American continent, could have triggered the full frantic fury of all-out war" (Schelling, 1966: 41). In other words, one might win simply by increasing the *risk* of war. Of course, what is true for one player is also true for the other in any symmetric game, which is why Schelling came to view intense interstate crises such as Cuba as risk-taking contests.

One of the analogies that Schelling (1960: 196; 1966: 123) used to make this point was of two men "fighting in a canoe." If the boat goes down, both players could drown. Worse still, once the canoe starts to wobble, neither might be able to stabilize it. Thus, in any crisis, there is an autonomous risk of war, a risk of things spiraling out of control. Schelling's (1960: ch. 8) intuition led him to argue that this was a type of risk that could be used to successfully manage an intense interstate conflict. Thus was conceived the "threat that leaves something to chance."[6]

Schelling (1966: 96) saw just such a threat implicit in the blockade even though "there was nothing about the blockade of Cuba that could have led straightforwardly into general war." But the blockade, as measured as it was, still carried with it the possibility of an inadvertent nuclear exchange. And, Schelling argued, it was precisely because President Kennedy

---

[5] Of course, no player makes the first move in a strategic form game where the assumption is that all players choose a strategy before the game begins. This is a feature of strategic form games that many first-wave strategic theorists, including Schelling, seemed not to recognize. For a discussion, see Rapoport (1964: 119). As will be seen in Chapter 8, however, there is a first-mover advantage when Chicken is played sequentially. Schelling most likely knew this intuitively.

[6] As Quackenbush (2001: 745) points out, the "threat that leaves something to chance" depends on the possibility of an accidental war. Empirical examples of accidental wars, including World War I, are nonexistent (Trachtenberg, 1990/1991; Zagare, 2011b).

successfully manipulated this autonomous risk that he won the war of nerves with Premier Nikita Khrushchev and was able to get the better of the Soviet Union in October 1962.

Schelling was not the only one to code the outcome of the crisis as a "win" for the United States and to attribute it to President Kennedy's adroit brinkmanship. Consider, for example, Arthur Schlesinger's (1965: 767) summary description of Kennedy's diplomatic performance:

> From the moment of challenge the American President never had a doubt about the need for a hard response. But throughout the crisis he coolly and exactly measured the level of force necessary to deal with the level of threat. Defining a clear and limited objective, he moved with mathematical precision to accomplish it. At every stage he gave his adversary time for reflection and reappraisal, taking care not to force him into "spasm" reactions or to cut off his retreat.

Schlesinger's interpretation of the crisis and, by extension, Schelling's, has not withstood the test of time and informed historical scrutiny. Michael Dobbs (2012), for example, recently noted that "the White House tapes demonstrate that Kennedy was a good deal more nuanced, and skeptical, about the value of 'red lines' than his political acolytes were. He saw the blockade—or 'quarantine' as he preferred to call it—as an opportunity to buy time for a *negotiated settlement*" (emphasis added; see also Dobbs, 2008). Similarly, after reviewing the transcripts of White House deliberations, Marc Trachtenberg (1985: 162) concluded that the documentary record does not support a view of "the crisis as a 'competition in risk-taking' à la Thomas Schelling." Speaking about US decision makers as a group, Trachtenberg noted that "no one wanted to keep upping the ante, to keep outbidding the Soviets in 'resolve,' as the way of triumphing in the confrontation."

Both Dobbs and Trachtenberg, then, find little evidence of the mathematically precise manipulation of threat levels that Schlesinger wrote about.[7] Of course, their empirical observations do not necessarily mean that Schelling's view of the crisis as a competition in risk taking can be cast aside. As the astrophysicist Carl Sagan once noted: "the absence of evidence is not the evidence of absence." Thus, before rejecting Schelling's explanation of the missile crisis, a more compelling reason would have to be given.

In the years immediately after Schelling wrote, a determinative assessment of his interpretation of the crisis was problematic. The main reason was that his approach, which Young (1975: 318) labels "manipulative bargaining theory," had "not yet yielded much in the way of deductively derived propositions that can be subjected to empirical validation." Some years later, Achen and Snidal (1989: 159) made the same point when they pointed out that "Schelling's 'threat that leaves something to chance' has yet to be given a coherent statement within rational choice theory."

---

[7] Snyder and Diesing (1977: 489–90) also find no example of the use of this and, for the most part related, coercive bargaining tactics in the sixteen major interstate crises they studied. See also Huth (1999) and Danilovic (2001, 2002).

At about the same time that Achen and Snidal wrote, however, Robert Powell (1987, 1990) developed a two-person sequential game model that nicely filled the theoretical void.[8] In Powell's model, one player begins play by deciding whether to accept the status quo, escalate the contest by challenging it, or attack. If that player chooses not to contest the status quo or not to attack, the game ends. But if escalation is chosen, the other player is faced with similar choices. Significantly, the four possible outcomes in this model are the same as in Chicken. If the player choosing first does not escalate or attack, the status quo prevails. If one player escalates and the other does not, the escalating player wins. If either player attacks, the game ends in disaster. And if both players escalate, the game continues until one player submits or until the game "gets out of control" and culminates in disaster. Powell assumes that, by choosing to escalate, a player unleashes an autonomous risk, beyond its control, of an all-out war. Thus, his model captures well Schelling's view of a nuclear crisis as a "competition in risk taking."

Given these assumptions, Powell shows that the existence of a crisis equilibrium, that is, a stable outcome that arises after a challenge by one player and resistance by the other, depends on incomplete information, that is, each side's lack of information about the values of its opponent.[9] But once a crisis occurs, the game can only end in one of three ways: a victory for the first player, a victory for the second, or a head-on collision.[10] The clear implication is that the political bargain that brought the Cuban crisis to a close cannot be adequately explained by a mutual fear that things might spiral out of control. If the Cuban crisis were truly a *competition* in risk taking, there could have been no compromise; there would have been either a clear "winner" or a thermonuclear war.

To put this in a slightly different way, Schelling's interpretation of the Cuban missile crisis and, arguably, of most other intense major power disputes,[11] is game-theoretically inconsistent with what is now a strong consensus among historians and foreign policy analysts that the resolution of the crisis was a political compromise or draw. When former US Secretary of State Dean Rusk learned on October 24 that several Soviet ships that had been approaching Cuba had either turned around or stopped dead in the water, he is said to have remarked, "We are eyeball to eyeball, and I think the other fellow just blinked" (May and Zelikow, 1997: 358). Rusk was probably correct. But, in this regard, the Soviets were not alone. The historical record shows that President Kennedy was not only more than eager to compromise but also willing to offer much more than he did to end the high-stakes stalemate that the Soviets referred to as "the Caribbean crisis."

The lack of fit between Schelling's theoretical explanation and the resolution of the crisis is indeed disturbing. But when this discrepancy is coupled with the absence of any compelling empirical trace that either President Kennedy or Premier Khrushchev carefully calibrated his threats in order to manipulate the other's behavior and induce the other's

---

[8] For a similar, albeit less general, model that is applied hypothetically to the missile crisis, see Dixit and Skeath (2004: ch. 14).

[9] Powell also shows that, under certain conditions, no challenge will be made and, hence, stable mutual deterrence can emerge. Zagare and Kilgour (2000: 54–7) challenge the adequacy of this deduction.

[10] The same is true of Dixit and Skeath's (2004) model. Powell's model, however, suggests that when deterrence breaks down, the connection between resolve and victory in a crisis does not always depend on a greater willingness to risk war.

[11] See footnote 7.

concession, it becomes difficult to sustain what has become the conventional interpretation of the crisis, that is, an explanation "à la Thomas Schelling."

4.4 Nigel Howard and the Theory of Metagames

After Schelling's, the next noteworthy attempt by a game theorist to explain the missile crisis was Nigel Howard's. Howard's analysis begins on Tuesday, October 16, the day that President Kennedy was told that the Soviets were installing missiles in Cuba. As Howard (1971) saw things, once the missiles were discovered, the United States had only two viable options: either to cooperate by blockading Cuba (B) or to defect from cooperation by attempting to remove the missiles with a "surgical" air strike (A). On October 22, after a full review of US options, Kennedy announced the blockade. According to Howard, once this announcement was made, the Soviet Union also had only two broad strategic choices: to either cooperate by Withdrawing (W) the missiles or not cooperate by Maintaining (M) them. These policy options give rise to a 2 × 2 normal form game. Like Schelling and most other strategic analysts of this era, Howard assumes a payoff structure that defines the crisis as a game of Chicken. Recall that, in Chicken, *Conflict* is the worst outcome, and *Compromise* is the second-best outcome, for both players.

In Howard's interpretation of the crisis, an all-out *Conflict* would ensue if the Soviet Union decided to maintain the missiles, and the United States launched an air strike to remove them. By contrast, a *Compromise* would be reached if the United States decided to blockade Cuba, and the Soviets responded by agreeing to withdraw their missiles.[12] Of course, this is what happened, so Howard, unlike Schelling, implicitly accepts what is now the standard understanding of the crisis's resolution, that is, that it was a draw.

As noted above, the existence of two nonequivalent and noninterchangeable Nash equilibria in Chicken, each with equal status as a solution candidate, confounds a game-theoretic analysis. But Howard's interpretation of the game's outcome creates an additional stumbling block: since the *Compromise* outcome is not a Nash equilibrium, how can its persistence be explained?[13] Why, in other words, didn't just one of the superpowers blink, as both Powell's model and a standard interpretation would suggest?

To answer this and related questions, Howard developed the *theory of metagames*. Building on an idea first suggested by von Neumann and Morgenstern (1944: 100–6), Howard altered the underlying game to reflect the possibility that the players might be able to anticipate each other's strategy choice. Presuming that each player bases its own strategy choice on the strategy it expects the other to select, a new game—the *metagame*—is

---

[12] A *Soviet Victory* is implied if the Soviets were to maintain their missiles and the United States took no aggressive action. Similarly, a *US Victory* would have occurred if the United States had used force to remove the missiles. Brams's interpretation of the possible outcomes of the crisis is the same as Howard's (see Figure 4.6). Brams, however, assigns different (ordinal) utilities.

[13] There is a Nash equilibrium in mixed strategies in Chicken. Under a mixed strategy equilibrium in a 2 × 2 game, all four outcomes occur with positive probability. Thus, one might expect the compromise outcome in Chicken to occur, but only sometimes, and not necessarily often. The mixed strategy equilibrium also has some anomalous normative properties (see O'Neill (1992) for a discussion). All of which suggests that it provides, at best, a very weak explanation for the resolution of the Cuban crisis; at worst, it provides no explanation at all.

rendered and played "in the heads" of the players prior to the play of the actual game. In the metagame, players choose *metastrategies* rather than strategies. Stable outcomes of the metagame are termed *metaequilibria*.[14]

One way to think about Howard's reformulation of classical game theory is as a theory of equilibrium selection. In the metagame, the metastrategies can be interpreted as signals that the players send to one another *before the game begins*.[15] These signals, verbal or otherwise, allow each of the players to anticipate the other's strategy choice. In other words, the theory of metagames attempts to model the impact of pre-play communication in a noncooperative game environment. Howard's goal is to identify those communication patterns that are consistent with rational choice, that is, are associated with the metaequilibria. Within the theoretical confines of the theory of metagames, an explanation is uncovered when a plausible connection is made between a communication pattern, a pair of metastrategies, and an observed metaequilibrium. Because pre-play communication allows the players to form a *common conjecture* about the way the game will be played, the problem of multiple noninterchangeable and nonequivalent equilibria is potentially rendered moot.[16]

To illustrate, assume that, as the crisis unfolded, the Soviet leadership believed that it could correctly anticipate the strategy choice of the United States. If this were the case, the range of choices available to the Soviet Union would expand. Rather than having just two strategies (i.e., (W) or (M)), it would now have $2 \times 2 = 4$ metastrategies:

1. Withdraw Regardless (W/W): choose (W) regardless of the US choice;
2. Maintain Regardless (M/M): choose (M) regardless of the US choice;
3. Tit for Tat (W/M): choose (W) if the United States chooses (B), and choose (M) if the United States chooses (A); and
4. Tat for Tit (M/W): choose (M) if the United States chooses (B), and choose (W) if the United States chooses (A).

These four metastrategies give rise to the 2×4 *first-level metagame* shown in Figure 4.2. Notice that the third metastrategy, Tit for Tat, is conditionally cooperative. It implies Soviet cooperation (W) if and only if Soviet leaders believe that the United States intends to cooperate (B).

There are three metaequilibria in the *first-level* metagame. Two correspond to the Nash equilibria in the original (simultaneous choice) Chicken game while the third—(A,W/M) —is strictly a product of the metagame. The additional metaequilibrium, however, does not materially expand the set of distinct rational strategic possibilities, that is, it is repetitive. In consequence, the central explanatory problem remains: the compromise outcome (3,3) continues to be (at least for now) an irrational event, that is, it is not a metaequilibrium of the first-level metagame.

But Howard continues. What if, he asks, the United States could predict the metastrategy of the Soviet Union and then based its choice on that prediction? If the United States were able to condition its strategy choice on the Soviet metastrategy, it could choose either (B) or (A) for

---

[14] A metaequilibrium is simply a Nash equilibrium in the metagame.

[15] For an insightful informal analysis of signaling, see Cohen (1987).

[16] A common conjecture is an agreement among the players about the way the game will be played.

Soviet Union

		Withdraw Regardless (W/W)	Maintain Regardless (M/M)	Tit for Tat (W/M)	Tat for Tit (M/W)
	B	(3,3)	(2,4)*	(3,3)	(2,4)
United States					
	A	(4,2)*	(1,1)	(1,1)	(4,2)*

Key: (x,y) = payoff to the United States, payoff to the Soviet Union
    4 = best; 3 = next best; 2 = next worst; 1 = worst
    * = metaequilibrium (Nash equilibrium)

Figure 4.2 First-level metagame of the Cuban missile crisis (Chicken)

each of the four Soviet metastrategies, which gives it $2 \times 2 \times 2 \times 2 = 16$ *second-level* metastrategies. For instance, the second-level metastrategy A/A/B/A requires the United States to

1. choose (A) if the Soviet Union chooses Withdraw Regardless;
2. choose (A) if the Soviet Union chooses Maintain Regardless;
3. choose (B) if the Soviet Union chooses Tit for Tat; and
4. choose (A) if the Soviet Union chooses Tat for Tit.

Notice that this metastrategy is also conditionally cooperative. It implies US cooperation (B) if and only if US leaders believe that the Soviet Union is itself conditionally cooperative (i.e., selects Tit for Tat).

The sixteen second-level metastrategies of the United States and the four first-level metastrategies of the Soviet Union imply a $16 \times 4 = 64$ outcome metagame. An abbreviated version of this matrix, listing only non-repetitive metaequilibria, is given in Figure 4.3. As before, a number of new metaequilibria appear. Among them is one that corresponds to the *Compromise* outcome (3,3). To support this outcome in equilibrium, however, each player must intend to choose conditionally cooperative metastrategies (Tit for Tat for the Soviet Union, and A/A/B/A for the United States) and convey this intention to the other. Specifically, the Soviet Union must intend to cooperate (by withdrawing the missiles) if (and only if) the United States cooperates by not using force to remove the missiles (i.e., by blockading Cuba). And the United States must intend to cooperate by blockading Cuba if (and only if) the Soviet Union cooperates by withdrawing the missiles.

This is an interesting and potentially important theoretical result. If it stands, a logically consistent rational choice explanation of the political bargain that ended the missile crisis will have been uncovered. Whether it stands, however, depends on the interpretation of the metaequilibria.

		Soviet Union			
		Withdraw Regardless (W/W)	Maintain Regardless (M/M)	Tit for Tat (W/M)	Tat for Tit (M/W)
	B/B/B/B	(3,3)	(2,4)*	(3,3)	(2,4)
	...	...	...	...	...
	A/B/B/A	(4,2)	(2,4)*	(3,3)	(4,2)
United States	...	...	...	...	...
	A/A/B/A	(4,2)	(1,1)	(3,3)*	(4,2)
	...	...	...	...	...
	A/A/A/A	(4,2)*	(1,1)	(1,1)	(4,2)*

Key: (x,y) = payoff to the United States, payoff to the Soviet Union
4 = best; 3 = next best; 2 = next worst; 1 = worst
* = metaequilibrium (Nash equilibrium)
... = unlisted metastrategies/outcomes

Figure 4.3 Second-level metagame of the Cuban missile crisis (Chicken)

Howard's construction is strictly *descriptive*: metaequilibria are established as theoretical possibilities only, and the metastrategies are theoretical statements about the content of the communication necessary to lead to some outcome in equilibrium. In Howard's view, no particular metaequilibrium has special status. Each, therefore, describes a logical possibility in a game between rational players. Which metaequilibrium eventually comes into play depends on what the players expect from one another, or what they communicate to each other, in pre-play bargaining and discussion.

In the present example, then, a bargain was struck not only because both players were prepared to cooperate but also because both indicated that they were willing to run the risk of all-out war (the Soviets by maintaining their missiles, and the United States by using an air strike to remove them) if the other was unwilling to cooperate. Of course, there were other rational strategic possibilities. As Howard notes, if only one of the players had been willing to risk war, that player would have won. For instance, if the Soviet Union limits itself to the two metastrategies that do not admit the possibility of the *Conflict* (1,1) outcome, Withdraw Regardless and Tat for Tit, the only remaining metaequilibria are associated with a victory for the United States.

Howard's explanation, however, merely begs the question, why that communication pattern—why that particular metaequilibrium? Worse still, there is a compelling strategic rationale that suggests that the metastrategies required to effect a compromise would *not* be selected by rational agents. Notice that the metastrategy A/B/B/A—or what Howard refers to as the "sure-thing" metastrategy—is weakly dominant for the United States, giving the Soviet Union good reason to suspect that the United States would choose it; and since Maintain Regardless is the Soviet Union's best response to A/B/B/A, the United States would have had a good reason to suspect that the Soviets were going to choose it. All of which suggests that the metaequilibrium associated with these two metastrategies, which would bring about a Soviet victory, might well evolve in an actual play of the metagame.

Howard, however, rejects this outcome as *the* solution to the metagame and denies that any particular reason exists for singling it out. In fact, he argues it would be *foolish* for the United States to select its sure-thing metastrategy because it induces a worse outcome than its "retaliatory" metastrategy A/A/B/A. Or, in Howard's (1974a: 730) own words, the sure-thing metastrategy is "the strategy of a 'sucker' who invites, and is ready to yield before, the most extreme ultimatum in the possession of his opponent, and is thus willing to surrender his position before any bargaining begins."

Harsanyi (1974b), however, argues convincingly that the use of any dominated metastrategy is irrational and, hence, not *credible*. Since a player with a dominant metastrategy always maximizes its expected utility by choosing it, there is no good reason for an opponent to believe that any other metastrategy would be chosen. This, in turn, implies that a player with a dominant metastrategy should choose it.[17] To do otherwise would be to invite calamity. Specifically, if the United States were to select its retaliatory metastrategy A/A/B/A, and the Soviet Union, anticipating the sure-thing metastrategy A/B/B/A, were then to select Maintain Regardless, each player's worst outcome, *Conflict*, would result.

It is difficult to ignore Harsanyi's admonition not to abandon the use of even a weakly dominant strategy. To be sure, Howard's metagame theory provides insight into the conditions that are both necessary and sufficient to effect a political compromise during a crisis. But, because it lacks a normative foundation, at least according to Howard, an explanation of the missile crisis within its theoretical confines can only be considered weak and incomplete.

4.5 Fraser, Hipel, and the Analysis of Options Technique

Niall Fraser and Keith Hipel's (1982–3) explanation of the Cuban missile crisis begins where Howard's leaves off. Like Howard's theory of metagames, Fraser and Hipel's *analysis of options* (or *improved metagame*) *technique* should be interpreted, at least in part, as a proto-theory of equilibrium selection. But, unlike Howard's attempt to reformulate classical game theory, Fraser and Hipel's subtle refinement of Nash's equilibrium concept adds a distinctly dynamic element to the analysis of complex conflicts.

As its name suggests, Fraser and Hipel's innovative methodology explores the stability characteristics of every feasible combination of strategy choices in a game and suggests a path leading to the selection of one equilibrium when multiple equilibria exist. Like Howard

---

[17] For the particulars of their debate, see Harsanyi (1973, 1974a, b) and Howard (1973, 1974a, b).

(see Section 4.4) and Brams (see Section 4.6), their analysis of the Cuban crisis begins with the discovery of the missiles on October 16. Stepping away from the narrow confines of a 2×2 normal form game and the standard Chicken analogy, however, they consider three options for the United States:

1. perform no aggressive action, either by doing nothing or by using normal diplomatic channels to try to induce the Soviets to remove the missiles,
2. destroy the missile sites with an air strike, or
3. blockade Cuba,

and three for the Soviet Union:

1. withdraw the missiles,
2. maintain the missiles, or
3. escalate the conflict.

Clearly, some of these options are mutually exclusive, while others could be selected concurrently. For instance, the Soviets could not both withdraw and maintain the missiles, although the United States could blockade and attack the missiles sites at the same time. After eliminating the mutually exclusive combinations of options, twelve feasible combinations and, hence, twelve different outcomes, remain. These combinations are listed as columns in Figure 4.4. To facilitate a computer implementation of their technique, Fraser and Hipel convert each outcome into a binary number, that is, the outcomes are "decimalized." These numbers, which range from 0 to 11 and which are arrayed in the last row of Figure 4.4, can also be considered numerical labels for the various outcomes.

Opposite two of the US and two of the Soviet options in Figure 4.4 are twelve columns that contain either a one (1) or a zero (0). A "1" indicates that the option has been selected, and a "0" indicates that it has not been selected. For instance, one option for the United States is to

Outcomes												
United States												
Air Strike	0	1	0	1	0	1	0	1	0	1	0	1
Blockade	0	0	1	1	0	0	1	1	0	0	1	1
Soviet Union												
Withdraw	0	0	0	0	1	1	1	1	0	0	0	0
Escalate	0	0	0	0	0	0	0	0	1	1	1	1
Decimalized	0	1	2	3	4	5	6	7	8	9	0	11

Figure 4.4 Players, options, and outcomes for the Cuban missile crisis (Source: Fraser and Hipel, 1982–3)

"do nothing," that is, neither blockading nor striking. This "do nothing" option is captured by the two zeros next to the two US options above the decimalized outcome 0 in Figure 4.4.

To explore the stability characteristics of the set of feasible outcomes, Fraser and Hipel start with preference assumptions.[18] In Figure 4.5, US preferences are given, from best to worst, in the second row of its (the upper) part of the array, and Soviet preferences are shown in the second row of its (the lower) part of the array. For instance, Fraser and Hipel assume that the United States most prefers outcome 4 (which is brought about when it does nothing and the Soviets withdraw their missiles) to outcome 6 (in which the Soviets withdraw their missiles in response to a blockade), and so on. Next, they ask whether any of the outcomes offer either player a "unilateral improvement," which is defined as a better outcome that a player can induce by unilaterally changing its strategy. Figure 4.5 lists all possible unilateral improvements, in descending order of preference, in the three rows beneath the preference vector for the United States and the two rows beneath the preference vector for the Soviet Union. The United States, for example, has a unilateral improvement from outcome 6, which is its second-best outcome, to outcome 4, which is its most preferred outcome. And since it can induce outcome 4 from 6 simply by switching from its blockade-only option to its "do nothing" option, a unilateral improvement from 6 to 4 is listed below outcome 6 in the preference vector for the United States.

To determine which of the twelve outcomes are in equilibrium, a multistep search procedure is necessary. These steps involve characterizing the stability, or lack thereof, of each outcome. In this regard, Fraser and Hipel identify three types of outcomes: *rational*, *sanctioned*, and *unstable*.[19] Rational outcomes are those for which a player has no unilateral improvements.[20] Rational outcomes are indicated by an "r" above each player's preference

	E	E	X	X	X	X	X	X	X	X	X	X	Overall Stability
United States	r	s	u	u	r	u	u	u	r	u	u	u	Player Stability
	4	6	5	7	2	1	3	0	11	9	10	8	Preference Vector
	4	4	4		2	2	2		11	11	11		UI
		6	6			1	1			9	9		UI
			5				3				10		UI
Soviet Union	r	s	r	u	r	u	r	u	u	u	u	u	Player Stability
	0	4	6	2	5	1	7	3	11	9	10	8	Preference Vector
	0		6		5		7	7	5	6	0		UI
								3	1	2	4		UI

Figure 4.5 Stability analysis tableau for the Cuban missile crisis (Source: Fraser and Hipel, 1982–3); UI = unilateral improvement

---

[18] Fraser and Hipel do not provide a detailed justification for their preference assumptions. Their assumptions, however, are not particularly unreasonable.

[19] There is a fourth type, an outcome that is rendered "stable by simultaneity" that does not come into play in their analysis of the missile crisis.

[20] Note that this is the sole criterion Nash uses to gauge the stability of a strategy combination.

vector in Figure 4.5. Clearly, the most preferred outcome of the United States (i.e., 4) and that of the Soviet Union (i.e., 0) are rational and so are labeled as such in Figure 4.5.

Sanctioned outcomes, designated by an "s" in the row above each player's preference vector, are those for which the *other* player can credibly induce a worse outcome for a player who acts on a unilateral improvement. A credible action is one which brings about a better outcome for the sanctioning player. For example, the unilateral improvement of the United States from outcome 6 to outcome 4 is sanctioned by the Soviet Union, since the Soviets can unilaterally induce outcome 0, which it prefers and which the United States does not prefer, to outcome 4. Unstable outcomes, designated by a "u," are those with an unsanctioned unilateral improvement. Any outcome that is not unstable for both players is in equilibrium. Only outcomes 4 and 6 pass this test and are denoted with an "E" in the first row of Figure 4.5. Nonequilibria are indicated by an "X."

To choose between these competing equilibria, Fraser and Hipel examine the game's status quo outcome, outcome 0, which was the state of the world on October 15, the day before the missiles were discovered. This is the outcome that is obtained when the Soviets maintain their missiles, and the United States does nothing. Although outcome 0 is rational for the Soviet Union, it is unstable for the United States, which can induce three other outcomes it prefers to outcome 0 simply by switching to another course of action. Of these three unilateral improvements, outcome 2 would be the US preference. Outcome 2 is induced by blockading Cuba. Hence, the US reaction.

The likely Soviet response to the blockade can be also deduced from its incentive structure. At outcome 2, the Soviets have a unilateral improvement to outcome 6, which it can bring about by withdrawing the missiles. Outcome 6, of course is an equilibrium. Since, by definition, neither player can do better at an equilibrium by unilaterally switching to another course of action, the game would rationally end there. Thus, Fraser and Hipel are able to successfully explain why the United States blockaded Cuba and why the Soviets responded by withdrawing the missiles.

What they are unable to explain, however, is why the dynamic process they explore did not stop at that point. Notice that the United States has a unilateral improvement from outcome 6 to outcome 4, which is also an equilibrium. Moreover, not only is outcome 4 *Pareto superior*[21] to outcome 6, but it is in fact a more accurate description of the flow of events that brought the crisis to a close. After all, once the Soviets withdrew their missiles, the United States dropped the blockade.

Fraser and Hipel are unable to reach this seemingly straightforward conclusion, and it is easy to understand why: their definition of an equilibrium is constrained by an arbitrary stopping rule. Recall that a unilateral improvement from an outcome is sanctioned for a player if the other player has a credible (i.e., rational) response that leaves the player acting on the unilateral improvement worse off. In effect, this criterion limits each player's foresight to a single move and countermove. Thus, in their view, the United States would not act on its unilateral improvement from outcome 6 to outcome 4 simply because it is sanctioned by the Soviet Union (one move and one countermove). But why would the

---

[21] An outcome is Pareto superior to another if both players prefer it to the other. In this case, both the United States and the Soviet Union prefer outcome 4 to outcome 6.

STEVEN J. BRAMS AND THE THEORY OF MOVES | 75

Soviets sanction that move by acting on its own unilateral improvement from outcome 4 back to outcome 0, the original status quo? The obvious answer is that they would not because a move from 4 to 0 would lead to a counter-countermove by the United States to outcome 2, which would be worse for the Soviets (assuming, for the sake of argument, that the process would stop there). But this answer (or any other) would require that the United States consider the consequences of three moves (or more) rather than just two. With laudable but nonetheless perplexing consistency, then, Fraser and Hipel fail to fully extend the strategic logic of their model. In consequence, their explanation of the missile crisis necessarily falls short of the mark.[22]

4.6 Steven J. Brams and the Theory of Moves

Perhaps sensing this, Steven J. Brams developed a more general dynamic modeling frame-work called the "theory of moves" and uses it to offer several (plausible) explanations of the crisis (Brams, 1985: 48–62; 2011: 226–40). Here, I concentrate on the theoretical core of all his explanations and ignore the subtle differences that set the various explanations apart.[23]

Moving away from both a normal form representation and the standard Chicken analogy, Brams begins by considering the "payoff matrix" given by Figure 4.6. The cells of this graphical

		Soviet Union	
		Withdraw (W)	Maintain (M)
United States	Blockade (B)	Compromise (3,3)*	Soviet Victory; US Capitulation (1,4)
	Air Strike (A)	"Dishonorable" US action; Soviets thwarted (2,2)	"Honorable" US action Soviets thwarted (4,1)

Key: (x,y) = payoff to the United States, payoff to the Soviet Union
4 = best; 3 = next best; 2 = next worst; 1 = worst
* = nonmyopic equilibrium

Figure 4.6 The Cuban missile crisis as seen by Brams

[22] Fraser and Hipel (1982–3: 8–15) also describe a *computational* model, called the *state transition model*, that conforms to the actual outcome of the crisis. While it uses the input of their improved metagame technique, it is not a game-theoretic model.
[23] Brams also develops a few explanations using standard game-theoretic concepts. For instance, he suggests that the compromise outcome can also be supported in equilibrium if Khrushchev deceived the United States by

summary of the crisis represent possible states of the world (or outcomes). The ordered pair in each cell of the matrix reflects Brams's understanding of the relative ranking of the four possible outcomes by the United States and the Soviet Union, respectively. Brams assumes that once a game begins, either player can move from whatever outcome is the initial state (or the status quo outcome), and if it does, the other can respond, the first can counter-respond, and so on. The process continues until one player decides not to respond, and the outcome that they are at is the final outcome. Any outcome from which neither player, looking ahead indefinitely, has an incentive to move to another state of the world, including the initial state, is said to be a *nonmyopic equilibrium* (Brams, 1994).

When Kennedy announced the blockade on October 22, the Soviet missiles were already being installed in Cuba. This, the initial state of the world, was the worst for the United States and best for the Soviet Union. Thus, this outcome is labeled *Soviet Victory; US Capitulation*. Brams suggests several reasons why the Soviet Union would then withdraw the missiles and induce its next-best outcome, the *Compromise* outcome, rather than stick with its initial choice. For example, if the United States had what he calls *moving power*, which is the ability to continue moving in a game when the other player cannot, it could induce the Soviet Union to end the game at (3,3) by forcing it to choose between (3,3) and (4,1). Or, if the United States possessed *threat power*, which is the ability to threaten a mutually disadvantageous outcome in the first play of a repeated game, it could similarly induce the Soviet Union to withdraw the missiles by threatening to remove them with an air strike if they did not.

But, regardless of the reason why the Soviets decided to withdraw the missiles, once they did, the game ended. Brams's explanation is that the *Compromise* outcome is a nonmyopic equilibrium, that is, neither player can do better by moving the game to another state of the world by changing its strategy choice, given that the other might then switch to another strategy, so that it might then be forced to also change its strategy, and so on.

To see this, consider Figure 4.7 which lists the sequence of moves and countermoves away from the *Compromise* outcome that would be touched off if the Soviet Union reversed its course and decided to maintain the missiles after all. At the last node of this tree, the United States has a choice of staying at its second-worst outcome or cycling back to the *Compromise* outcome, its second-best outcome. As indicated by the arrow, the United States would rationally move.

Similarly, at the previous node, the Soviet Union would move away from (4,1), its least preferred outcome, to its next-worst outcome (2,2). But it would do so not because it prefers (2,2) to (4,1)—which it does—but because it would anticipate the outcome induced by the American choice at the subsequent node, which is better still. Counterintuitively, perhaps, the United States would, at the second node of the tree, also rationally move to its best outcome, not because it is its best outcome but because it prefers to cycle back to

suggesting that the compromise outcome was his most preferred (when it was not) or if his preferences "deteriorated" as the crisis progressed and the compromise outcome actually was his most preferred. Additionally, Brams constructed an extensive form game model in which the compromise outcome is a (subgame perfect) Nash equilibrium.

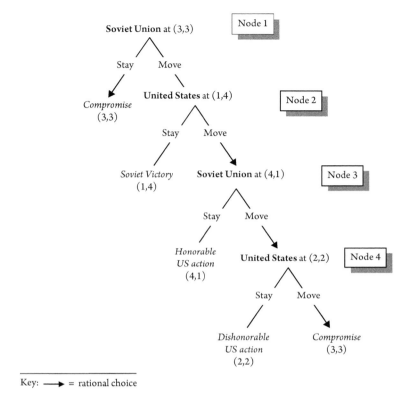

Figure 4.7 Moves and countermoves during the Cuban missile crisis, starting with the Soviet Union at the *Compromise* outcome (3,3)

Compromise, which is its second-best outcome, to terminating the game at (1,4), its least preferred outcome.

Finally, Brams argues that the Soviet Union would stick with its choice (at the top of the tree) to withdraw the missiles simply because, by not doing so, it would only cycle back to the *Compromise* outcome. As Brams (2001) asks, "What is the point of initiating the move–countermove process if play simply returns to 'square one'?"

It would be a straightforward exercise to show that there is also no incentive for the United States to precipitate the move–countermove process once the *Compromise* outcome is reached. And since neither player has a long-term incentive to move away from this outcome if and when it is reached, it constitutes a nonmyopic equilibrium.[24]

_____

[24] Interestingly, Brams (2001, 2011: 235) shows that it was in the interest of the United States to allow the Soviet Union to precipitate the move–countermove process rather than initiate the process itself. Thus, the United States could afford to be "magnanimous" in 1962.

Brams's explanation proceeds from this fact. As an equilibrium, the *Compromise* outcome is, by definition, consistent with the rationality postulate. And since it is the only nonmyopic equilibrium in the outcome matrix of Figure 4.6, it is the only outcome that can persist under rational play. All of which is another way of saying that the Soviet Union withdrew the missiles because, looking ahead, they believed that maintaining them would not be in their long-term interest. Given the American blockade and the incentive structure of the game, they simply could not win.

There is obviously much to like about the theory-of-moves framework. Indeed, I have used it myself to explore the dynamics of both one-sided and two-sided deterrence relationships and to analyze a number of acute interstate crises, including the Berlin crisis of 1948, the Middle East crisis of 1967, the Cease-Fire Alert crisis of 1973, and the Falkland–Malvinas crisis of 1982 (Zagare, 1987). The theory of moves is a simple and extremely intuitive methodology for analyzing complex interstate conflicts like the missile crisis; it is also more general than the analysis of options technique developed by Fraser and Hipel; its major solution concepts are easy to calculate and interpret; and since it is based only on ordinal utilities, it requires fewer "heroic assumptions" than many other game-theoretic frameworks.

In the case of the missile crisis, however, its strengths are also it weaknesses. Since the concept of a nonmyopic equilibrium has not as yet been successfully defined in games in which the players have more than two strategies each, it can only be used to assess the rationality of four outcomes at a time.[25] This limitation explains why Brams starts his analysis on October 22, the day that Kennedy announced the blockade. In consequence, its theoretical range is restricted. Why did the Soviets precipitate the crisis by installing the missiles in the first place, and why did the United States then respond by blockading Cuba? These are questions that Brams does not address and, in fact, cannot address within the confines of a single integrated game model using the theory-of-moves framework. Thus, while his analysis of the endgame of the crisis is both insightful and penetrating, his explanation remains incomplete.[26]

4.7 R. Harrison Wagner and Games with Incomplete Information

By the mid 1980s, it was apparent that a sea change was underway in the game-theoretic literature of international relations. First, there was a distinct shift away from the static environment of strategic form games, where the central equilibrium concept is Nash's, toward the dynamic framework provided by the extensive form, where the accepted measure of rational behavior is Selten's (1975) concept of a *subgame perfect equilibrium*. And, second, the until-then-standard assumption of complete information fell by the wayside, and the analysis of games of incomplete information became the norm.

---

[25] But see Brams (1994: 11–17) for an analysis of a 3 × 3 game that illustrates how the theory of moves could be extended to larger games. There has, however, been no systematic development of the theory for such games.
[26] It is not, however, inconsistent with the more general explanation developed in Chapter 5.

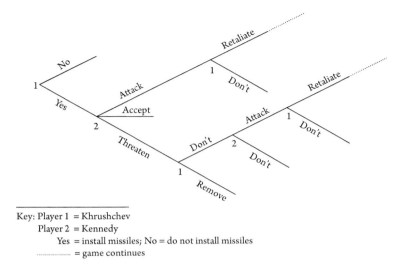

Key: Player 1 = Khrushchev
　　　Player 2 = Kennedy
　　　Yes = install missiles; No = do not install missiles
　　　················ = game continues

Figure 4.8 The Cuban missile crisis as seen by Wagner

Both of these developments are clearly reflected in the work of R. Harrison Wagner (1989), who was the first to make use of the modern theory of games with incomplete information to analyze the Cuban crisis. Starting with a straightforward crisis bargaining model (see Figure 4.8) that he claims broadly mirrors both the choices and the sequence of moves made by the United States and the Soviet Union, Wagner deduces what must have been true, game theoretically, (1) for Khrushchev to introduce the missiles in Cuba; (2) for Kennedy to have implemented a blockade (or some other ultimatum) rather than respond in a more aggressive way; and (3) for the bargain that ended the crisis to have come about. For example, in terms of preferences, Wagner asserts, persuasively, albeit predictably, that Khrushchev could not have anticipated the actual resolution of the crisis since, by Wagner's reckoning, it was "extremely unlikely that Khrushchev preferred this outcome to the one that would have resulted from an initial decision not to place the missiles in Cuba" (Wagner, 1989: 181). Along the same lines he suggests, in terms of beliefs or probabilities, that "Khrushchev must not have believed that…[the probability that Kennedy was prepared to take military action]…was high enough that he would choose voluntarily to remove the missiles if Kennedy demanded it, since otherwise he would not have decided to put them in Cuba" (Wagner, 1989: 184–5).

Proceeding backward through the game tree of his model, Wagner then constructs a (sequential) equilibrium[27] (Kreps and Wilson, 1982), or a set of strategies, that is consistent with the rational choices of both Kennedy and Khrushchev, given their inferred (required) preferences and beliefs. The formal analysis yields some interesting insights. For example, the characteristics of the equilibrium that Wagner discovers suggests that Khrushchev's

---

[27] A sequential equilibrium is a refinement of perfect Bayesian equilibrium. For a discussion, see Morrow (1994: ch. 7).

refusal to halt work on the missile sites provided him with valuable information about Kennedy's preferences, information that eventually made a compromise possible. Thus, Wagner argues persuasively that the crisis was not a "*competition* in risk taking," as Schelling (1966) suggests; rather, it was an exercise in bargaining and information gathering that is more than consistent with the fact that both Kennedy and Khrushchev lacked complete information about one another's preferences.[28]

Wagner's study provides a plausible and insightful description of the strategic dynamic that characterized the missile crisis. At the same time, it falls short of providing a compelling explanation. As Kaplan (1964: 339) notes, "to explain something is to exhibit it as a special case of what is known in general." But since Wagner's method is to deduce preferences from action choices, it cannot but fail to fully explain actual behavior (Morrow, 1994: 22).[29] For that, a general model that not only identifies a plausible equilibrium, but also specifies the general conditions under which it exists, is necessary. Of course, the identification of a plausible explanatory equilibrium in a general model is a necessary but not sufficient condition for a fully satisfying explanation of any real world event. Ideally, any such descriptor of actual play would be unique. But since Wagner's approach does not allow him to identify, let alone eliminate, coexisting equilibria, his explanation of the missile crisis must, ultimately, be judged deficient.

4.8 Coda

This chapter surveys and evaluates previous attempts to use game theory to explain the strategic dynamic of the Cuban missile crisis, a crisis that has been characterized, without exaggeration, as the "defining event of the nuclear age" (Allison and Zelikow, 1999: ix). All of the attempts are judged to be either incomplete or deficient in some other way. The earliest, most well-known, and most enduring interpretation of the crisis drew its inspiration from the pioneering work of Thomas Schelling. Schelling's "threat that leaves something to chance," however, fails as an explanation, on both theoretical and empirical grounds. Formal analyses of this bargaining tactic reveal it to be inconsistent with the consensus interpretation of the crisis's outcome. Equally troubling is the scant empirical evidence that the Kennedy administration either manipulated the risk of war during the crisis with "mathematical precision," as Schlesinger (1965: 767) and some other insider accounts have claimed, or successfully made use of any related brinkmanship tactics that resulted in a clear US victory. The crisis ended only when *both* sides "blinked."

---

[28] Wagner was right to dismiss Schelling's characterization of the Cuban crisis, but for the wrong reasons. Wagner's conclusion was based, in part, on his belief that the Soviets never put their bombers on alert. Wagner's source for his belief was Betts (1987). But Betts was mistaken. By October 26, "Soviet armed forces were on full alert" (Fursenko and Naftali, 1997: 266). Nonetheless, the fact that the Soviets did not make their alert known to US decision makers is inconsistent with Schelling's interpretation of the crisis.

[29] As discussed in Chapter 3, Wagner's approach, called *revealed preferences*, should be contrasted with the procedure of *posited preferences* that assumes an actor's goals rather than deducing them from actual behavior. For a discussion of the implications and differences, see Riker and Ordeshook (1973: 14–16). See also Hausman (2012: ch. 3).

Nigel Howard's metagame analysis of the missile crisis also fails to provide a compelling explanation. To be sure, Howard shows the superpower compromise that settled the crisis is a metaequilibrium in his game model and, consequently, consistent with a minimal definition of rational behavior. However, since the normative foundation of the compromise metaequilibrium is more than suspect, Howard is unable to explain why rational agents would settle on it, other than by observing that one of the players in his model (i.e., the United States) would lose if it were "perfectly" rational. Needless to say, rational choice explanations that reject even some of the logical imperatives of the rationality postulate are less than satisfying.

Similarly, the improved metagame technique of Fraser and Hipel falls short of the explanatory mark. Like Howard, Fraser and Hipel find that the compromise outcome is an equilibrium in their dynamic model but are unable to explain, at least game theoretically, why it, and not another coexisting equilibrium, ended the crisis. Making matters worse, the coexisting equilibrium is in fact a more accurate descriptor of the eventual resolution of the crisis than is the equilibrium implied by their modeling technique. Finally, since their description of the crisis begins after the missiles were installed, they fail to address one of the three critical questions that a complete explanation of the crisis requires.

Brams's theory-of-moves framework avoids the arbitrary foresight limitation that Fraser and Hipel assume of the players in their model. It also mitigates some of the problems associated with the existence of multiple, coexisting equilibria. But Brams takes as his starting point the imposition of the blockade on October 24. As a result, his explanation, like Fraser and Hipel's, is incomplete. Specifically, Brams is unable to explain why the Soviet Union decided to challenge the status quo by installing the missiles in the first place, and why the "initial" step taken by the United States was not an escalatory choice.

Wagner's model, which makes use of the modern methodology of games of incomplete information, does, in fact, address all of the central questions about the crisis. Wagner's model is carefully constructed from the facts of the crisis as they were known to him at the time. As such, it can be considered a concise and even a more than plausible *description* of the events of October 1962. But the explanatory power of his model is suspect. Theories and models that are constructed from facts cannot but fail to explain those facts. In consequence, explanations like Wagner's that verge on the tautological, are, ultimately, unconvincing (King, Keohane, and Verba, 1994: 19–23; Morrow, 1994: 22).

5

. . . . .

A General Explanation of the Cuban Missile Crisis

5.1 Introduction

Explanations of the Cuban missile crisis are common in the strategic literature. Like the competing game-theoretic explanations which were discussed in detail in Chapter 4 and were shown to be empirically suspect, incomplete, or tautological, more conventional explanations also fall short of the explanatory mark. Allison and Zelikow (1999: 78–109), for example, develop four different rational actor explanations of the Soviet decision to install the missiles. But the various explanations they describe, which are representative of the conventional wisdom, are all couched in terms of the strategic situation that existed in early 1962, when the decision to install the missiles was actually made. It should be clear, however, that case-specific explanations such as theirs do not completely suffice. As Abraham Kaplan (1964: 339) reminds us, "to explain something is to exhibit it as a special case of what is known in general."[1]

The absence of a general explanation of the missile crisis in the mainstream strategic literature is most likely due to way the foundational questions have been posed: why did the Soviet Union install the missiles in Cuba, why did the United States respond with a blockade and not an air strike or an invasion, and why did the Soviet Union remove the missiles? In this chapter, I address these questions more generally. My purpose, therefore, is to develop an explanation of the crisis that is not only consistent with the documentary record as it is known today but also more general than idiosyncratic explanations like those summarized by Allison and Zelikow (1999).

To this end, I explore the strategic dynamic of the missile crisis in the context of a single integrated game-theoretic model of interstate conflict initiation, limitation, and escalation. As will be seen, this model brings with it a clear set of theoretical expectations about the conditions under which intense interstate disputes occur and, if and when they do, are successfully resolved (or not). Thus, the explanation I offer is neither ad hoc nor post hoc. Rather, it is a logically implied consequence of an explicit set of assumptions applied to a

[1] See also Morrow (1994: 52). This chapter is based on Zagare (2016).

Game Theory, Diplomatic History and Security Studies. Frank C. Zagare. Oxford University Press (2019).
© Frank C. Zagare 2019. DOI: 10.1093/oso/9780198831587.001.0001

transparent theoretical model and not, as are most existing explanations, an after-the-fact rationalization of US and Soviet action choices. Put in a slightly different way, the explanation that is derived from the model applies not only to the Cuban case but to other interstate conflicts as well. This is as it should be; as King, Keohane, and Verba (1994: 43) point out, "where possible, social science research should be both general and specific: it should tell us something about classes of events as well as about specific events at particular places."

5.2 The Asymmetric Escalation Game

In developing this explanation, I explore the equilibrium structure of the *Asymmetric Escalation Game* (see Figure 5.1).[2] The Asymmetric Escalation Game provides a rich theoretical context in which to explore the missile crisis. For one, it is a general model that admits a range of conflict possibilities. Beyond its generality, the internal structure of Asymmetric Escalation Game model also closely tracks the decision-making environment that conditioned the Cuban crisis. This is, perhaps, its most attractive feature and the principal reason it provides a compelling theoretical context for explaining why the United States and the Soviet Union were able to settle their dispute short of war. Of course, closeness of fit is not a sufficient condition for rendering an abstract model suitable for empirical application. Also required is a set of theoretically derived and empirically supported preference and information assumptions. Below, I show this to be the case as well.

As Figure 5.1 shows, there are two players, the Challenger and the Defender, and six outcomes in the Asymmetric Escalation Game. The Challenger (i.e., the Soviet Union) begins play at Node 1 by deciding whether to contest the status quo. If the Challenger makes no demand (by choosing (C)), the outcome *Status Quo* (SQ) is obtained. But if the Challenger initiates conflict and demands a change to the existing order (by choosing (D)), the Defender (i.e., the United States) decides (at Node 2) whether to capitulate or concede (by choosing (C)) or to respond, and if the latter, whether to respond in kind (by choosing (D)) or to escalate the conflict (by choosing (E)).

Capitulation ends the game at *Defender Concedes* (DC). If the Defender responds, the Challenger can escalate or not at Nodes 3a or 3b. If the Challenger is the first to escalate (at Node 3a), the Defender is afforded an opportunity at Node 4 to counter-escalate. *Limited Conflict* (DD) occurs if the Defender responds in kind and the Challenger chooses not to escalate at Node 3a. The outcome *Challenger Escalates (Wins)* (ED) occurs if, at Node 4, the Defender chooses not to counter-escalate. Similarly, the outcome is *Defender Escalates (Wins)* (DE) if the Challenger chooses not to counter-escalate at Node 3b. *All-Out Conflict* (EE) results whenever both players escalate.

Because the Asymmetric Escalation Game is a general model, it is applicable to a wide range of empirical circumstances and is in no way restricted by the terms used to denote its component parts. For example, the outcome *Limited Conflict*, which occurs if and only if both players defect but neither escalates, could easily and quite naturally be associated with an ongoing real world dispute that lingers on short of war. But the outcome could also be

---

[2] The Asymmetric Escalation Game model will be used in Chapter 6 to explore the July Crisis of 1914. It has also been used to study NATO'S 1998 war with Serbia over Kosovo (Quackenbush and Zagare, 2006).

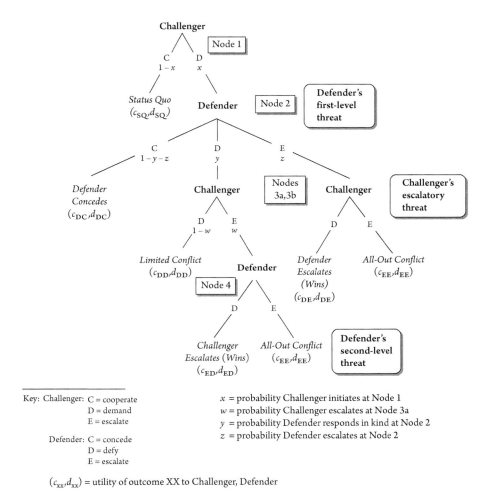

Figure 5.1 The Asymmetric Escalation Game with incomplete information

used to capture the denouement of an intense crisis that is resolved, possibly in short order, by political negotiations, as was the case in 1962. It would be a straightforward exercise to change the nomenclature of the Asymmetric Escalation Game to conform to the specifics of the Cuban crisis, in which case the outcome *Limited Conflict* could be variously labeled "brokered settlement," "negotiated outcome," or "compromise," but that would change nothing. Thus, to maximize generality and to facilitate cross-case comparisons, the labels originally used to construct the model will be retained.

All of which raises the issue of how closely the Asymmetric Escalation Game model conforms to the actual situation in 1962. It is not a reach at all to suggest that the Challenger's choice at Node 1 captures the dilemma facing Khrushchev early that year. Had Khrushchev done nothing, that is, chosen (C), the lopsided strategic relationship of the superpowers would have remained unbalanced, and a status quo in which Cuba was open to an American attack would have continued indefinitely (Khrushchev, 1970: ch. 20). On the other hand, the Soviet Premier was obviously contesting the status quo, that is, choosing (D), when he decided to ship the missiles to Cuba. US decision makers saw it this way as well (May and Zelikow, 1997: 235).

But what about the Defender's tripartite choice at Node 2? It is widely accepted that the Kennedy administration considered a bewildering number of nuanced responses, including but not limited to doing nothing, blockading Cuba, removing the missiles with a "surgical" air strike, and removing the missiles *and* Cuban prime minister Fidel Castro with a massive air strike and a subsequent invasion (Allison and Zelikow, 1999). Nonetheless, all were tactical options that correspond roughly to the choices available to the Defender at Node 2. Of course, doing nothing (i.e., choosing (C)) was always an available response; but, as Secretary of Defense Robert McNamara put it on October 18, there were only two additional alternatives: "one is a minimum action, a blockade approach, with a slow buildup to subsequent action. The other is a very forceful military action with a series of variances as to how you enter it" (May and Zelikow, 1997: 162). In terms of the model, the United States could either measure its response (i.e., choose (D)) or escalate the conflict (i.e., choose (E)).

It is noteworthy that McNamara went on to recommend to those advising the president that they "consider how the Soviets are going to respond" to whatever course of action the United States took. Significantly, at that point the State Department's Soviet expert Llewellyn Thompson chimed in: "Well, not only the Soviet response, but what the response to the response will be" (May and Zelikow, 1997: 162). Nodes 3a and 3b reflect the possible Soviet counter-response to an American action choice: to escalate (i.e., to choose (E)) or not (i.e., to choose (D)). Of course, if the Soviets were the first to escalate (at Node 3a), the United States always had the option of counter-escalating. Node 4 takes account of this possibility, which is the one that Thompson had in mind.

To summarize briefly, the Asymmetric Escalation Game is a general model of interstate interaction that bears a prima facie connection to the broad outlines of the Cuban missile crisis. It admits two distinct conflict possibilities: one limited and one all out. As well, there are two distinct paths to *All-Out Conflict* in the model. One occurs when the Defender escalates first at Node 2 and the Challenger counter-escalates at Node 3b. The second, which corresponds to the classic escalation spiral, results when the Challenger initiates a conflict, the Defender resists at Node 2, the Challenger is the first to escalate at Node 3a, and the Defender counter-escalates at Node 4. To a large extent, it was a fear of this potentially disastrous sequence of moves and countermoves that conditioned the choices made by both Kennedy and Khrushchev during the crisis.

5.3 Preferences and Type Designations

The preference assumptions that will be used to explore the underlying strategic dynamic of the Cuban missile crisis are summarized in Table 5.1. The Challenger's preferences are listed in the first column, from best to worst; the Defender's are given in the second. For example,

Table 5.1 Preference assumptions for the Asymmetric Escalation Game.

Challenger	Defender
Defender Concedes	*Status Quo*
Status Quo	*Defender Escalates*
Challenger Escalates	*Defender Concedes* or *Limited Conflict*
Limited Conflict	*Challenger Escalates* or *All-Out Conflict*
Defender Escalates or *All-Out Conflict*	

the assumption is that the Challenger most prefers *Defender Concedes*, then the *Status Quo*, and so on. No fixed preference assumption is made for outcomes contained in the same cell of Table 5.1. Thus, in what follows, the Challenger could prefer *Defender Escalates* to *All-Out Conflict* or the reverse. Similarly, the Defender's preference between *Defender Concedes* and *Limited Conflict* and between *Challenger Escalates* and *All-Out Conflict* is left open. The Challenger's and the Defender's relative preferences for these three sets of paired outcomes are the crucial explanatory variables of the version of the model described herein.

The three pairs of unspecified preference relationships represent threats that the players may or may not prefer to execute. The Challenger has only one threat: to escalate (i.e., choose (E)) or not (i.e., choose (D)) at Nodes 3a and 3b. The Defender, however, has two threats: a *tactical level* threat to respond in kind (i.e., choose (D)) at Node 2, and a *strategic level* threat to escalate (i.e., choose (E)) at Nodes 2 and 4.

Each player's willingness, or lack thereof, to execute its threat(s) determines its *type*. The Challenger, having only one threat, may be one of two types: *Hard* Challengers are those that prefer *All-Out Conflict* to *Defender Escalates*; Challengers with the opposite preference are called *Soft*. Defenders, by contrast, are more difficult to type cast. A Defender that prefers *Limited Conflict* to *Defender Concedes* is said to be *Hard* at the first (or tactical level), while a Defender with the opposite preference is said to be *Soft* at the first level. Similarly, a Defender that prefers *All-Out Conflict* to *Challenger Escalates* is said to be *Hard* at the second (or strategic level), while a Defender with the opposite preference is said to be *Soft* at the second level. Thus, a Defender may be one of four types: Hard at the first level but Soft at the second (i.e., type HS), Soft at the first level but Hard at the second (i.e., type SH), Hard at both levels (i.e., type HH), or Soft at both levels (i.e., type SS).

The assumption is that each player knows its own type (preferences) but is unsure of its opponent's. The Defender's lack of information about the Challenger's type, and the Challenger's about the Defender's, constitutes the principal source of uncertainty in the model. Specifically, the Defender believes the Challenger to be *Hard* with probability p_{Ch}, and *Soft* with probability $1 - p_{Ch}$. Likewise, the Challenger believes the Defender to be of type HH with probability p_{HH}, of type HS with probability p_{HS}, of type SH with probability p_{SH}, and of type SS with probability p_{SS}. These beliefs and all other elements of the model, including the choices available to the players at each decision point, the outcomes of the game, and the preference relationships, as specified in Table 5.1, are assumed to be common knowledge.

For the most part, the (postulated) preferences that will be used to analyze the Asymmetric Escalation Game are both straightforward and transparent. Underlying the arrayed preferences given in Table 5.1 is the standard assumption that the players prefer winning to losing. To reflect the costs of conflict, the players are also presumed to prefer to win or, if it comes to it, to lose at the lowest level of conflict. Thus, the Challenger prefers *Defender Concedes* (outcome DC) to *Challenger Wins* (outcome ED)—and so does the Defender.[3]

There is, however, one assumption that may not be so obvious. Specifically, the assumption is that neither player prefers that the other execute any threat it may possess. In terms of preferences, this means that both players prefer the *Status Quo* to *Limited Conflict*, and *Limited Conflict* to *All-Out Conflict*. In other words, all threats, when executed, hurt. Threats that hurt are called *capable* (Schelling, 1966: 7; Zagare, 1987: 34).

Capable threats should be distinguished from threats that are *credible*. Credible threats are defined as threats that are believable precisely because they are rational to execute, that is, threats that a player prefers to carry out. Clearly, perfectly credible threats require complete information about preferences. Such is not the case, however, in the present analysis of the Asymmetric Escalation Game, where the players are assumed to have only probabilistic knowledge (i.e., subjective beliefs) about one another's type. In the analysis that follows, these beliefs are taken as a measure of each player's credibility.

For instance, the greater the value of p_{Ch} (i.e., the Defender's belief that the Challenger is Hard), the greater is the perceived credibility of the Challenger's threat. Similarly, the greater the value of p_{HH}, the greater is the perceived credibility of the Defender's tactical and strategic level threats. The overall probability that the Defender prefers conflict to capitulation at the first (or *tac*tical) level is the perceived credibility of the Defender's first-level threat. This probability, that the Defender is of type HH *or* type HS, is denoted $p_{Tac} = p_{HH} + p_{HS}$; similarly, the perceived credibility of the Defender's second-level (or *str*ategic) threat is $p_{Str} = p_{HH} + p_{SH}$.

5.4 Equilibria

In this section, I briefly describe the perfect Bayesian equilibria of the Asymmetric Escalation Game with incomplete information. In this game, a perfect Bayesian equilibrium must specify the action choice of both a Hard and a Soft Challenger at Nodes 1, 3a, and 3b and for all four types of Defenders at Nodes 2 and 4. It must also indicate how each player updates its beliefs rationally about other players' types in the light of new information obtained as the game is played out. For instance, should the Challenger instigate a crisis by choosing (D) at Node 1, the Defender will have an opportunity to reevaluate its initial beliefs about the Challenger's type before it makes a choice at Node 2. Similarly, if and when the Challenger is faced with a decision at Node 3a, it will have observed the Defender's choice of (D) at Node 2. The assumption is that the Challenger will rationally reassess its

---

[3] Note that these preferences are entirely consistent with the preferences used to analyze the Tripartite Crisis Game in Chapter 3. In other words, they are not theoretically ad hoc.

beliefs about the Defender's type and, therefore, the Defender's likely response at Node 4, based on that observation.

The equilibrium structure of the Asymmetric Escalation Game is more than complex. There are eighteen perfect Bayesian equilibria in the Asymmetric Escalation Game with incomplete information (Zagare and Kilgour, 2000: app. 8; Kilgour and Zagare, 2007). Making matters worse, many of the equilibria are distinguished only by minor technical differences that are of little theoretical interest or import. Clearly, a straightforward description of these equilibria would not only be tedious but unproductive as well.

A special case analysis, however, alleviates the problem. The number of perfect Bayesian equilibria in the Asymmetric Escalation Game is dramatically reduced when the assumption is made that the Challenger is *likely* Hard. Moreover, the perfect Bayesian equilibria in the special case fully exemplify the existence conditions in the general case. In other words, although this assumption about the Challenger's type simplifies the analysis of the Asymmetric Escalation Game, it does so without any serious loss of information. Little is to be gained, therefore, by examining its strategic structure in the absence of this simplifying assumption.

The assumption that the Challenger is likely Hard, however, is not just convenient; it is also consistent with the beliefs and the expectations of the Kennedy administration throughout the crisis. Both the president and the vast majority of his advisors firmly believed that the Soviets would respond forcefully, regardless of the course of action taken.[4] Kennedy himself thought that the most probable Soviet target would be Berlin. Others, however, feared an attack on the missile sites in Turkey. All of which is to say that the special case analysis is both theoretically and empirically justified.[5]

As Table 5.2 reveals, there are six rational behavioral possibilities when it is highly likely that the Challenger is Hard. These six perfect Bayesian equilibria can be conveniently placed into three major groups. The first is a family of several equilibria called the *Escalatory Deterrence Equilibria*. But since all members of this family are based on beliefs that are implausible, they will be ignored.[6] This leaves only five possible solutions of the game: the *No-Response Equilibrium* (NRE), which always exists, and the four members of the *Spiral Family*, of which precisely one will always coexist with the No-Response Equilibrium.

Of the five plausible perfect Bayesian equilibria that exist in the special case (see Table 5.2), two are deterrence equilibria (Det_2, and Det_3). These two closely related equilibria, which are in the Spiral Family, are called the *Limited-Response Deterrence Equilibria*. Under either equilibrium form, the Challenger never initiates at Node 1 (i.e., $x_H = x_S = 0$), and the outcome of the game is always *Status Quo*. As their name implies, equilibria of this

---

[4] The belief was accurate (Fursenko and Naftali, 2006: 472).
[5] Secretary of State Dean Rusk even put it in the terms of the model. Speaking at a meeting in the Cabinet Room of the White House just before the president's televised address, he remarked that it was "clear now that the hard-line boys have moved into the ascendancy" in the Kremlin (May and Zelikow, 1997: 255).
[6] For a member of this family to exist, the Defender must believe that any demand for a change in the *Status Quo* would be a mistake made by a genuinely Soft Challenger (i.e., the Defender's updated belief, probability r, that the Challenger is Hard given that the Challenger chooses D at Node 1 must be relatively small). There may be situations when this kind of belief is, in fact, plausible. But the special case analysis, which is based on the Defender's a priori belief that the Challenger is very likely hard, is patently not one of them. For a detailed explanation, see Zagare and Kilgour (1998: 73–4).

Table 5.2 Equilibria of the Asymmetric Escalation Game when the Challenger has high credibility.*

	Challenger					Defender						
	x		w		q_{HH}	y		z				r
	x_H	x_S	w_H	w_S		y_{HH}	y_{HS}	z_{HH}	z_{HS}	z_{HH}	z_{SS}	
Escalatory Deterrence Equilibria (typical)												
Det$_1$	0	0	1	1	Small	0	0	1	1	1	1	$\leq d_1$
No-Response Equilibrium												
NRE	1	1	Large		Small	0	0	0	0	0	0	p_{Ch}
Spiral Family of Equilibria												
Det$_2$	0	0	0	0	$p_{Str\|Tac}$	1	1	0	0	0	0	$\geq d_2$
Det$_3$	0	0	d^*/r	0	c_q	1	v	0	0	0	0	$\geq d_2$
CLRE$_1$	1	1	0	0	$p_{Str\|Tac}$	1	1	0	0	0	0	p_{Ch}
ELRE$_3$	1	1	d^*/p_{Ch}	0	c_q	1	v	0	0	0	0	p_{Ch}

* Table 5.2 is excerpted from Table A8.1 in Zagare and Kilgour (2000: app. 8), which should be consulted for details of definitions and interpretations. Definitions of the strategic and belief variables appearing in Table 5.2 are summarized here for convenience.

The probability that the Challenger initiates at Node 1 of the Asymmetric Escalation Game is denoted x. In fact, this probability can depend on the Challenger's type—if the Challenger is Hard, the initiation probability is x_H; if Soft, x_S. Likewise, w_H and w_S are the probabilities that Hard and Soft Challengers, respectively, escalate at Node 3a. At Node 3b, the Challenger always chooses (E) if Hard, and (D) if Soft.

Similarly, the Defender chooses (D) at Node 2 with probability y, (E) with probability z, and (C) with probability $1 - y - z$. Again, these probabilities can depend on the Defender's type, so they are denoted y_{HH}, z_{HS}, and so on. It can be proven that $y_{SH} = y_{SS} = 0$ at any perfect Bayesian equilibrium. At Node 4, the Defender chooses (E) if strategically Hard (type HH or SH); otherwise, it chooses (D).

Finally, players revise their initial probabilities about their opponent's type as they observe the opponent's actions. Of these revised probabilities, the only two that are important to the equilibria are shown in Table 5.2. The Defender's revised probability that the Challenger is Hard, given that the Challenger initiates, is denoted r. The Challenger's revised probability that the Defender is of type HH, given that the Defender chooses (D) (response in kind) at Node 2, is denoted q_{HH}.

category do not require the Defender to escalate first. In fact, the form of deterrence that emerges under either Det$_2$ or Det$_3$ rests *entirely* on the more limited threat of responding in kind at Node 2.

This characteristic alone sets the Limited-Response Deterrence Equilibria apart from Det$_1$ and all other members of the Escalatory Deterrence Equilibria group. Additionally, since the Limited-Response Deterrence Equilibria are based on plausible beliefs, they are not so easy to dismiss, at least as theoretical possibilities. Indeed, as will be seen in Chapter 6, the fact that the two Limited-Response Deterrence Equilibria exist as plausible rational strategic possibilities casts a significant light on the way an important question about the onset of war in 1914 is answered.

The existence of a Limited-Response Deterrence Equilibrium depends solely on the Challenger's beliefs about the Defender's type (the Defender's a priori beliefs are completely immaterial to their existence). Specifically, for Det_2 or Det_3 to exist, the Defender's first- and second-level threats must both be highly credible: the Challenger must believe it quite likely that the Defender is tactically Hard, and, given that the Defender is tactically Hard, the Challenger must place a fairly high probability on the Defender being strategically Hard also.[7] Given these beliefs, the Challenger generally intends not to escalate at Node 3a because it believes that the Defender will likely counter-escalate at Node 4; and because the Challenger believes that the Defender will almost certainly respond in kind at Node 2—most likely forcing the Challenger to back down at Node 3a—the Challenger decides *not* to initiate at Node 1.

Clearly, all three deterrence equilibria are inconsistent with Soviet behavior during the crisis and can, therefore, be immediately eliminated as possible descriptors of the striking events of October 1962. There are, therefore, only three other rational strategic possibilities: the No-Response Equilibrium, one representative of the *Constrained Limited-Response Equilibrium* group, $CLRE_1$; and one member of the *Escalatory Limited-Response Equilibrium* group, $ELRE_3$. Under any of these equilibria, the Challenger *always* chooses (D) at Node 1 (i.e., $x_H = x_S = 1$). At minimum, then, each is consistent with the Soviet decision to install ballistic missiles in Cuba. In addition, the most likely outcome under each equilibrium is *Defender Concedes*. Thus, we have a compelling theoretical reason, rather than an empirical inference, to explain why Khrushchev was taken aback when Kennedy announced the blockade in a televised speech on Monday, October 22.

Of these three remaining rational strategic possibilities, the No-Response Equilibrium can also be eliminated on empirical grounds. As its name suggests, under this equilibrium form, the Defender always concedes and never responds, either in kind or by escalating at Node 2 (i.e., the strategic variables y and z always equal zero for all four types of Defender), which is why the only outcome that is consistent with rational choice under the No-Response Equilibrium is *Defender Concedes*.[8] The same cannot be said, however, about either the Constrained Limited-Response Equilibrium $CLRE_1$ or the Escalatory Limited-Response Equilibrium $ELRE_3$. In fact, since a *Limited Conflict* is a theoretical possibility under either equilibrium form, both remain potential descriptors of actual play during the Cuban missile crisis. What remains to be shown, therefore, is not whether the action choices of the United States and the Soviet Union are consistent with these two equilibria but whether the beliefs of President Kennedy, and especially of Premier Khrushchev, are consistent with those that are necessary to support these choices in equilibrium.

$CLRE_1$ is the only form of the Constrained Limited-Response Equilibrium that exists when the Challenger is likely Hard. Under this equilibrium, the outcome *Status Quo* is never obtained; the Challenger *always* initiates. For its part, the Defender responds in kind if it is tactically Hard (i.e., of type HH or HS). Otherwise, the Defender capitulates. Since this member of the Spiral Group of perfect Bayesian equilibria exists only when the Defender is likely Soft at the first level, the most likely outcome of play under $CLRE_1$ is *Defender Concedes*. Thus, when the Challenger chooses (D) at Node 1, it does so with the expectation that its demands will almost certainly be met.

---

[7] More technically, $p_{Str|Tac}$ must be large.
[8] The Defender gives in because the Challenger is very likely Hard and, therefore, prone to escalate first at Node 3a or counter-escalate at Node 3b. To support its choice at Node 3a, however, the Challenger must believe that a Defender who unexpectedly *responds in kind* at Node 2 is more likely to be of type HS than of type HH. This is a plausible belief since, ceteris paribus, type HH Defenders are more likely to *escalate* than type HS Defenders.

Put in another way, should the Defender respond in kind, the Challenger will be surprised. In this unlikely event, the Challenger will be forced to update its beliefs about the Defender's type. Clearly, the Challenger will conclude that the Defender is tactically Hard, since only tactically Hard Defenders can rationally choose (D) at Node 2. Moreover, under any Constrained Limited-Response Equilibrium, *if* the Defender is Hard at the first level, then it is also likely Hard at the second level as well, that is, it is more likely to be of type HH than of type HS. Fearing this possibility, the Challenger is, understandably, deterred from escalating at Node 3a; instead, it always chooses (D) at Node 3a, settling for a *Limited Conflict*.

The Constrained Limited-Response Equilibrium group is strategically significant, especially for the purposes of this chapter, if only because a *Limited Conflict* is most likely when a member of this group is in play. This is why particular attention is paid to the conditions under which $CLRE_1$ exists. But, more generally, the conditions that give rise to the existence of a Constrained Limited-Response Equilibrium may help to explain why, at times, states abruptly shift gears and adjust their behavior mid crisis.

Limited Conflict is also possible under $ELRE_3$, the only form of Escalatory Limited-Response Equilibrium that exists when the Challenger is likely Hard, but the possibility is remote, at best. In fact, the most likely outcome of a game played under this equilibrium form is, once again, *Defender Concedes*. Whenever $ELRE_3$ is in play, the Challenger, whatever its type, always chooses (D) at Node 1, thereby upsetting the *Status Quo*. What happens next depends on the Defender's type. Under $ELRE_3$, the Defender is likely to be tactically Soft (i.e., of type SS or SH). Such Defenders *always* concede at Node 2, which is why *Defender Concedes* is the most likely outcome under any Escalatory Limited-Response Equilibrium. In the less likely event that the Defender is Hard at the first level, it would respond in kind, with certainty if it is also Hard at the second level (i.e., of type HH), and probabilistically if it is Soft at the second level (i.e., of type HS). Given the probabilities, however, a response in kind will once again surprise the Challenger.

Under either $CLRE_1$ or $ELRE_3$, then, the Challenger always initiates, so the *Status Quo* never survives. The Defender responds in kind, either with certainty or probabilistically, if it is tactically Hard (i.e., of type HH or HS). Otherwise, it simply capitulates, and the outcome is *Defender Concedes*. Since both $CLRE_1$ and $ELRE_3$ exist only when the Defender is seen to be likely Soft at the first level, (i.e., when p_{Tac} is low), a response in kind will always come as a surprise to the Challenger. Of course, when this happens, the Challenger is forced to update its beliefs about the Defender's type. Clearly, the Challenger will now know that the Defender is, in fact, tactically Hard, since only tactically Hard Defenders can rationally choose (D) at Node 2.

Up to this point of surprise and reevaluation, behavior and expectations are similar under $CLRE_1$ and $ELRE_3$. What separates these two equilibria are the Challenger's expectations should the Defender unexpectedly choose (D) at Node 2. Under $CLRE_1$, if the Defender is Hard at the first level, then it is also likely Hard at the second level, which is why Challengers never escalate first under a Constrained Limited-Response Equilibrium. It is also why a *Limited Conflict* is a distinct theoretical possibility whenever $CLRE_1$ is in play.

While a *Limited Conflict* is also a theoretical possibility under $ELRE_3$, that possibility is just that. $ELRE_3$ exists only when a tactically Hard Defender is much more likely to be of type HS than of type HH. It is for this reason that Hard Challengers tend to escalate first at Node 3a. At this point, the Defender will most likely back off, and the outcome will be *Challenger Escalates (Wins)*. But, from time to time, the Challenger's belief about the Defender's type will be wrong. When this happens, the Defender will counter-escalate and an *All-Out Conflict* will take place. As will be seen in Chapter 6, the escalation spiral that brought about World War I is a case in point.

In the Asymmetric Escalation Game with incomplete information, therefore, *Limited Conflict* can only take place when either $CLRE_1$ or $ELRE_3$ is in play. For either equilibrium to exist, however, the Challenger must, at minimum, believe that the Defender is likely to capitulate immediately (because it is thought to be tactically Soft). Clearly, this theoretical requirement was met during the Cuban crisis and helps to explain why the missiles were placed in Cuba in the first place. But, for the crisis to have been resolved as it was, additional conditions would have to be met as well. Obviously, the Defender would also have to respond unexpectedly, and its response would have to be in kind rather than escalatory—precisely because the Defender believes that an escalatory response would lead to an all-out conflict. Again, this belief is consistent with the expectations of the Kennedy administration about the likely Soviet response to either an air strike and/or an invasion. None of this is in the least bit surprising. What would be surprising, however, is for *Limited Conflict* to actually occur under $ELRE_3$. But if and until it can be eliminated on other than probabilistic grounds, it must be considered a rational strategic possibility.

5.5 Explanation

Up to this point, I have shown that a *Limited Conflict* is most likely to occur in the Asymmetric Escalation Game with incomplete information when play takes place under the Constrained Limited-Response Equilibrium $CLRE_1$, and that the key to its existence is the Challenger's initial and updated beliefs about the Defender's type. Thus, the hypothesis is that crises that are resolved politically will most likely occur when a Challenger, expecting an easy victory, meets unexpected resistance and then concludes, perhaps reluctantly, that discretion is the better part of valor. In this section, I explain the political compromise that resolved the Cuban missile crisis by demonstrating a strict correspondence between these behavioral expectations and Soviet action choices *and* beliefs. Since this is, fortunately, a straightforward exercise, the explanation that I offer is at once natural and intuitive. But this is as it should be, at least most of the time. Moreover, a theoretically derived explanation that is in accord with the facts on the ground is at once more compelling and more satisfying than ad hoc explanations and ex post rationalizations, but especially when many of the relevant facts are no longer in dispute (Gaddis, 1997; Stern, 2012). Facts do not necessarily speak for themselves. Theories are required to give them both meaning and context.

In the Cuban case, many of the undisputed facts involve Soviet action choices: although Khrushchev's motivation is unclear (Allison and Zelikow, 1999: 107–9), the missile crisis was precipitated when US decision makers became aware that the Soviet Union was in the process of installing medium- and intermediate-range ballistic missiles in Cuba in mid

October 1962. Khrushchev was surprised not only that the missiles were discovered but also that the Kennedy administration reacted by clamping a blockade around Cuba. We also know that, eventually, a settlement was brokered: in exchange for removing the missiles, Khrushchev received a public assurance from the United States that it would not invade Cuba, and a secret assurance that it would, in due course, remove American-controlled Jupiter missiles from Turkey. The clear theoretical expectation is that the brokered agreement (i.e., a limited conflict) would have had to been preceded by a series of events that led Khrushchev to reevaluate his initial beliefs about the likely consequences of his actions. Otherwise, the crisis's resolution is simply inexplicable.

The reevaluation process, which began even before a personal letter from the president and a copy of his televised address were delivered to the Kremlin on October 22, did not take very long. What explains Khrushchev's dramatic policy reversal? It was not, as many have concluded, the thinly veiled threat that the president's brother, Robert F. Kennedy (1969: 108), delivered when he met with the Soviet ambassador, Anatoly Dobrynin (1995: 88), on Saturday, October 27; nor was it the attorney general's pledge to remove the Jupiters from Turkey, made the same evening; it was also none of the "seven things" that happened during the day of October 27 and that Secretary of State Dean Rusk thought might induce the Soviets to reverse course (May and Zelikow, 1997: 616). It wasn't even just the blockade (Allison and Zelikow, 1999: 128). With the exception of the blockade, all of these supposed signals were sent after Khrushchev (1970: 553) changed his mind and decided "to look for a dignified way out of this conflict."

"No single piece of information seems to have moved Khrushchev to his new position" (Fursenko and Naftali, 1997: 260). And while "there is little evidence to explain exactly why Khrushchev reversed his assessment of American intentions" (Allison and Zelikow, 1999: 125), there is no doubt and very little dispute that, for one reason or another, he became "convinced that the Soviet Union could not keep ballistic missiles in Cuba without going to war" (Fursenko and Naftali, 1997: 259). And it was war that Khrushchev (1990: 176) "didn't want."

Khrushchev's strong belief that war was likely should the Soviets "inflame the situation" and escalate the conflict by running the blockade and pushing forward with the installation of the missiles is consistent with the beliefs necessary to support $CLRE_1$ in equilibrium but is inconsistent with the beliefs associated with the existence of $ELRE_3$ (Malin and Khrushchev, 1962: 2). Recall that, under $CLRE_1$, the Challenger believes that a Defender who is tactically Hard is likely to be strategically Hard as well. This belief leads logically to the expectation that escalation at Node 3a will result in an *All-Out Conflict*. By contrast, under $ELRE_3$, the Challenger believes that a Defender who is tactically Hard is more likely to be strategically Soft and that, therefore, an escalatory choice at Node 3a will most likely bring about the outcome *Challenger Escalates (Wins)*. All of which is to say that $ELRE_3$ can now be eliminated on empirical grounds as a viable rational strategic alternative, so that $CLRE_1$ is the *only* perfect Bayesian equilibrium in the Asymmetric Escalation Game that is consistent with both the beliefs and the action choices of US and Soviet decision makers throughout the crisis.

Consistent with his beliefs about the consequences of an escalatory choice, Khrushchev did a strategic about-face and decided to "conduct a reasonable policy" (Malin and Khrushchev, 1962: 2). "The decision to end the crisis through diplomatic means was made on the night of Wednesday October 25" (Fursenko and Naftali, 2006: 616, fn. 69) at a meeting of the Soviet presidium. Khrushchev began that meeting by explaining why he thought that the missiles should be withdrawn: "The Americans say that the missile installations in Cuba must be dismantled. Perhaps this will need to be done. This is not capitulation on our part. Because if we fire, they will also fire."

But he did not back down entirely. He wanted to bargain: "We have to give the opponent a sense of calm and, in return, receive assurances concerning Cuba." Then he suggested his terms: "Kennedy says to us: take your missiles out of Cuba. We respond: 'Give firm guarantees and pledges that the Americans will not attack Cuba.' That is not a bad [trade]" (Malin and Khrushchev, 1962: 2).

Not surprisingly, his proposal was unanimously supported by the presidium. But it was left up to Khrushchev to decide when and how to seal the deal. That moment came soon after the presidium met. "Early on Friday, October 26, Khrushchev received a stream of information indicating the likelihood that the Americans were readying an attack for October 27" (Fursenko and Naftali, 2006: 486). Time was obviously running out, or so he believed. Hence, on October 26, he wrote Kennedy a long rambling letter that outlined the bargain that, eventually, ended the crisis (May and Zelikow, 1997: 485–91; Stern, 2012: 139). Most of what occurred afterwards, including Khrushchev's infamous second letter, which was written on October 27 and in which he roiled Kennedy by publically demanding the removal of the missiles in Turkey, was simply diplomatic haggling, even if no one recognized it at the time. It would take a few more days to work out the details.[9]

5.6 The United States and North Korea: A Cuban Missile Crisis in Slow Motion?

Game-theoretic models can be applied in a variety of ways. For example, we just used a generic escalation model to construct an explanation for an acute interstate crisis that took place over fifty years ago. Could such a model also be used in real time to predict an ongoing event or one that is expected to occur sometime in the near future? The short answer is a qualified "yes." To understand why, consider for a moment the long-simmering dispute between the United States and North Korea, the basic parameters of which have remained unchanged for some time and across several US administrations.

According to Graham Allison (2017), North Korea's ongoing policy push to develop a nuclear-capable intercontinental ballistic missile is "a Cuban Missile Crisis in slow motion." If Allison is correct, then the Asymmetric Escalation Game model would, arguably, be an appropriate context in which to explore the current "crisis."[10] It is not clear, however, that Allison is correct.

---

[9] This is not to say that the deal could not have fallen apart. But that was unlikely. Khrushchev, who did not bring up the missiles in Turkey in his first letter, was prepared to settle the crisis even without their removal. And Kennedy was prepared to sweeten the pot by including a promise to remove US-controlled missiles in Italy and, perhaps, Great Britain.

[10] And, possibly, the Joint Comprehensive Plan of Action that the P5+1 (China, France, Germany, Russia, the United Kingdom, and the United States), and the European Union reached with Iran in 2015.

Of course, there are similarities. It is an established fact that the ballistic missiles that were installed in Cuba were capable of delivering a nuclear blow to much of the continental United States; recent improvements in North Korea's missile program are believed to have given Pyongyang the same capability—if not presently, then in the not too distant future. And, from the point of view of American policy makers, the leaders of both nations— Castro in Cuba, and Kim Jong-un in North Korea, are seen as mercurial and borderline irrational. It is easy to understand why Allison might see these two situations as similar and offer policy advice based on that assessment.

But there are also important differences. In 1962, the Cuba was not a player; the missiles of October were fully controlled by the Soviet Union, not by the Cuban government. And while North Korea is generally considered a Chinese client state, China's ability to influence Kim's policy choices has its limits. As well, while the Cuban crisis was a clearly a face-off between two nuclear superpowers, the slow-moving crisis developing on the Korean penin- sula involves a nuclear superpower and a minor power, albeit one with a nascent nuclear capability but whose supposed protector, China, packs a considerably more powerful nuclear punch. The distinct circumstances surrounding these two cases, one past and one ongoing, immediately raise a question about the appropriateness of the Asymmetric Escalation Game model for analyzing the political dynamic of the North Korean situation.

If, on reflection, it is determined that it is not a good fit, an appropriate model would have to be either identified or constructed. One distinct possibility is that the current crisis is better analyzed as a one-sided deterrence relationship. In that case the Unilateral Deterrence Game (see Chapters 2 and 7) could be used. On the other hand, if the relationship between the United States and North Korea is seen as one of mutual deterrence, the Generalized Mutual Deterrence Game, to be discussed in Chapter 8, would be a far better choice. In either case, the underlying model would have to be carefully calibrated to reflect the relative power imbalance and unique circumstances that define the relationship. Since the strategic characteristics of these two models, not to mention of the Asymmetric Escalation Game, are significantly different, it should be clear that the choice of an applicable model for any ongoing event is not always a straightforward exercise. It is very easy to go wrong whenever a model is used to predict behavior in real time, especially since significant data limitations will undoubtedly exist. And, as the United States found out in 2003, overconfidence in a strategic assessment can easily lead to a foreign policy debacle.

But let us further assume, just for the sake of discussion, that the Asymmetric Escalation Game model does, in fact, capture the salient characteristics of the "Cuban Missile Crisis in slow motion" and that North Korea is one of the two players in the game? What, then, is its role? Is North Korea the Challenger, and the United States the Defender, or is it the other way around? If North Korea's pursuit of an intercontinental nuclear striking force is simply a tactic to deter an invasion and protect Kim's regime, then North Korea might properly be thought of as the Defender in the model. In that case, the United States could only play the role of the Challenger. Some analysts, however, believe that Kim's goal is nuclear blackmail. In this case, the roles of North Korea and the United States would be reversed. Or, like the Germans in 1905 and 1906, North Korea may simply want to separate the United States from its allies in the region, Japan and South Korea (Rich and Sanger, 2017) (in which case, the Tripartite Crisis Game could be the model of choice), What, in other words, does

North Korea want—what are its preferences? As noted in Chapter 2, since preferences cannot be directly observed, their determination is oftentimes the most intractable problem that one must overcome before using a game model to either construct an analytic narrative or scrutinize an ongoing event. The problem is particularly acute when one of the players operates in a closed society. Again, it is very easy to go wrong here, too.

All of which is to say that point predictions drawn from any model based on inputs that are speculative at best is a dangerous game, especially since many models have multiple equilibria, the existence of which further confounds both explanations and predictions. Even within the confines of the Asymmetric Escalation Game, a range of rational strategic possibilities exist. As will be seen in Chapter 6, under the conditions that existed in 1914, the great power war that eventually broke out was always a distinct, albeit somewhat remote, theoretical possibility; by contrast, in 1962, slightly different parameter conditions, fortunately, enabled the United States and the Soviet Union to settle on an equilibrium that did not escalate to the highest level. Neither denouement, however, was predetermined. Since the strategic dynamic of any particular future crisis or event will likely be highly path dependent, expectations about the flow of these events can only be seen as exceedingly contingent—contingent not only on the specific path taken but also on the reliability of the model's inputs. When all the stars are properly aligned, game models can and have been fruitfully used to anticipate conflict behavior (Bueno de Mesquita, 2009). But it is not always easy to align all the stars, and lining them up is as much an art as it is a science.

5.7 Coda

This chapter developed a new explanation of the 1962 missile crisis, one that was constructed without reference to the facts of the Cuban case. Specifically, the equilibrium structure of the Asymmetric Escalation Game was used to explain the initiation, development, and resolution of the crisis. One and only one of the model's several equilibria, $CLRE_1$, a member of the Constrained Limited-Response family, was shown to be consistent with the beliefs and the action choices of US and Soviet decision makers and, significantly, with the way the crisis was eventually resolved. Answers to all three of the foundational questions associated with the crisis were derived from an examination of its strategic characteristics. These answers are neither ad hoc nor ex post; rather, they are the clear a priori theoretical expectations of a single integrated game-theoretic model of interstate conflict initiation, escalation, and resolution.

For example, why did the Soviet Union precipitate a crisis by installing nuclear-capable missiles in Cuba? Under $CLRE_1$, the answer is manifest: Soviet actions were motivated, at least in part, by the clear expectation that the United States would not respond, either because it would have been too late to do so if and when the missiles were discovered, or because Khrushchev thought that the Kennedy administration would be unwilling to respond forcefully. Whether the Soviet decision was further motivated by a strong desire to redress an unfavorable strategic balance, protect a well-placed ally, or some combination of these and other factors is a secondary question that will most likely never be definitively answered (Allison and Zelikow, 1999: 77–109).

Why was the response of the United States measured and not escalatory? Again, the strategic characteristics of $CLRE_1$ provide an unambiguous answer. The blockade was seen as an "initial step" that carried with it a message: stop or we will shoot. As well, US decision makers believed that it was the course of action "most likely to secure our limited objective—removal of the missiles—at the lowest cost" (Sorensen, 1965: 782). At the same time, an air strike and/or an invasion carried with it an unacceptably high risk of a superpower war—a risk which Kennedy famously estimated to be "somewhere between one out of three and even" (Sorensen, 1965: 795). Needless to say, both of these beliefs are required for a limited conflict to occur under any Constrained Limited-Response Equilibrium, including $CLRE_1$.

Finally, why was the settlement of the crisis a political compromise under which the Soviets withdrew their missiles in exchange for a public US promise and a private US assurance? The short answer is that the Soviets got the message implicit in the blockade and the other signals, intended or not, sent by the United States. Or as Snyder and Diesing (1977: 397) would put it, the Soviet Union underwent a "strategy revision ... [that was] ... initiated when a massive input of new information [broke] through the barrier of the image and [made Soviet decision makers] realize that [their] diagnosis and expectations were somehow radically wrong and must be corrected." All of which is to say that the Soviet decision to withdraw the missiles was the rational response to the additional information they acquired about US preferences while the crisis was playing out. For both the United States and the Soviet Union, then, an escalatory move was simply too risky. Hence, their bargain.

6

· · • · ·

Explaining the 1914 War in Europe

6.1 Introduction

The outbreak of World War I remains one of the most perplexing events of international history. It should be no surprise, then, that rationalist interpretations of the July Crisis are a diverse lot, ranging from the sinister, on the one hand, to the benign, on the other. The dark view is that German leaders simply wanted a war in 1914; the less baleful interpretation is that the war was unintended and, at least in some sense, accidental. Somewhere in-between are those intentionalist accounts that attempt to show how instrumentally rational agents were led to the choices that gave rise to an escalation spiral.[1]

Levy's (1990/1991) attempt at explanation, with preference assumptions that shade toward the sinister but with conclusions that approach those of the inadvertent war school, is a good example of an in-between interpretation of the events that led, eventually, to World War I. Levy (1990/1991: 184) locates the cause of the war in those "economic, military, diplomatic, political, and social forces ... [that] ... shaped the policy preferences of states-men and the strategic and political constraints within which they had to make extraordinarily difficult decisions."

But Levy's rational choice explanation is, as are most explanations of the July Crisis, theoretically ad hoc. It contains no formal structure, game theoretic or otherwise, to demonstrate a direct mapping between postulated preferences and the eventual denouement of the crisis. As well, it suffers from the absence of the inclusion among the possible outcomes of the crisis of an outcome labeled "status quo"—or an equivalent.[2] The latter is a particularly problematic deficiency, since the heart of any explanation of the start of the Great War must

---

[1] This chapter is based on Zagare (2009b, 2015a).

[2] In his discussion of German preferences, Levy appears to consider the status quo to have been a possible outcome. Yet, he does not include it in his summarizing array. The exclusion and temporary inclusion of this important explanatory variable is symptomatic of ad hoc theorizing, a certain amount of which is almost unavoidable and, therefore, to some extent, forgivable. Nonetheless, the casual way this important variable is considered remains an important deficiency of Levy's analysis of the July Crisis. The reason for this is that, in a crisis with two conflict outcomes, one limited and the other all-out, the Challenger's relative preference between the status quo and each conflict outcome is strategically determinative. Depending on the specifics and on information constraints, deterrence might succeed, succeed partially, or fail altogether (Zagare and Kilgour, 2000).

Game Theory, Diplomatic History and Security Studies. Frank C. Zagare. Oxford University Press (2019).
© Frank C. Zagare 2019. DOI: 10.1093/oso/9780198831587.001.0001

contain an explicit statement of exactly why the prevailing European state system failed to withstand a challenge in early August 1914.

One purpose of this chapter is to overcome some of the deficiencies of extant "in-between" intentionalist accounts of the war's outbreak, and to do so without taking the sinister view that Germany was simply an evil empire seeking expansion. This explanation, normally associated with the German historian Fritz Fischer (1967, 1975), is theoretically uncomplicated and logically straightforward. It suffers, however, from an almost exclusive focus on German policies and decisions. Today, most historians and political scientists regard Fischer's argument as incomplete at best, and misleading at worst (Langdon, 1991; Mombauer, 2002).

It is not my intention in this chapter to revisit the debate that the so-called Fischer thesis set off in the 1960s. Rather, my goal is more modest: I hope to demonstrate, formally, using an incomplete information game model, that Trachtenberg's (1991: 57) contention that "one does not have to take a particularly dark view of German intentions" to explain the onset of war in 1914 (and "question the 'inadvertent war' theory") is logically correct, and to do this in a theoretically rigorous way. Along the way, I also hope to answer a number of related questions about the July Crisis.

Before proceeding, however, there is one important proviso. In what follows, I expressly do not attempt to offer an explanation of Britain's entry into the war and its failure to deter a German attack on Belgium and France. This question, which is addressed elsewhere (Zagare and Kilgour, 2006), is largely immaterial to my immediate purpose of constructing an explanation of the escalation of the local contest between Austria–Hungary and Serbia to a strictly continental war that also involved Germany, Russia, and France.

6.2 Background

Archduke Franz Ferdinand, heir apparent to the Austro-Hungarian throne, was assassinated in Sarajevo on June 28, 1914. By the end of the first week of July, German leaders had issued the so-called Blank Check, pledging unconditional support of Austria's reactive decision to deal harshly with Serbia, in effect agreeing to a coordinated strategy that ceded control of some critical aspects of German foreign policy to decision makers in Vienna.[3]

Even with Germany's backing, though, Austria was slow to move. It was not until the 23rd of July that Vienna delivered its humiliating ultimatum to Belgrade, an ultimatum which, according to Farrar (1972: 8), signaled the beginning of the "European stage" of the crisis. The next day, the details of the ultimatum were formally conveyed to other European leaders, including Russia's foreign minister, Serge Sazonov, who, after consulting with both the French and British ambassadors, convened a meeting of the Russian council of ministers. As Spring (1988: 57) notes, this meeting was "the critical point for Russia in the July crisis." The decisions reached that afternoon established the type of player Russia would be in the days that followed: Sazonov's inclination toward a hard-line policy was supported by the rest of the government and, on the following day, ratified by the tsar. The agreed upon

---

[3] For an analysis of the Blank Check, see Zagare (2009a).

strategy was multifaceted; it covered a number of contingencies and revealed a clear hierarchy of objectives.

More than anything, the Russians wanted to defuse the crisis before it could further intensify. Accordingly, the council of ministers approved Sazonov's proposal to ask Austria for an extension of the ultimatum's deadline and to encourage Serbia to accede to as many of Vienna's terms as possible. Of course, these measures could always fail. In this eventuality, Serbia was also to be urged not to resist an Austrian invasion. As one might expect, this latter suggestion was not well received by the Serbs (Stokes, 1976: 70).

The most important decision made on July 24, though, concerned what Russia would do *if* Austrian troops marched against Belgrade. According to an internal foreign ministry memorandum, "it was decided in principle" to implement a *partial* mobilization of the Russian army and navy "and to take other military measures should the circumstances so require" (Geiss, 1967: 190). The next day, the tsar formally endorsed this recommendation and agreed "to enforce throughout the entire Empire the order for the period preparatory to war" (Geiss, 1967: 207). As Trachtenberg (1991: 76) notes, with this decision "the crisis had moved into its military phase." Depending on circumstances, Russia was now prepared for either a limited response (i.e., a partial mobilization against Austria) or an escalatory response (i.e., a full mobilization against both Austria and Germany), should the need occur.

Even still, as late as July 27, no irrevocable choices had been made by any European government. True, Austria–Hungary and Germany had decided on a joint course of action, and Austria had issued a demanding ultimatum, but no significant military steps had actually been taken. Similarly, Russia and France had decided to stand together, and Russia had developed a strategy that took account of various contingencies, but no overt military plan had in fact been implemented. In other words, neither side had as yet mobilized, either fully or partially, for war. But the status quo would not long endure. After all, the Austrian ultimatum had already expired, Serbia had rejected its most humiliating conditions, the German foreign office was still urging immediate action, and Vienna remained intent on crushing the Serbian "viper."

The Austrian intent was realized at noon on July 28 when a telegram declaring war was sent to the Serbian government. By the next morning (July 29), Austro-Hungarian gunboats opened fire on Belgrade. This was the first of four critical moves in a game that would, in short order, lead to a European war that pitted the Dual Alliance of Austria–Hungary and Germany against Russia and France.

6.3 Asymmetric Escalation Game Redux

To understand the dynamic process that eventually, albeit briefly, involved four of the European great powers in a continental conflict, consider once again the Asymmetric Escalation Game, which is reproduced here for convenience as Figure 6.1. As noted in Chapter 5, this model was specifically designed to analyze intense interstate disputes, such as the one under investigation, that involve at least two distinct levels of conflict, one limited and another unlimited. The empirical fit between this model and the events that led to the

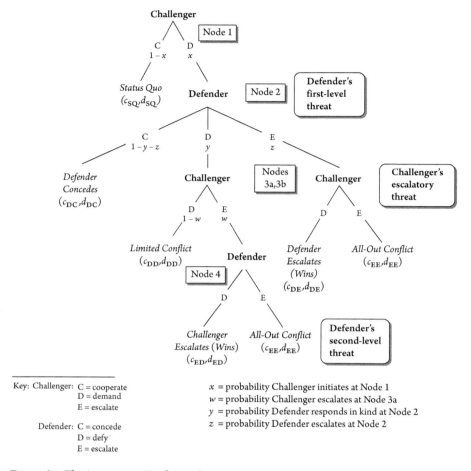

Figure 6.1 The Asymmetric Escalation Game

European phase of the war is especially close. This model is, therefore, a powerful tool for understanding the escalatory process that led, eventually, to the initiation of armed conflict by Germany in early August 1914.

The Asymmetric Escalation Game is a two-player noncooperative game. In what follows, I associate Austria–Hungary and Germany with the Challenger,[4] and Russia and France with the Defender. The assignment of these roles to the two players is not difficult to justify. Clearly,

---

[4] Notice that Austria–Hungary is listed as the first (principal) member of the coalition. This is no accident. For all intents and purposes, the German government played a subordinate role in actual decision making prior to August 4, when it invaded Belgium. And, by that time, the die had already been cast. As Williamson (1991: 196) notes, "the steps that pushed Europe toward war were taken Vienna. The support given by Berlin simply confirmed and assured that the Hapsburg decision to settle accounts would this time be a military solution rather than a diplomatic one."

the Austrian declaration of war against Serbia was seen, at least in St. Petersburg, as a direct challenge to Russia's standing as a great power. As well, the first significant Russian decision was doubtless a defensive reaction to the Austrian declaration of war and the bombardment of Belgrade.

More difficult to justify, however, is the assumption that (1) Austria–Hungary and Germany constituted a single decision-making unit and (2) that Russia and France made concerted choices from July 28 and afterwards. Indeed, neither assumption is entirely consistent with the known facts. Decision makers in Vienna and Berlin, for instance, had slightly different objectives,[5] sometimes possessed private information that they did not share with each other, and were not always operating under the same set of constraints.[6] Similarly, in the actual play of the game I examine presently, all of the critical choices made by the Defender were made in St. Petersburg, not in Paris. In other words, France, which Remak (1971: 354) judges to be the nation "least responsible for the outbreak of the war," was a relatively minor player during the most critical stages of the crisis.[7]

Nonetheless, there are a number of good reasons for treating the two coalitions as a single player. One important justification is that members of both alliances had agreed on a joint strategy well before Austria's declaration of war. As noted earlier, given German support, Austria–Hungary had decided to take a hard-line approach toward Serbia. Similarly, with France's backing, Russia was determined to stand firm and resist an Austrian challenge to Serbia's political integrity. Additionally, by treating the two alliances as a unit, the analysis that follows is simplified considerably. It should also be noted that there is ample precedent for this particular player assignment. In their classic study of interstate crises, for instance, Snyder and Diesing (1977: 94) view each coalition as a unified actor. And, in most empirical studies of interstate disputes, Germany and Austria–Hungary are grouped together as a single entity during the July Crisis.[8]

On the other hand, there will be times when we will have to stand back from this simplifying assumption—just a bit—to get a firmer picture of how and why the game terminated as it did. As will be seen, this will especially be the case after July 29, when decision makers in Vienna and Berlin began to interpret the world differently and, in consequence, work at cross-purposes.[9] It was at this point in the game that Germany lost control of its ally and that critical decision making was thoroughly monopolized by the Hapsburg monarchy.

[5] Berlin's goal was to preserve its alliance with Austria; Vienna's to preserve its standing as a great power. Nonetheless, in the game they played with Russia and France, their preferences converged (see Section 6.4).

[6] Equally problematic is the assumption that the German government itself acted as a single unit throughout the crisis. But since the divergent tendencies and underlying preferences of Kaiser Wilhelm II, Chancellor Theobald von Bethmann Hollweg, Foreign Secretary Gottlieb von Jagow, Chief of the General Staff Helmuth von Moltke (the Younger), and others surfaced when decision making was still centered in Vienna, the lack of coordination among German leaders is not particularly material to the present discussion.

[7] But see Clark (2012) for the argument that France played a major role in fomenting this crisis.

[8] See, for example, Huth (1988) or Danilovic (2002).

[9] This is not to say that their preferences diverged. In my opinion, Levy is correct (1990/1991: 162) when he concludes that Austria–Hungary and Germany had identical (ordinal) utility functions. It is clear, however, that despite a commonality of purpose, German leaders placed a much higher probability on the likelihood that Russia would intervene than did Austria–Hungary after Russia ordered partial mobilization on July 29 (see Section 6.6.3). More technically, the (updated) beliefs of Germany's leaders about Russia and France's type differed from those held by Austria's foreign minister, Count Leopold Berchtold, and other key decision makers in Vienna.

These reservations aside, the Asymmetric Escalation Game is particularly well suited for examining situations where more than one type of conflict outcome is possible, as was the case at the end of July 1914. It is, therefore, a much more refined model than most other game-theoretic models of the escalation process.

Recall from Chapter 5 that, in the Asymmetric Escalation Game, the Challenger begins play at Node 1 by whether deciding to initiate a crisis. If the Challenger turns away from a confrontation (by choosing (C)), the *Status Quo* (SQ) obtains. But if the Challenger initiates conflict (by choosing (D)), the Defender decides (at Node 2) whether to capitulate (by choosing (C)) or to respond, and if the latter, whether to respond in kind (by choosing (D)), or to escalate (by choosing (E)).

Capitulation ends the game at *Defender Concedes* (DC). If the Defender responds, the Challenger can escalate (or not) at Nodes 3a or 3b. If the Challenger is the first to escalate (at Node 3a), the Defender is afforded an opportunity at Node 4 to counter-escalate. *Limited Conflict* (DD) occurs if the Defender responds in kind and the Challenger chooses not to escalate at Node 3a. The outcome *Challenger Escalates (Wins)* (ED) occurs if, at Node 4, the Defender chooses not to counter-escalate. Similarly, the outcome is *Defender Escalates (Wins)* (DE) if the Challenger chooses not to counter-escalate at Node 3b. *All-Out Conflict* (EE) results whenever both players escalate.

The connection between the initial choice facing Austria–Hungary and Germany in late July 1914 and the Challenger in the Asymmetric Escalation Game should be obvious. After Sarajevo, the Dual Alliance could have either accepted a humiliating status quo by doing nothing (i.e., by choosing (C)) or sought to modify it by choosing (D) and demanding its alteration. Austria's intent to contest the status quo was clearly signaled on July 23, when it issued its ultimatum to Serbia, and realized on July 28, when it finally declared war.

It is noteworthy that once Austria began to shell Belgrade on July 29, decision makers in St. Petersburg faced a set of options that closely parallel the choices available to the Defender at Node 2 of the Asymmetric Escalation Game: Russia could have stood aside and accepted (C) the Austrian attempt as a fait accompli, it could have measured its response (D) with a *partial* mobilization directed only against Austria, or it could have significantly escalated (E) the conflict with a full mobilization of its army directed against both Austria and Germany.[10] It was generally understood in St. Petersburg that an escalatory choice (i.e., full mobilization) was likely to lead to war with Germany (Trachtenberg, 1990/1991: 126), whether that escalatory choice was made initially at Node 2 or subsequently at Node 4.

Unlike a full mobilization, however, a partial mobilization did not *necessarily* imply a war between Russia and Germany. On July 27, Germany's foreign secretary, Gottlieb von Jagow, told both the French ambassador, Jules Cambon, and the British ambassador, Sir Edward Goschen, that Germany would not mobilize against Russia if the Russian mobilization were directed only against Austria–Hungary, but that it would counter-mobilize if subsequently Russia mobilized against Germany as well (Geiss, 1967: 245, 253). In other words, the implications of a partial mobilization were unclear. What is clear, however, is that, in the event of a partial mobilization, the Austrian response would be critical, which helps to

---

[10] Jannen (1996: 279) characterizes the Russian general mobilization as "a major escalation of the crisis." Langdon (1991: 60) and Trachtenberg (1990/1991: 146) offer similar descriptions.

explain why the locus of decision making shifted from Berlin to Vienna after the tsar backed away from full mobilization on Wednesday, July 29.

We know that, empirically, the partial mobilization decision was implemented shortly after news reached the Russian capital that Austria–Hungary had declared war, and that there was a cause and effect relationship between these two events (Geiss, 1967: 262). Indeed, as indicated in Section 6.2, the decision to react in this way had been made in advance, at the meeting of the Russian council of ministers on July 24. At this point, the choice facing Austria–Hungary was similar to the choice facing the Challenger at Node 3a of the Asymmetric Escalation Game. One option was simply to step back. For example, Austria–Hungary could have accepted the proposal made by the British foreign secretary, Sir Edward Grey, on July 24 that "the four nations not immediately concerned—England, Germany, France, and Italy—should undertake to mediate between Russia and Austria" (Kautsky, 1924: no. 157). Or Austrian leaders could have accepted the "Halt-in-Belgrade" proposal, which was made by the kaiser on July 27 and stipulated that Austria announce that it intended to occupy the Serbian capital, but only temporarily until Serbia carried out the promises it made in its response to the Austrian ultimatum (Kautsky, 1924: no. 293). Or Austria could have taken the German chancellor Theobald von Bethmann Hollweg's strong hint on July 29 to defuse the crisis by entering into direct discussions with St. Petersburg (Kautsky, 1924: no. 396). All of which indicates that, as late as July 29, a *Limited Conflict* was a distinct possibility—there was still a way out of the crisis if Austria–Hungary had wanted one.

Of course, the Hapsburg Empire was not looking for an escape clause. Disregarding last-minute pleas from Berlin to accept mediation (Kautsky, 1924: no. 441), Austria–Hungary decided to plow on. After learning of the Russian partial mobilization, it decided to mobilize against Russia. Meanwhile, it continued its advance toward Serbia. By resisting mediation and pushing forward militarily, Austria–Hungary clearly intensified the conflict. It did not take long for Russia to respond (at Node 4). On July 30, the tsar consented to full mobilization of Russian forces against both Germany and Russia. As Trachtenberg (1990/1991) shows, the Russian mobilization decision was a war choice. Before long, German troops were marching into Belgium.[11]

---

[11] Notice that the game tree of the Asymmetric Escalation Game provides the Challenger with an opportunity to counter-escalate (at Node 3b) should the Defender escalate first at Node 2. In the context of the events of July 1914, this decision node models a possible response to a full mobilization by Russia immediately after the Austro-Hungarian declaration of war against Serbia. By contrast, there is no analogous decision node for the Challenger after the Defender's escalatory choice at Node 4. The implicit assumption in the latter instance is that a full Russian mobilization simply implies a (German) counter-mobilization and, hence, an *All-Out Conflict* between the Dual Alliance and the Russian—French alliance. This is clearly an inconsistency, but, given the assumption (see Section 6.6) that the Challenger's threat to counter-escalate is highly credible, it is an inconsistency that is of no analytic import. The inconsistency could easily be eliminated either by adding another decision node after the Defender's escalatory choice at Node 4 or by eliminating the Challenger's Node 3b decision. There are a number of reasons why I decided not to make these changes to the underlying game form. First, as noted, the inconsistency is in no way material. Second, eliminating the inconsistency would likely complicate the narrative and the associated analysis. And third, redrawing the tree to eliminate the inconsistency would obscure (1) the model's generality and (2) the relationship of the Asymmetric Escalation Game to the family of models that delineate perfect deterrence theory (see Chapter 7).

For generations historians and political scientists have attempted to explain why. If a limited conflict was a distinct possibility, why did it fail to materialize? Was there an inevitable slide to an all-out war, as Britain's chancellor of the exchequer, David Lloyd George, suggested after the war, or was the general European conflict preventable? What role did perceptions play in the way the game played out? In what follows, I attempt to answer these and related questions by examining the equilibrium structure of the Asymmetric Escalation Game with incomplete information.

6.4 Preferences

Depending on the choices of the players, there are six possible outcomes of the Asymmetric Escalation Game. These outcomes, and the choices that lead to them, are indicated both verbally and symbolically on the game tree shown in Figure 6.1. Note once again that the model admits two distinct conflict outcomes: *Limited Conflict* occurs only when the Challenger defects at Node 1, the Defender responds in kind at Node 2, and the Challenger chooses not to escalate at Node 3a; *All-Out Conflict* occurs whenever both players escalate.

It is also worth pointing out once more that there are two distinct paths to *All-Out Conflict* in the Asymmetric Escalation Game. The first results when the Defender escalates immediately at Node 2 and the Challenger retaliates at Node 3b. The second path conforms to a classic escalation spiral: the Challenger initiates at Node 1, the Defender responds in kind at Node 2, the Challenger escalates first at Node 3a, and the Defender counter-escalates at Node 4.

In the discussion of the Asymmetric Escalation Game that follows, all previous assumptions about preferences are retained. Table 6.1 restates the particulars. The preferences arrayed in Table 6.1 presume that both players prefer winning to losing. To reflect the costs of conflict, the players are also assumed to prefer to win or, if it comes to it, to lose at the lowest level of conflict. Thus, the Challenger prefers *Defender Concedes* (outcome DC) to *Challenger Wins* (outcome ED)—and so does the Defender.

As before, several critical preference relationships are left open in Table 6.1. These relationships are associated with threats that the players may or may not prefer to execute. Recall that the preferences associated with these threats define each player's type. Since the Challenger has only one threat, to escalate or not at Node 3b, it may be one of two types:

Table 6.1 Preference assumptions for the Asymmetric Escalation Game.

Challenger	Defender
Defender Concedes	*Status Quo*
Status Quo	*Defender Escalates*
Challenger Escalates	*Defender Concedes* or *Limited Conflict*
Limited Conflict	*Challenger Escalates* or *All-Out Conflict*
Defender Escalates or *All-Out Conflict*	

Hard Challengers are those that prefer *All-Out Conflict* to *Defender Escalates*; Challengers with the opposite preference are called Soft.

The Defender, by contrast, has two threats in the Asymmetric Escalation Game: a *tactical level* threat to respond in kind at Node 2, and a *strategic level* threat to escalate at Node 4. A Defender that prefers *Limited Conflict* to *Defender Concedes* is said to be Hard at the first, or tactical, level, while a Defender with the opposite preference is said to be Soft at the first, or tactical, level. Similarly, a Defender that prefers *All-Out Conflict* to *Challenger Escalates* is said to be Hard at the second, or strategic, level, while a Defender with the opposite preference is said to be Soft at the second, or strategic, level. Thus, the Defender may be one of four types: Hard at the first level but Soft at the second (i.e., type HS), Soft at the first level but Hard at the second (i.e., type SH), Hard at both levels (i.e., type HH), or Soft at both levels (i.e., type SS).

In the analysis that follows, all retaliatory threats are taken to be capable in the sense that the player who initiates conflict ends up worse off if and when the other player retaliates. In terms of preferences, this means that both players prefer the outcome *Status Quo* to *Limited Conflict*, and *Limited Conflict* to *All-Out Conflict*. This final assumption about preferences, however, is neither innocuous nor noncontroversial—as I explain in Section 6.5.

6.5 Some Caveats

There are a few devils in the details of the preference assumptions I make that, perhaps, require exorcism. Before proceeding, however, it will be useful to comment, briefly, on the connection between the six theoretical outcomes of the Asymmetric Escalation Game and the real world events they are meant to represent.

The outcome with the clearest meaning is the one labeled *Status Quo*, which I take to be the existing European order as of July 1914. As things stood shortly after Sarajevo, neither Austria–Hungary nor Germany placed a high value on this outcome, which provides further justification for the identification of the governments in Vienna and Berlin with the player called Challenger. Specifically, German leaders looked around and saw both a dominating Great Britain and a rising Russia—which was tied closely to France, a long-time rival. For their part, Austria's policy makers feared that their polyglot empire would soon implode if Serbian subversives were not soon eradicated. Clearly, both Austria–Hungary and Germany were dissatisfied powers as the July Crisis unfolded.

Another outcome whose meaning should be clear is *Defender Concedes*. *Defender Concedes* is simply a more generic term for the outcome that Levy (1990/1991) calls "localized war." *Defender Concedes*, therefore, is intended to capture the denouement of a war that takes place in the Balkans and pits Austria–Hungary against "tiny Serbia" (Geiss, 1967: 363).

Defender Escalates (Wins) and *Challenger Escalates (Wins)* refer to one-sided victories for the Defender and the Challenger, respectively, that come about after an escalatory move by one player and capitulation by the other. In the context of the July Crisis, Russia (i.e., the Defender) would clearly have gained a political and diplomatic advantage had it implemented a full mobilization of its army and forced Austria–Hungary and Germany to back off. Similarly, Austria–Hungary and Germany (i.e., the Challenger) would have gained the upper hand, and probably split the entente, had the partially mobilized Russian army stood down as Belgrade was leveled and Serbia dismembered.

The remaining two outcomes of the Asymmetric Escalation Game have names that may be misleading. Within the context of this chapter, *All-Out Conflict* corresponds to the outcome that Levy (1990/1991) refers to as a "continental war," that is, a war in which only the four major continental European powers and Serbia are involved. This, of course, is the war that actually broke out in Europe after Germany declared war on Russia on August 1.

The final outcome of the Asymmetric Escalation Game, *Limited Conflict*, requires a most careful exegesis. Normally, the term "limited conflict" is reserved for an actual war in which two or more nations fight but at least one of the involved nations either has a limited objective or fails to use all the weapons or resources at its disposal. But this is not the sense of the term here, where *Limited Conflict* refers to any outcome that comes about after the Challenger contests the status quo, the Defender measures its response, and the Challenger decides not to escalate. In 1914, for example, a limited conflict would have evolved had Austria–Hungary agreed to mediate its dispute with Serbia after Russian troops had been mobilized in the Balkans. The broad outlines of this outcome, therefore, correspond closely to the outcome Levy (1990/1991) identifies as "negotiated peace."

All this said, one might well ask whether the preference assumptions summarized in Table 6.1 stand up to empirical scrutiny. After all, these are generic preferences that were developed to represent an interesting and important general case. It is more than possible, therefore, that there are some critical differences between these (posited) preferences and those of the actual players during the July Crisis.

For example, the Defender in our model strictly prefers the outcome *Status Quo* to any other outcome. But was Russia truly a satiated power in 1914? Schroeder (1972: 335) makes a compelling case that it was not,[12] that it had designs on large swaths of Austrian territory, and that it was simply waiting for the aging emperor's death to annex Galicia and other parts of Franz Joseph's sure-to-disintegrate empire. Clearly, Russia was also a dissatisfied power on the eve of World War I.

On the other hand, for our purposes, Russia's territorial ambitions are largely immaterial. One reason is that, in the Asymmetric Escalation Game, the Defender never makes a choice between the *Status Quo* and any other outcome, so that its relative ranking is theoretically irrelevant. Additionally, unless one wishes to argue that Russia and France deliberately provoked the crisis in order to humiliate Austria–Hungary and Germany by forcing the Dual Alliance to back down, implying that the Defender preferred *Defender Escalates (Wins)* to the *Status Quo*, the assumption that Russia and France's highest ranked outcome was the *Status Quo* is entirely defensible within the context of the Asymmetric Escalation Game and the set of outcomes associated with it.[13]

Fischer (1967, 1975), though, would most certainly object to the relatively low ranking of *All-Out Conflict* (i.e., a continental war) by the Challenger (see Table 6.1). In Fischer's opinion, Germany deliberately instigated a war in 1914 in a bid for world power. If Fischer is correct, and

[12] See also Butterfield (1965).
[13] The argument is not entirely without merit. For the role of France, see Clark (2012); for Russia, see McMeekin (2013).

there are some who believe that he is, *All-Out Conflict* was Germany's highest ranked outcome.[14] In consequence, deterrence by Russia and France (without the assistance of Great Britain) would have been impossible. In other words, Fischer's assumption about German preferences, in and of itself, constitutes a sufficient condition for the outbreak of war in Europe on August 1.

Both Schroeder (1972: 336–7) and Remak (1971: 361), however, agree that Fischer's argument is not necessary for an explanation of the escalation spiral that led, eventually, to World War I. I hope to demonstrate in Section 6.6 why they are correct. But for now, in the tradition of William of Ockham, I simply adopt the less demanding assumption.[15]

Levy's (1990/1991) contention that both Austria–Hungary and Germany preferred *All-Out Conflict* (i.e., a continental war) to *Limited Conflict* (i.e., a negotiated peace) also runs counter to the preference assumptions arrayed in Table 6.1.[16] With respect to Austria–Hungary, Levy's conclusions are debatable but difficult to establish. To be sure, as the crisis intensified, Vienna did everything it could to avoid mediation. But was this because it preferred a continental war or, as suggested in Section 6.6.2 and as Jannen (1996) forcefully argues, because it did not believe that Russia would intervene? With respect to Germany, Levy's conclusions are even more problematic—unless one is willing to discount completely the sincerity of Bethmann Hollweg's frantic, last-ditch effort to moderate Austria–Hungary's behavior (discussed in Section 6.6.3). Even Immanuel Geiss (1967: 88), Fischer's student and disciple, is unwilling to go that far.

Some readers may also find fault with the fact that the Asymmetric Escalation Game, and the outcomes associated with it, do not include Great Britain as a player and the possibility of the wider, world war that eventually broke out when Britain declared war on Germany. One reason why I have chosen not to model Great Britain's choices in this chapter is that my purpose is to explain, in the simplest possible way, the escalation spiral that led to the outbreak of war on the continent. Including Great Britain among the players would only unnecessarily complicate matters. Additionally, the game played between Great Britain and Germany in 1914 is analyzed elsewhere (Zagare and Kilgour, 2006). It would be a straightforward exercise to include that game as proper subgames of the Asymmetric Escalation Game.[17] But doing so would not alter the argument I make here.

Finally, the Challenger's postulated preference for *Status Quo* over *Limited Conflict* (i.e., a negotiated peace) also runs counter to both Fischer's and Levy's assessments of Austro-Hungarian and German preferences. As noted in Section 6.4, in the analysis of the Asymmetric Escalation Game, I assume that both the Challenger and the Defender possess capable threats at every level of play. Consistency with this assumption requires that the Challenger prefer *Status Quo* to *Limited Conflict*.

---

[14] Copeland's (2000: 57) argument that Germany "preferred major war to even a localized war or a negotiated solution" is even more extreme. With respect to Moltke and other German military leaders, he may well be correct. But as Copeland (2000: 59–60) and others (e.g., Williamson and May, 2007: 363) note, the military was not in control of German policy in 1914. Had they been, the war most likely would have come sooner, perhaps as soon as 1875 when Moltke (the elder) first proposed a preemptive war against France (Förster, 1999: 351).

[15] Stone (2009: 23–5) also argues that Germany wanted a European war in 1914.

[16] Levy (1990/1991: 162) claims that Fischer would also argue that both Austria–Hungary and Germany preferred *All-Out Conflict* to *Limited Conflict*.

[17] Notice the plural. To completely extend the Asymmetric Escalation Game, the game developed in Zagare and Kilgour (2006) would have to be substituted for the Challenger's Node 3b decision *and* appended subsequent to the Defender's decision at Node 4.

How critical is this assumption? In the analysis that follows, it plays a relatively minor role. After all, in the end the crisis escalated to the highest level, suggesting that the Challenger's *relative* ranking of these two outcomes was of little moment. On the other hand, it has important implications for how a hypothetical question about the inevitability of conflict in 1914 is answered. This question will be addressed in Section 6.6.2.

6.6 Special Case Analysis and Discussion

In analyzing the Asymmetric Escalation Game with incomplete information, I focus once again on the special case in which the Challenger is likely Hard, that is, when the Challenger's threat to counter-escalate at Node 3b is highly credible. The special case analysis is not difficult to justify. First, as noted in Chapter 5, although this assumption vastly simplifies the analysis of the Asymmetric Escalation Game, it does so with no serious loss of information. The more important reason, however, is empirical. The assumption that the Challenger is likely Hard is entirely consistent with the facts on the ground at the end of July in 1914. According to Berghahn (1993: 197), the "hard-liners" were in control in Germany. And, as the crisis unfolded, both the Russians and French took it for granted that (1) Austria–Hungary had Germany's backing and (2) a full Russian mobilization implied a general European war. Neither would be reasonable inferences if the Challenger (i.e., Austria–Hungary/Germany) were seen as likely to be Soft.

The six perfect Bayesian equilibria in the special case analysis of the Asymmetric Escalation Game with incomplete information were described in detail in Chapter 5. Recall that there were two *plausible* deterrence equilibria under which the outcome *Status Quo* always survived rational play;[18] *Defender Concedes* is the only outcome that is consistent with rational choice when the No-Response Equilibrium exists. Under $CLRE_1$, a member of the Constrained Limited-Response group, escalation spirals are precluded; *All-Out Conflict*, however, remains a distinct rational strategic possibility whenever an Escalatory Limited-Response Equilibrium, such as $ELRE_3$, is in play. The theoretical characteristics of these six perfect Bayesian equilibria (see Table 6.2 for the particulars) that exist when the Challenger is likely Hard enables us to address a number of interesting questions about the events that led to the outbreak of war in Europe in early August 1914.[19]

6.6.1 What Were They Thinking?

One question that arises immediately concerns the expectations of both Germany and Austria during the period following the Hoyos mission[20] on July 6 (when the Blank Check

[18] Det_1, along with all other members of the family of several equilibria called the Escalatory Deterrence Equilibria, is based on implausible beliefs.
[19] The definitions of the strategic and belief variables appearing in Table 6.2 are given in Chapter 5, in the footnote to Table 5.2.
[20] Shortly after the archduke's assassination on June 28, an Austrian delegation was sent to Berlin to represent the Austrian position and to sound out the likely German response to a Russian attack, either actual or threatened, should Austria move against Serbia. The delegation was led by the Austrian foreign minister's *chef de cabinet*, Count Alexander Hoyos.

Table 6.2 Equilibria of the Asymmetric Escalation Game when the Challenger has high credibility.

	Challenger				Defender						
	x		w		q_{HH}	y		z				r
	x_H	x_S	w_H	w_S		y_{HH}	y_{HS}	z_{HH}	z_{HS}	z_{SH}	z_{SS}	
Escalatory Deterrence Equilibria (typical)												
Det_1	0	0	1	1	Small	0	0	1	1	1	1	$\leq d_1$
No-Response Equilibrium												
NRE	1	1	Large		Small	0	0	0	0	0	0	p_{Ch}
Spiral Family of Equilibria												
Det_2	0	0	0	0	$p_{Str\vert Tac}$	1	1	0	0	0	0	$\geq d_2$
Det_3	0	0	d^*/r	0	c_q	1	v	0	0	0	0	$\geq d_2$
$CLRE_1$	1	1	0	0	$p_{Str\vert Tac}$	1	1	0	0	0	0	p_{Ch}
$ELRE_3$	1	1	d^*/p_{Ch}	0	c_q	1	v	0	0	0	0	p_{Ch}

was issued) and the delivery of the Austrian ultimatum on July 23. What, in other words, were the leaders in Vienna and Berlin thinking during the so-called silent period of the crisis? The equilibrium structure of the Asymmetric Escalation Game with incomplete information gives a very strong suggestion about the likely content of their thoughts.

We know that, in the wake of the archduke's assassination, Austria requested and received a strong commitment of support from Germany. We also know for a fact that the German promises played a major role in Vienna's decision, which was reached at a meeting of the Austro-Hungarian common ministerial council on July 7, to pursue a hard-line policy toward Serbia. Thus, it seems safe to conclude that, once the Blank Check had been issued, there was little or no chance that Vienna would not cash it. All of which implies that, after Sarajevo, neither of the two plausible deterrence equilibria were likely to come into play.

Once the deterrence equilibria are eliminated as empirically unlikely, there are only three theoretical options, one of which is the No-Response Equilibrium. There is a strong possibility that leaders at both the Ballhausplatz and the Wilhelmstrasse anticipated that play would occur under this equilibrium form. Had that been the case, they would have fully expected a one-sided victory. Recall that *Defender Concedes* is the only possible outcome under the No-Response Equilibrium. But much the same could be said for the two remaining theoretical possibilities, $CLRE_1$ and $ELRE_3$. Under either of these two perfect Bayesian equilibria, *Defender Concedes*, although not certain, is the most likely outcome.

To put all this in a slightly different way, regardless of which of the three non-deterrence equilibria Austria–Hungary and Germany believed to be in play, one would expect, theoretically, that they had the clear expectation that Russia and France were unlikely to offer any meaningful resistance. Of course, empirically, we know that this was indeed their expectation, at least initially, so our answer should come as no surprise. Still, it is encouraging to find out that the equilibrium structure of the Asymmetric Escalation Game with incomplete information is fully consistent with the facts as they existed in early July 1914. Were this not the case, the explanatory and predictive power of the model would be more than suspect.

6.6.2 Was War Avoidable?

A second important question is whether or not the crisis in Europe was inevitable, that is, whether Austria–Hungary and Germany could have been deterred from instigating a crisis in Europe toward the end of July 1914. This is a difficult question to answer, although, like many others, I shall attempt to do so.

One answer is that, after the Blank Check was issued, the die had been cast—the prevailing status quo was no longer sustainable. But, to accept this answer, one must also hold to the view that, in the period before Austria–Hungary finally issued its ultimatum, no new information about Russia's, France's, and perhaps Great Britain's attitude could have stayed the Dual Alliance from its appointed rounds. Needless to say, this is a difficult position to sustain.

Assuming, then, that Austro-Hungarian and German perceptions were subject to revision, the answer is clear. The existence of two distinct Limited-Response Deterrence Equilibria in the Asymmetric Escalation Game attests to the theoretical possibility that the crisis could have been averted. Of course, what is theoretically possible is not necessarily likely. Such is the case in the Asymmetric Escalation Game when the Challenger is likely Hard.

To understand why, consider for now Figure 6.2 which depicts in three-dimensional space the existence conditions associated with the Spiral Family of four perfect Bayesian equilibria. Along with the No-Response Equilibrium, one and only one member of this family will exist at any one time (Zagare and Kilgour, 2000: 272).

The Defender's credibilities determine which Spiral Family equilibrium exists. In Figure 6.2, every possible combination of the Defender's credibilities is represented as a point in the tetrahedron shown in the center of this figure. The right horizontal axis represents the probability that the Defender is of type HH, the lower-left (horizontal) axis the probability that the Defender is of type SH, and the vertical axis the probability that the Defender is of type HS. Thus, any point c in the three-dimensional triangle, or simplex, has a combination of non-negative coordinates, (p_{HH}, p_{HS}, p_{SH}) whose sum is less than or equal to 1. The fourth credibility, p_{SS}, equals the difference between this sum and 1; this amount is also the (perpendicular) distance between the point (p_{HH}, p_{HS}, p_{SH}) and the front face of the tetrahedron. For example, the point $(0,0,0)$ represents the combination $p_{HH} = p_{SH} = p_{HS} = 0, p_{SS} = 1$.

Speaking more informally, Figure 6.2 can be visualized as a corner of a room with two walls and a floor, all at right angles—the fourth face of the simplex is the downward sloping plane. The side wall is light gray, the back wall is medium gray, and the floor is dark gray. Of course, to enable us to peer into this corner, the front face must remain transparent.

Observe that the two Limited-Response Deterrence Equilibria occupy a relatively small area of the simplex. Thus, ceteris paribus, it is not all that likely that either would come into play in the Asymmetric Escalation Game.[21] Of course, not all things were equal in 1914, so

---

[21] To see this, simply compare the relative size of the cutouts associated with the two Limited-Response Deterrence Equilibria with those of $CLRE_1$ and $ELRE_3$.

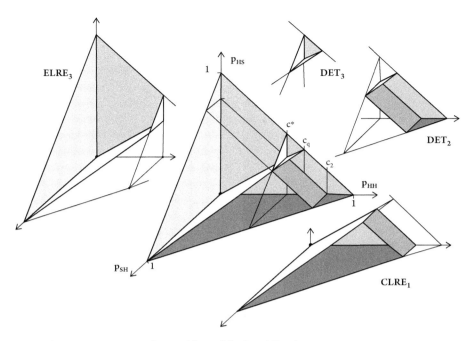

Figure 6.2 Existence regions for equilibria of the Spiral Family
(Source: Zagare and Kilgour, 1998)

we are led to follow up. Given that deterrence was theoretically possible in July, what would have had to occur for this possibility to become a reality? The conditions associated with the existence of Det_2 and Det_3 provide a succinct answer.

Notice from Figure 6.2 that the two closely related Limited-Response Deterrence Equilibria occupy a small region in the right-hand side of the tetrahedron, where p_{HH} is large, p_{HS} is not too large, and p_{SH} and p_{SS} are small. In this region, the Defender is likely tactically Hard; this explains its propensity under either Det_2 or Det_3 to respond in kind at Node 2, whatever its actual type. But this tendency alone is not sufficient to deter the Challenger. The Defender's willingness to respond in kind also rests on its ability to dissuade the Challenger from escalating at Node 3a. For this to occur, the Defender's second-level threat must be highly credible as well; in other words, for deterrence to succeed under either Det_2 or Det_3, the Defender must likely be both strategically and tactically Hard—that is, p_{HH} must be large, that is, *both* of the Defender's threats must be fairly credible.

It is clear, however, that this condition, which is necessary for deterrence success in the Asymmetric Escalation Game, was not satisfied during the July Crisis. There is ample documentary evidence that both German and Austro-Hungarian leaders formulated their policies with the expectation that Russia, even with the backing of France, would not offer anything but token resistance if and when Austria moved against Serbia. The kaiser, for example, initially believed that the risk of a war was minimal, since neither Russia nor France was prepared for one (Geiss, 1967: 71, 77; Massie, 1991: 862). Bethmann Hollweg (1920: 126) shared the kaiser's opinion.

Vienna's views were similar. According to the Italian ambassador in St. Petersburg, by mid July, Austria "was capable of taking an irrevocable step with regard to Serbia based on the belief that, although Russia would make a verbal protest, she would not adopt forcible measures for the protection of Serbia against any Austrian attempts" (Albertini, 1952: vol. 2, 184). Astonishingly, not even the full mobilization order issued by the tsar on July 30 would alter Vienna's expectations. In a telegram that Franz Conrad von Hötzendorf, the chief of Austria's general staff, sent to his German counterpart, Helmuth von Moltke, on the night of July 31, Conrad matter-of-factly observed that the Austrian leadership was "not sure yet whether Russia is merely threatening, therefore we could not allow ourselves to be diverted from the action against Serbia" (quoted in Fischer, 1975: 507).

In summary, deterrence was a distinct, but not highly likely, possibility in July 1914. An overly aggressive act against Serbia could have been averted, at least in theory. But, and this is a big "but," given that Russian and French credibility was negligible, the theoretical possibility could not be realized. Thus, while a European war was certainly not inevitable, the status quo was not very likely to survive, even in the short run.

Parenthetically, it should be noted that this conclusion rests directly on the assumption, discussed in Section 6.5, that each member of the Dual Alliance preferred the *Status Quo* to *Limited Conflict* and *All-Out Conflict*, that is, that both threats of Russia and France were capable. Of the two, Austria–Hungary's and Germany's preference for the *Status Quo* over *Limited Conflict* is the more problematic. Would Austria–Hungary and Germany still have provoked a crisis if they had (1) believed that they could avoid a continental war but (2) thought that the issues that separated Vienna from Belgrade would be subject to multilateral mediation? If one concludes, as does Fischer (1967) and, by extension, Levy (1990/1991), that a negotiated peace (*Limited Conflict*) was ranked relatively low by both Austria–Hungary and Germany, the answer would have to be "yes." Most revisionist historians, including those who hold that World War I was in some sense inadvertent, would strongly disagree.

6.6.3 Why Did It Happen?

Finally, why did the policies of Germany and Austria–Hungary, coordinated at the onset of the crisis, diverge so dramatically after the Russian partial mobilization on July 29? To address this question, I refer once again to the evolving beliefs of the leadership groups in Vienna and Berlin and the equilibrium structure of the Asymmetric Escalation Game with incomplete information.

The common policy, reached early in July after the kaiser, with Bethmann Hollweg's concurrence, issued the Blank Check, is easy to explain. As noted in Section 6.6.2, both the Wilhelmstrasse and the Ballhausplatz initially believed that Russia was unlikely to respond if Austria moved aggressively against Serbia. To be sure, German leaders preferred that Austria–Hungary act quickly (in the kaiser's words, it was "now or never"). But they did so precisely because they also believed that any delay would decrease the probability that a fait accompli could be pulled off successfully. In consequence, as the days passed in July, Berlin

SPECIAL CASE ANALYSIS AND DISCUSSION | 115

continued to press Vienna to act—not that it mattered. While Austria was determined to strike, it was also determined to strike at a time and at a place of its own choosing. Timing aside, however, both members of the Dual Alliance preferred action to inaction and were clearly intent on taking it. Until mid July, the policies and expectations of Germany and Austria–Hungary were, for all intents and purposes, identical. All of which helps to explain why, in the theoretical and empirical literature of international relations, these two closely aligned nations are generally considered to have comprised a single unit in 1914.

The commonality of purpose and expectations, however, would not hold forever. As rumors of Austria's intentions began to circulate in European capitals, German officials started to fear that Austria might not be able to have its way with Serbia before Russia could react (Berghahn, 1993: 210). The equilibrium structure of the Asymmetric Escalation Game with incomplete information helps to explain what happened next.

In Section 6.6.1, it was suggested that of the five possible perfect Bayesian equilibria in the Asymmetric Escalation Game, the two Limited-Response Deterrence Equilibria were inconsistent with German and Austrian expectations and, therefore, were not likely to come into play. The same, however, cannot be said of the remaining three rational strategic possibilities.

The three still viable perfect Bayesian equilibria share a number of important characteristics. Whether play takes place under the No-Response Equilibrium, $CLRE_1$, or $ELRE_3$, the Challenger always begins play by initiating at Node 1. In other words, under any of these three equilibria, the Challenger's action choice is always the same. As well, there is also no chance that the Defender will respond by escalating immediately (i.e., $z_{HH} = z_{HS} = z_{SH} = z_{SS} = 0$) at Node 2. And, finally, a one-sided victory (i.e., *Defender Concedes*) is the most likely outcome under each equilibrium form. This means that, under most real world circumstances, it may not be possible to determine, empirically, which of the three equilibria had actually come to define play.

There is, however, one critical difference that may sometimes provide a clue—even before the game is actually played out. Recall that, under the No-Response Equilibrium, the Defender never intends to respond, either in kind or by escalating, regardless of its type. By contrast, under either $CLRE_1$ or $ELRE_3$, tactically Hard Defenders always respond in kind when they are strategically Hard, and sometimes or always respond in kind when they are strategically Soft.[22] All of which means that, up to the point at which German leaders came to fear that Russia *might* act to protect Serbia, the No-Response Equilibrium was the only perfect Bayesian equilibrium consistent with the expectations of the decision makers in Berlin (and Vienna). And, had the events of July 1914 unfolded as both Germany and Austria–Hungary initially hoped, World War I would have never occurred; rather, the third Balkan war would have been a localized conflict between Austria–Hungary and Serbia.

Of course, everything did not go according to plan. As Austria fiddled about, German beliefs about Russian intentions changed. These expectations are clearly inconsistent with the existence of the No-Response Equilibrium—which can now confidently be eliminated as a potential descriptor of the events of late July. This leaves (for now) just two theoretical possibilities: $CLRE_1$ and $ELRE_3$. As shall be seen, until July 30, each provides a plausible explanation of the unfolding crisis. In fact, until push came to shove, German leaders acted as if $CLRE_1$ was

---

[22] Tactically, Soft Defenders never respond in kind.

actually in play. Unfortunately, the beliefs and action choices of Austria's leaders were consistent with the existence of ELRE$_3$. The contention here is that, had this not been the case, the crisis would have been resolved differently. I explain why next.

Once German leaders began to become concerned about the possible involvement of other powers, they intensified their pressure on Austria to act. It should also be no surprise to learn that, at the same time, they also made a concerted effort to deter Russian interference. For example, on July 19, Germany's foreign minister, Gottlieb von Jagow, placed a notice in the *North German Gazette*, a quasi-official publication, that expressed his government's position that "the settlement of differences which may arise between Austria–Hungary and Serbia should remain localized" (Geiss, 1967: 142). The notice should be interpreted as a thinly veiled threat directed at the entente: Germany would back Austria in a war with Russia and France.

The note in the *North German Gazette* was followed up on July 21 with a cable that was sent by Bethmann Hollweg to Germany's ambassadors in Russia, France, and Great Britain that instructed them to convey the same message officially. The cable reiterated the chancellor's desire for "localization of the conflict," which was a euphemism for the *Defender Concedes* outcome, that is, a bilateral war between Austria–Hungary and Serbia.

The dispatch was quite revealing. Implicit in it was the notion that Vienna was finally about to take an irrevocable step. As the chancellor explained to his ambassadors, Austria "had no other course than to enforce its demands upon the Serbian Government by strong pressure, and if necessary, to take military measures." But a localized conflict also required that other governments remain uninvolved. To make this more likely, Bethmann Hollweg instructed his ambassadors to warn, this time less subtly, that "the intervention of any other Power would, as the result of the various alliance obligations, bring about inestimable consequences" (Kautsky, 1924: no. 100). Clearly, German policy now focused on precluding intervention in, and deterring escalation of, the dispute. It would remain so for the remainder of the month.

Two days after this broadside, at 6:00 p.m. on July 23, the ultimatum that was designed to be rejected was delivered to authorities in Belgrade by Austria's ambassador to Serbia. Not only were the terms of the ultimatum harsh, but its deadline was exceedingly short. Serbia would have two days—just 48 hours—to respond. It is noteworthy that, on July 25, the day the ultimatum was set to expire, Germany's foreign secretary was still of the opinion that "neither London, nor Paris, nor St. Petersburg wants war" (Pogge von Strandmann, 1988: 102). Jagow's strong belief helps to explain the continuing pressure the German foreign office put on Vienna to take decisive action.

Diplomatic and political foreplay came to an end on July 28 when Austria finally declared war—much to the relief and surprise of the German government. Whether it was the result of wishful thinking or of a sober calculation that simply turned out to be misguided, the leaders of both countries still believed that localization of the conflict remained the most likely outcome, that is, that a fait accompli could still be accomplished. Their illusion, however, would not last very much longer. Later that day, Russia informed Germany and the other powers that "in consequence of Austria's declaration of war on Serbia," it would declare a partial mobilization "in the military districts of Odessa, Kiev, Moscow and Kazan tomorrow," July 29. Significantly, the partial mobilization announcement also underscored "the absence of any intentions of a Russian attack on Germany" (Geiss, 1967: 262).

The Russian response, while a direct consequence of Austria–Hungary's challenge to the status quo, was clearly measured. As Fromkin (2004: 190–1) explains:

> "partial mobilization" consisted of a number of measures, some feasible and others not, none of which would have significantly helped to defend Russia and most of which put Russia in a less advantageous position than before. It was an essentially political concept, muddled and unclear, intended to convey the message that Russia was resolved to act if necessary, but did not wish to alarm or provoke Germany or Austria as a full mobilization—a real mobilization—would have done.

To put this in a slightly different way, the intent of the Russian decision to respond in kind (i.e., to choose (D) at Node 2) was deterrence (Fay, 1930: 439; Trachtenberg, 1990/1991: 130; Williamson and May, 2007: 348). The Russian leadership wanted to send the message that, if necessary, it was prepared to escalate the conflict—that it was not only tactically but also strategically Hard. The message was received loud and clear, at least in Berlin, where the Russian decision came as a shock (Williamson, 1991: 208). As Massie (1991: 870) concludes,

> Germany now faced the growing likelihood of war with Russia. German policy had been to encourage a localized Balkan war, punish a regicide state, and restore the fortunes of a crumbling ally. Russian intervention had been discounted. The Tsar's army was considered unready and the Kaiser and his advisors had expected Russia to give way, as she had five years earlier in the Bosnian Crisis. The prospect was glittering: localization accomplished; general war avoided; Serbia crushed; Austria reborn; Russia stripped of her status as a Great Power; the balance of power in the Balkans and Europe realigned. Russian mobilization against Austria demolished this dream.

Not surprisingly, the Russian partial mobilization brought about a stunning reversal of Germany's approach to the crisis.[23] Whereas, previously, Bethmann Hollweg had sought to localize the conflict by (1) encouraging Austria–Hungary to move against Serbia and (2) discouraging other powers from intervening, he now began to urge restraint on his ally and to encourage a political (i.e., a negotiated) resolution of the crisis.

To this end, the first in a series of what Fischer (1975: 495) calls the "world-on-fire" telegrams was sent to Count Heinrich von Tschirschky, the German ambassador to Austria, on July 28 at 10:15 p.m. Berlin time. In this message, Bethmann Hollweg urged Austria to moderate its policy lest it "incur the odium of having been responsible for a world war." He specifically asked that Vienna accept the kaiser's Halt-in-Belgrade proposal, which had been made earlier in the day, by announcing (1) that it had no interest in acquiring Serbia territory and (2) that its occupation of Belgrade was temporary and contingent on Serbian compliance with the terms of the ultimatum. At the same time, the German chancellor wanted it made clear, probably because he was in the process of attempting to make it clear himself,

---

[23] Fay (1930: 402–16) traces this turnaround to the late afternoon of July 27.

that Tschirschky was "to avoid very carefully giving rise to the impression that we wish to hold Austria back" (Kautsky, 1924: no. 323).[24]

Albertini (1952: vol. 2, 477–9), Fischer (1975: 72), Geiss (1967: 223), and Schmitt (1930: vol. 2, 171) interpret this telegram (no. 174) in the worst possible light, claiming that Bethmann Hollweg's motivation was simply to place the blame for war, should it come, on Russia.[25] They argue that the chancellor had delayed transmitting the kaiser's proposal to Vienna and that he had intentionally undermined it by subtly altering its substance. But as Kagan (1995: 200) notes, "that judgment seems unduly harsh." Although Bethmann Hollweg's injunction to Tschirschky may have lacked a sense of urgency, most historians now hold that his efforts on the night of the 28th to moderate Austria's behavior were sincere (Lebow, 1981: 136; Langdon, 1991: 180).[26]

In any event, shortly thereafter Bethmann Hollweg followed up with another telegram (no. 192) that was less ambiguous. This telegram, transmitted on 2:55 a.m. Berlin time, "urgently and impressively" urged Vienna to enter into direct negotiations with Russia. Five minutes later, after learning that this suggestion had once before been rejected, he again telegrammed Tschirschky, in apparent desperation, reiterating that a "refusal to hold any exchange of opinion with St. Petersburg" would be a "serious error" and a "direct provocation" of Russia. He concluded by directing his ambassador to "talk to [Austria's foreign minister] Count [Leopold] Berchtold at once with all impressiveness and great seriousness" (Kautsky, 1924: nos. 395, 396). As Nomikos and North (1976: 156–62) observe, "taken in sum, [the telegrams] represent a by no means inconsiderable effort to slow down, if not alter, the course of events now unfolding."

Some have attributed this dramatic policy shift to a strong warning from the German ambassador in London, Prince Karl von Lichnowsky, that Great Britain was very unlikely to stand aside in any war that involved France (e.g., see Albertini, 1952: vol. 2, 520–2, and Massie: 1991: 871). But Trachtenberg (1990/1991: 136) argues persuasively that "it was the news from Russia about partial mobilization that played the key role in bringing about the shift in Bethmann's attitude." As Trachtenberg goes on to note, "the evidence strongly suggests that the decisive change took place *before* the chancellor learned of Grey's warning, but *after* he had found out about Russia's partial mobilization" (emphasis in original).

In the terms of the model, Germany had, on July 28, updated its prior estimate of Russia's type. Until Russia announced its partial mobilization, both Berlin and Vienna were operating on the premise that Russia would almost certainly capitulate, as it had previously, that is,

---

[24] Some (e.g., Geiss, 1967: 223) have pointed to this injunction as conclusive evidence of Bethmann Hollweg's disingenuousness. But, in a memorandum to Grey on July 30, the British ambassador in Berlin reported that the German chancellor was concerned that too much pressure on Vienna might make matters worse (Gooch and Temperley, 1926: vol. 9, no. 329).

[25] By contrast, Williamson and May's (2007: 361) more benign interpretation is that the German chancellor "scuttled" Wilhelm's proposal because he continued "to believe that he could keep the war local." Similarly, Clark (2000: 209) concludes that "the view that [Bethmann] had already begun to harness his diplomacy to a policy of preventive war cannot be supported from the documents. It is more probable that he was simply already committed to an alternative strategy that focused on working with Vienna to persuade Russia not to overreact to Austrian action."

[26] See also Nomikos and North (1976: 156–7). Mombauer (2001: 286, 185) tacitly accepts the sincerity of Bethmann Hollweg's efforts. She claims that while Moltke and other German military leaders pushed for a preventive war at the end of July, both the kaiser and the chancellor lost their courage and got "cold feet."

that it was Soft at both the tactical and the strategic level. The Russian partial mobilization, however, is inconsistent with this assessment. In the Asymmetric Escalation Game with incomplete information, only tactically Hard Defenders rationally respond in kind. It should come as no surprise, then, that both Germany and Austria would revise their initial beliefs in light of the new information obtained.

At this point, it should be emphasized that this pattern of surprise and reevaluation is consistent with the existence of either of the two remaining solution candidates, $CLRE_1$ and $ELRE_3$. As Figure 6.2 suggests, both $CLRE_1$ and $ELRE_3$ exist only when the credibility of the Defender's first-level threat is insufficient to sustain either of the two Limited-Response Deterrence Equilibria. In both cases, this reduction in the Defender's credibility gives even a Soft Challenger an incentive to initiate at Node 1. After all, since the Defender believes that the Challenger is likely Hard, the Defender is completely deterred from escalating first (under any perfect Bayesian equilibrium of the Spiral Family, including $CLRE_1$ and $ELRE_3$). The Challenger, therefore, calculating that the Defender is likely to concede at Node 2, takes decisive action. It is in this sense that a response in kind will come as a surprise to the Challenger, regardless of which of these two perfect Bayesian equilibria are actually in play.

What distinguish $CLRE_1$ from $ELRE_3$, however, are the inferences that are made when the Defender unexpectedly responds. Notice from Figure 6.2 that the upper face of the $CLRE_1$ region slopes upward, away from the bottom edge of the left side wall. This sloping "ceiling" means that, under this equilibrium form, the probability that the Defender is of type HS is always small relative to the probability that it is of type HH. By contrast, under $ELRE_3$, the Defender is less likely to be of type HH, and much more likely to be of type HS than of type HH, than under $CLRE_1$. These critical differences lead to different behavioral patterns whenever the Defender unexpectedly responds in kind.

Under $CLRE_1$ the Challenger believes it more likely than not that the Defender will counter-escalate at Node 4. This explains why the Challenger never escalates first under this equilibrium form and why the outcome *Limited Conflict*, if it occurs at all, is most likely to do so under the conditions associated with the existence of $CLRE_1$. Play under $ELRE_3$, however, is another story. Here, since the Challenger believes that it is unlikely that the Defender will counter-escalate at Node 4, it may rationally decide to escalate at Node 3a.[27] If the Challenger's belief is incorrect, the result will be tragic, and the escalation spiral complete. In the Asymmetric Escalation Game with incomplete information, then, the conditions that support the existence of $ELRE_3$ uniquely describe the path to *All-Out Conflict*.

What is striking about the reevaluation process in 1914 is that German and Austrian leaders drew diametrically opposite conclusions from the measured Russian response. German leaders looked into the abyss and did not like what they saw. The July 29 warning from London may have had an impact here. But it was likely the Russian response in kind that was critical. After all, unless the Russians further pressed the issue, the question of what Britain would do was moot. In any event, after the Russian partial mobilization, Bethmann Hollweg came to believe that Russia would not back down if Austria proceeded with its invasion. He also realized that, to protect its western flank, Russia would also have to implement a general mobilization, which would clearly threaten Germany. And if Russia mobilized against

---

[27] Unless it is Soft, which is unlikely.

Germany, Germany would be compelled to mobilize as well, and any German mobilization implied a two-front war against both Russia and France. Of course, an attack against France might bring Great Britain into the conflict too and, as Grey had just warned, this would bring about "the greatest catastrophe that the world has ever seen" (Kautsky, 1924: no. 368). Bethmann Hollweg considered the latter eventuality as nothing less than a "leap into the dark."

All of which is to say that, after the partial mobilization, German political leaders concluded that Russia was not only tactically Hard but likely strategically Hard as well, and that they drew the proper inferences from this new assessment of Russia's type. Accordingly, consistent with the defining characteristics of a Constrained Limited-Response Equilibrium, Bethmann Hollweg quickly reversed course and urged moderation on Germany's only real ally. His purpose was clearly to avoid the consequences of the escalation spiral that is implied by play under an Escalatory Limited-Response Equilibrium such as $ELRE_3$.

It is unfortunate, indeed, that, on July 29, the critical decisions were not being made in Berlin. At this time, the locus of decision making was in Vienna, and Bethmann Hollweg obviously realized it—hence the desperate tone of his telegrams to Tschirschky. Significantly, the reaction in Vienna to the Russian partial mobilization was starkly different than that in Berlin. The Austrian leadership in general, but Berchtold in particular, did not believe that Russia would further escalate the conflict. As Vienna saw things, while the partial mobilization revealed that Russia was tactically Hard, it did not follow that it was strategically Hard as well. In fact, the Austro-Hungarian leadership drew exactly the opposite conclusion (Albertini, 1952: vol. 2, 388). According to Jannen (1996: 263, 249), "the Austrians simply did not take the threat of Russian intervention seriously." At the height of the crisis, "Berchtold continued to believe that he could keep Russia talking while Conrad crushed Serbia." In consequence, policymakers in Vienna

acted as if Russia did not exist. Possibly they were overconfident about the deterrent effect of Berlin's "blank check"; possibly they exaggerated Romanov adherence to the principle of monarchical solidarity and the need to avenge the Sarajevo murders. Certainly, they failed to pay even elementary attention to the danger signals of [a] Russian military response. Until late in the whole process, the senior leadership blissfully directed its attention only southward. . . . Berchtold, Conrad von Hötzendorf and the others, now programmed for action against Serbia, disregarded any information that might require them to modify their plans—and ambitions. They would do what they wanted and, of course, preferred to do: fight Serbia (Williamson, 1983: 27).[28]

Jannen (1983: 74) offers a slightly different perspective on why Austria's policy makers seemed to ignore the possible reaction of Russia. In his opinion, Berchtold and others in the inner circle simply succumbed to psychological stress:

It has been argued here that Austro-Hungarian decision-makers were responding to real problems and real threats, but that they were responding to them unrealistically.

---

[28] Albertini (1952: vol. 2, 686) expresses a similar view.

They had been subject to accumulating stress and fears from a wide range of sources long before the assassination and were seeking to reduce stress through a variety of psychological mechanisms. After the assassination, particularly given the symbolic and emotional importance of such an event, they could not tolerate the further stress entailed by uncertain negotiations over uncertain solutions. The assassination therefore acted as an immensely powerful catalyst that both raised their fears and anxieties to levels that burst the restraints that had hitherto contained them, and presented an external enemy, Serbia, upon whom such fears and their resultant aggression could be discharged. In the face of the psychological needs thus generated, war with Russia did not matter.

It is difficult to take issue with Jannen's assessment that decision makers in both Vienna and Berlin operated in a highly charged psychological environment (Holsti, North and Brody, 1968). But it does not follow that the mere existence of stress can explain the apparent oversight in Austrian preparation. For one, Jannen's argument is seemingly at odds with the detailed planning at the Ballhausplatz in early July that took account of not only the impact of mobilization on the Dual Monarchy's harvest but also the whereabouts of both the president and the prime minister of France (Zagare, 2009a). Additionally, Jannen's explanation also ignores the very purpose of the so-called Hoyos mission, that is, to evaluate Germany's reliability as an ally.

In any event, based on his belief that Russia would stand aside, Berchtold decided to deflect all of Bethmann Hollweg's frantic last-minute injunctions. When beseeched by Tschirschky to "be satisfied by the occupation of Serbian territory" (Kautsky, 1924: no. 465), he delayed by claiming that he would have to consult with Franz Joseph before replying. And, with respect to Grey's proposal for four-power mediation, he accepted the suggestion of Hungary's minister–president, Count István Tisza, that he should say that he was "ready to approach it in principle but only on the condition that [Austrian] operations in Serbia be continued and the Russian mobilization stopped" (Geiss, 1967: 321). As Albertini (1952: vol. 2, 677) observes, "this was tantamount to outright rejection."

Berchtold was clearly drawing a line in the sand. On July 30, he approached the emperor for permission to proceed toward general mobilization which, when carried out, would directly threaten Russia. This, according to Albertini (1952: vol. 2, 659), was "another big step in the direction of war." All the while, he insisted that all Austrian "demands must be accepted integrally and we cannot negotiate about them in any way" (Geiss, 1967: 320). In essence, by failing to moderate his government's policy, Berchtold escalated the conflict (Schmitt, 1930: vol. 2, 155–6).

Of course, since Berchtold's beliefs about Russia's likely response were incorrect, the results of his hard-line policy were entirely predictable.[29] Throughout the crisis, Russia had been on the verge of a general mobilization. On July 29, for example, the same day that the partial mobilization was implemented, the tsar had, in fact, agreed to a full mobilization, only to rescind his order after learning that the kaiser was attempting to mediate the dispute. But by the next day, Nicholas could no longer resist the pressure from his foreign minister.

---

[29] In one sense, Berchtold was, in fact, correct. Discussions with St. Petersburg would continue until August 6, when Austria finally declared war on Russia.

It is very likely, as Turner (1968: 85–6) contends, that it was (1) the news that Austria was shelling Belgrade and (2) Bethmann Hollweg's warning on July 29 to stop mobilizing that turned Russia's foreign minister, Sazonov, around and prompted him to push for full mobilization. But, in his telegram sent to the kaiser the next day to justify the action (Kautsky, 1924: no. 487), the tsar, whose consent was critical, tied his decision to sign the general mobilization ukase directly to his mistaken belief that Austria had already mobilized against Russia (Albertini, 1952: vol. 2, 576). Berchtold's inflexible hard-line policy, lacking as it did any conciliatory gestures, had made it all too easy for the tsar to believe the worst about Austrian behavior.[30]

As the saying goes, the rest is history. At 5:00 p.m. on July 30, an hour after the tsar had given his consent, the orders calling for a general mobilization were issued. At noon the next day, Austria–Hungary, following suit, mobilized against Russia. Shortly thereafter, Germany issued ultimata to Russia and France, demanding that all of their mobilization efforts be canceled. Of course, neither St. Petersburg nor Paris responded in the affirmative. Consequently, on August 1, 1914, Germany declared war on Russia and, two days later, on France. Continental Europe was now at war. The theoretical characteristics of $ELRE_3$ help to explain why.

War broke out in Europe in 1914 because both Austria–Hungary and Germany believed that Russia would stand aside when Austria moved aggressively against Serbia. Localization was not only their objective; it was also their firm hope and expectation. Of course, both members of the Dual Alliance were mistaken. Russian policy makers had already decided that Russia could not abandon Serbia and still survive as a great power. But, fearing war, they declined to escalate the crisis. Instead, Russian leaders settled on a limited response, a partial mobilization against Austria, which was intended as a warning shot across the bow.

Decision makers in Berlin quite clearly got the message. Unfortunately, their counterparts in Vienna did not. And it was in Vienna that the crucial choice not to pull back was made. By refusing to compromise, Austrian leaders escalated the crisis. Russia responded by mobilizing the rest of its forces against Germany. Of course, it was well understood in St. Petersburg that this act of counter-escalation "almost certainly meant war" (Fay, 1930: 479). Sadly, this was just about the only belief confirmed by events.

To conclude, it should be noted that $ELRE_3$ is the *only* perfect Bayesian equilibrium of the Asymmetric Escalation Game that is consistent with both the expectations and the action choices of the key players in July 1914. Several outcomes, including a one-sided victory (i.e., localization) and a limited conflict (i.e., a negotiated settlement) are possible under this equilibrium form. Unfortunately, escalation spirals are also real and distinct possibilities. Testimony to this distressing fact is the continental war that broke out in early August 1914.

---

[30] Sazonov's (1928: 203) memoirs confirm the importance the tsar placed on the untenable position he believed Russia to be in as a consequence of Austria's (as yet still undeclared) general mobilization. It is unclear exactly how the tsar came to be misinformed. But there is no indication that Sazonov went out of his way to set the record straight. In his foreword to the diary of Baron Schilling (1925: 9), who was the head of the chancellery in the Russian foreign ministry, Sazonov cites the fact that "Austria's mobilization was in full swing" as one of several factors that led to Russia's general mobilization. It is clear that here Sazonov was carefully parsing his words. Russia was the first major power to fully mobilize.

6.7 Coda

The war was no accident. In 1915 Bethmann Hollweg explained it by noting that a number of factors had forced Germany "to adopt *a policy of utmost risk*, a risk that increased with each repetition, in the Moroccan quarrel, in the Bosnian crisis, and then again in the Moroccan question" (Jarausch, 1969: 48). Clearly, the German chancellor had come to realize that he had rolled the dice one time too many. As Joachim Remak (1971: 366) has insightfully observed, sometimes, "it happens." The laws of probability almost guarantee it.

To say that the war was no accident, however, is not the same thing as saying that it was inevitable. As demonstrated in Section 6.6.2, there are compelling theoretical reasons to conclude that the crisis could have been avoided. But there are a number of empirical reasons as well (Mulligan, 2010). Consider counterfactually what would have occurred if, inter alia,

- the Serbians had been less strategic in their response to the Austrian ultimatum and capitulated entirely;
- the Russians had backed off of their support of Serbia;
- the French had not stood by the Russians;
- the Germans had been able to convince the Austrians not to invade Serbia;
- the Austrians had been less demanding of the Serbs;
- the British had early on made clear to the Germans their commitment to France;
- the Germans had avoided violating Belgium's neutrality;
- the tsar had not inadvertently revealed to the kaiser that Russia was in the process of mobilizing; or
- the British offer of neutrality had not been rescinded at the last moment.

This short list could easily be extended.[31] Clearly, it is very difficult to agree with those who see the war as inevitable. But if it was not, who was at fault? Suspects are not hard to find. In fact, the blame game actually started prior to the outbreak of hostilities, as each of the major powers and a few of the minor powers released a highly selective collection of official documents—some of which were fraudulent and many of which were intentionally misleading—that were all designed to deflect culpability for the war.

After the war, the blame game continued, and, not surprisingly, all five of the European powers were fingered. Both the kaiser and the Russian foreign minister blamed the British, as does Niall Ferguson (1998). Many historians, including Fritz Fischer (1967, 1975), Annika Mombauer (2013), Max Hastings (2013), and the political scientist Dale Copeland (2000), point to the Germans. The important work of Samuel Williamson (1991:196), however, clearly shows that many of "the steps that pushed Europe toward war were taken in Vienna." At the same time, Christopher Clark's (2012) book demonstrates convincingly that the French were more highly involved than is generally understood. And Sean McMeekin's

---

[31] See, for example, Crawford (2014).

(2011) penetrating analysis of *The Russian Origins of the First World War* most certainly implicates the Russians. So, who should be held accountable?

In my view, all of these answers, and none of these answers, are correct. In *The Sleepwalkers*, Clark assiduously tried to avoid answering this probably irresolvable question. In the end, however, I believe that his (2012: 561) conclusion that the July Crisis was "genuinely inter-active" is the most persuasive. In other words, had the policies and decisions of *any* of the five major powers and of Serbia been other than they were, the nature of the war would have been much different. Indeed, the war might not have occurred at all. This is an argument that is more than consistent with the explanation of the July Crisis developed in this chapter and also with the most recent historiography on the origins of the Great War (Zagare, 2015a).[32]

It is not, however, consistent with a number of other explanations of the onset of a major power war in early August 1914, including,

- Barbara Tuchman's (1962) accidental war thesis, which reappears in a slightly different guise in the work of Thomas Schelling (1960, 1966) and a few other rational choice theorists of the early 1960s;
- the related black swan argument of Bernadotte Schmitt (1944) and Richard Ned Lebow (1981) that the war was the result of a highly unlikely singular event (i.e., the assassination of the archduke);
- the "war is inevitable" hypothesis of William Thompson (2003), Paul Schroeder (1972, 2007), and several others;
- the empirically dubious argument of nuclear realists like Kenneth Waltz (1993) that the war occurred because its perceived costs were too low;[33]
- the cult of the offensive argument (Van Evera, 1999), which falls apart logically and empirically; and
- any explanation that singles out Germany, Austria, Russia, France, or Great Britain as the causal villain (Zagare, 2011b).

---

[32] See, inter alia, Clark (2012), Hastings (2013), MacMillan (2013), McMeekin (2013), Mombauer (2013), and Otte (2014b).
[33] See Förster (1999) for a factual rebuttal.

PART III
Security Studies

7

Perfect Deterrence Theory

7.1 Introduction

In the aftermath of World War II, a large, oftentimes contradictory, but nonetheless influential literature emerged in the field of security studies that is commonly referred to as (classical) deterrence theory.[1] In this chapter, I summarize this important strand of theory and compare it to an alternative (game-theoretic) specification called perfect deterrence theory. It is my contention that classical deterrence theory is both logically deficient and empirically inaccurate (Zagare, 1996a). I briefly explain why. Perfect deterrence theory, in contrast, provides a powerful perspective from which to view the dynamics of interstate conflict avoidance, initiation, limitation, escalation, and resolution. It is, in fact, the theoretical framework that was used to organize the analytic narratives developed in Part II of this book. Unlike classical deterrence theory, it makes consistent use of the rationality postulate, is prima facie in accord with the empirical record, and makes common-sense policy prescriptions that are grounded in strict logic.[2]

7.2 Classical Deterrence Theory

Although the roots of classical deterrence theory can be traced to the contentious debate between realists and liberals during the interwar period about the causes of great power wars in general, and of World War I in particular (Carr, 1939), it is generally understood that Bernard Brodie's (1946, 1959a, b) work is seminal. Shortly after atomic bombs were dropped, first on Hiroshima and then on Nagasaki, Brodie realized that war prevention would become the primary goal of the American security apparatus in what was sure to become a world of nuclear plenty. How and under what conditions that goal might be achieved is the principal question addressed in this vast literature.

---

[1] Classical deterrence theory is also sometimes referred to as rational deterrence theory. However, there are other theories of deterrence that are based on rational choice and would therefore seem to fall under the umbrella of rational deterrence theory as well.

[2] This chapter is based on Zagare (2018b).

Game Theory, Diplomatic History and Security Studies. Frank C. Zagare. Oxford University Press (2019).
© Frank C. Zagare 2019. DOI: 10.1093/oso/9780198831587.001.0001

The earliest attempts to answer this question relied heavily on extant realist theory. Since the conventional wisdom at the time was that a power asymmetry was an all-but-sufficient condition for major power conflicts, it followed that the key to peace was a carefully calibrated strategic balance. Balance of power theory, however, did not hold up to intense logical and empirical scrutiny. As evidence began to accumulate that both world wars were actually contested under parity conditions (Organski, 1958: ch. 11; Organski and Kugler, 1980), a new variable was added to explain away the glaring discrepancy between fact and theory: the cost of war. What eventually emerged was a part of the literature I have elsewhere referred to as *structural deterrence theory* (Zagare, 1996a). Structural deterrence theorists, of whom Kenneth Waltz (1964, 1979, 1981) is the exemplar, focus on the interplay of strategic structure and the cost of conflict to explain the absence of a superpower conflict during the Cold War period. In their view, parity, when coupled with sufficiently high war costs, all but guarantees peace (Intriligator and Brito, 1984). By contrast, structural deterrence theorists contend that peace may break down—even under parity—when the costs of conflict are, or are believed to be, low. For Waltz (1993: 77), the breakdown of the international system in 1914 occurred precisely because decision makers in Berlin, Vienna, St. Petersburg, Paris, and London thought that a war could be waged on the cheap.[3]

While Waltz and other structural theorists (e.g., Mearsheimer, 1990) explored the systemic conditions that they believed explained the "long peace" (Gaddis, 1986, 1987) of the Cold War period, a second group of strategic thinkers approached the question from a choice theoretic perspective. Pitched at the micro level of analysis, this strand of the literature, which I label *decision-theoretic deterrence theory*, includes the work of psychological choice, expected utility, and game theorists. The exemplar of this part of the deterrence landscape is Thomas Schelling (1960, 1966). Schelling, whose work on deterrence will be discussed more fully in Chapter 8, dabbled in all three genres. Zakaria (2001) credits Schelling as the "inventor" of modern deterrence theory. Others, however, point to William Kaufmann (1956).

These two strands in the strategic studies literature, which together constitute the dominant paradigmatic (i.e., realist) articulation of deterrence theory, have much in common. For example, structural deterrence theorists conclude that bilateral nuclear relations are exceeding stable. In fact, they argue, that as the costs of conflict rise, the probability of a war between nuclear rivals "approaches zero" (Waltz, 1990: 740; see also Intriligator and Brito, 1981: 256). Since rational conflicts are believed to be unlikely, the gravest threat to peace in the nuclear era is seen to be accidental war.

Decision-theoretic deterrence theorists take these conclusions as axiomatic. Indeed, they are built into the very structure of the game these theorists use as the underlying metaphor of a bilateral nuclear relationship: Chicken (Jervis, 1979: 291). In Chicken (see Figure 7.1), the outcome *Conflict* is a mutually worst outcome. In consequence, rational conflict is precluded. In a strict 2×2 ordinal game like Chicken, a mutually worst outcome can never be part of a Nash equilibrium. Thus, *Conflict* can only occur when rational players miscalculate, that is, when an accident occurs.

---

[3] The empirical basis for this claim is weak. See, for example, Förster (1999).

State B

	Cooperate (C)	Defect (D)
Cooperate (C)	*Status Quo* (3,3)	*B Wins* (2,4)*
Defect (D)	*A Wins* (4,2)*	*Conflict* (1,1)

State A (Cooperate (C), Defect (D) on left)

Key: (x,y) = payoff to State A, payoff to State B
    4 = best; 3 = next best; 2 = next worst; 1 = worst
    * = Nash equilibrium

Figure 7.1 Chicken

Structural deterrence theory and decision-theoretic deterrence theory share many of the same logical and empirical flaws.[4] For example, structural deterrence theorists are unable to explain the absence of a superpower conflict during the period of American nuclear superiority unless they make an ad hoc adjustment to the theory, as Waltz himself points out (1993: 47). States generally do not jump through "windows of opportunity," even when the motivation to do so exists (Lebow, 1984; Jervis, 1985: 6). The abstract version of the theory clearly implies that the United States should have exploited the obvious strategic advantage it enjoyed throughout the 1950s and the early 1960s, not only against the Soviet Union, but against lesser powers as well. As Jervis (1988a: 342) notes, structural deterrence theorists are unable to "explain the fact that the United States did not conquer Canada sometime in the past hundred years." Or as Gaddis (1997: 88) more tactfully puts it, "the actions the United States took [during the early Cold War years] failed to fit traditional patterns of great power behavior."

For their part, decision-theoretic deterrence theorists are also hard put to explain the absence of a superpower conflict during the most intense periods of the Cold War era without logical contradiction. And it is easy to understand why. *Conflict* is not the only outcome in Chicken that is inconsistent with rational play. The outcome *Status Quo*, which results when both states cooperate and neither attacks the other, is also not a Nash equilibrium. As a result, the long peace can only be explained within the confines of decision-theoretic deterrence theory by assuming, simultaneously, that the players are rational when they are being deterred but irrational when they are deterring and threatening mutual destruction.[5] All of which helps to explain why Achen (1987: 92) has observed that "far from leaning too heavily

---

[4] For a full discussion, see Zagare (1996a). For the empirical deficiencies of decision-theoretic deterrence theory in particular, and structural realism more generally, see also Vasquez (1998: 163–4, and ch. 9).

[5] For a discussion of several unsuccessful attempts to eliminate this logical inconsistency, see Zagare and Kilgour (2000: ch. 2).

on rational choice postulates, 'rational deterrence theory' necessarily assumes that nations are not always self-interestedly rational."

7.3 Perfect Deterrence Theory

Perfect deterrence theory was developed to overcome the empirical and logical problems that plague classical deterrence theory.[6] As developed by Zagare and Kilgour (2000), this theoretical structure is composed of a number of interrelated game models that are analyzed under a common set of preference assumptions. The assumptions are both intuitively obvious and, for the most part, innocuous. Most of them should already be familiar to the reader.

In each of these models the players are assumed to prefer winning to losing and to do so at the lowest possible cost. Most other preference relationships are taken as strategic variables. For example, some players might prefer *Conflict* to losing. Players with such a preference are assumed to have *credible* threats, that is, threats that are rational to execute. Other players, with the opposite preference, have threats that lack credibility. The players also may or may not prefer the *Status Quo* to *Conflict*. A player whose opponent prefers the *Status Quo* to *Conflict* is said to have a capable threat, that is, a threat that hurts (Schelling, 1966: 7). Threats that do not hurt are considered incapable.

Finally, in perfect deterrence theory the players are not necessarily undifferentiated as they are in classical deterrence theory (Legro and Moravcsik, 1999: 13). Specifically, in perfect deterrence theory, *Defenders* are players who prefer the *Status Quo* to all other outcomes, while *Challengers* are those who are motivated to upset it. A bilateral relationship wherein both players are Challengers is said to constitute a *Mutual Deterrence Game*. *Unilateral Deterrence Games* are those in which a satisfied Defender and a dissatisfied Challenger are involved.[7] Perfect deterrence theory examines the strategic impact of various configurations of credible and capable threats under conditions of complete and incomplete information in each of its four constituent games.

The variable nature of these critical preference relationships is, perhaps, the most important way that perfect deterrence theory is distinguished from decision-theoretic deterrence theory (Quackenbush, 2011: 747). In either formal or informal studies that take as their starting point the game of Chicken, all preferences are necessarily fixed. But another important difference between these two theoretically distinct variants of deterrence theory is perfect deterrence theory's strict adherence to Selten's perfectness criterion.[8] As noted in

---

[6] Sörenson (2017: 198), who uses modeling criteria to evaluate these two theoretical competitors, concludes that perfect deterrence theory is "a clear advancement" over its classical rival.

[7] The assumption of differentiated actors is not ad hoc in perfect deterrence theory, as it is in most manifestations of classical deterrence theory. Perfect deterrence theory is connected, theoretically, with power transition theory (Organski and Kugler, 1980), which sees the international system as hierarchical rather than anarchistic. In a hierarchical system, the dominant state and its allies are generally content with the status quo. For a discussion of the linkage between power transition theory and perfect deterrence theory, see Zagare (1996b, 2007).

[8] Powell (1990), whose work is in the tradition of Schelling and other decision-theoretic deterrence theorists, is an exception to the rule. His analysis of nuclear deterrence is consistent with Selten's criterion. But there are other problems with his explanation of the long peace. For a discussion, see Zagare and Kilgour (2000: 54–7).

Section 1.4, some Nash equilibria are supported by threats that are not rational (or credible) to execute. Selten's perfectness criteria eliminates these equilibria as rational strategic possibilities and, not incidentally, the logical inconsistency that is associated with them.

Perfect deterrence theory's name reflects its reliance on Selten's definition of rational strategic behavior. Adherence to the perfectness criterion assures logical consistency which is, arguably, perfect deterrence theory's most important characteristic. Walt (1999) notwithstanding, logical inconsistencies are clearly fatal to the health and well-being of any theory. As pointed out in Chapter 1, logically inconsistent theories are incapable of being falsified. By contrast, because it has clear empirical implications, perfect deterrence theory can, and has, been rigorously tested (Quackenbush, 2010a, 2011). Many of its theoretical propositions are also consistent with the empirical record, as I discuss later in this section.

7.3.1 Strategic Variables

Perfect deterrence theory is a general theory of conflict initiation and resolution. Unlike classical deterrence theory, perfect deterrence theory makes no particular assumption about the cost of conflict. It is, therefore, applicable to a much wider range of strategic relationships. Perfect deterrence theory is simply not a divergent theory of nuclear war avoidance. Rather, it is a universal theory, applicable to both nuclear and nonnuclear interactions. As such, it can be used to help explain why crises occur, why some conflicts escalate and others do not, and when and why limited and all-out wars are waged. Perfect deterrence theory's empirical domain is not even restricted to interstate interactions. As a general theory of strategic interaction, it is potentially applicable to intergroup or interpersonal conflict of interest situations whenever and wherever they may occur.

In perfect deterrence theory, the cost of conflict is, nonetheless, a critical strategic variable. Its value relative to other variables determines both the *capability* and the *credibility* of a deterrent threat. As noted, capable threats are threats that hurt. Threats that hurt are those that leave a player worse off than if the prohibited action was not taken. Operationally, this means that one player's deterrent threat is capable only if the other player, the threatened player, prefers the status quo to the outcome that would result if the threat were executed. Conversely, a deterrent threat will lack capability whenever the threatened player prefers to act, even if the threat is carried out (Zagare, 1987: 34).

The threat of nuclear retaliation is clearly a capable threat. Classical deterrence theorists contend that capability constitutes a *sufficient* condition for deterrence success, which is why they believe that the mere possession of nuclear weapons all but guarantees a lasting peace (e.g., Bundy, 1983; Levy, 1988: 489–90; Waltz, 1993: 53–54). In perfect deterrence theory, by contrast, deterrence may fail even when all threats are capable. For instance, by definition, both players have a capable threat in the game of Chicken (see Figure 7.1). Yet the status quo does not survive rational play in this game.

On the other hand, within the theoretical confines of perfect deterrence theory, a capable threat constitutes a *necessary* condition for deterrence success, that is, deterrence will always fail whenever a deterrent threat lacks capability (Zagare and Kilgour, 2000: 81–4). To illustrate, consider for now the Unilateral Deterrence Game given in Figure 7.2. This game, which is one of perfect deterrence theory's four constituent games, depicts a one-sided

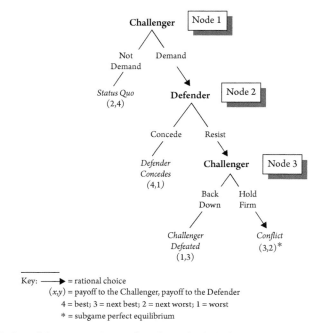

Key: ──▶ = rational choice
(*x,y*) = payoff to the Challenger, payoff to the Defender
4 = best; 3 = next best; 2 = next worst; 1 = worst
* = subgame perfect equilibrium

Figure 7.2 Unilateral Deterrence Game when the Defender's threat is credible but not capable

(or asymmetric) deterrence relationship in which a Defender seeks to deter a Challenger, but not the other way around.[9] On its face it would appear to be an appropriate context in which to explore the strategic relationship of China and the United States, Russia and Ukraine, and North and South Korea, inter alia.

In the Unilateral Deterrence Game, the Challenger begins play (at Node 1) by deciding whether to contest the existing order. If no demand is made, the *Status Quo* prevails. But if a demand is made, the Defender must decide (at Node 2) whether to resist it or concede the issue. Concession results in the outcome *Defender Concedes*. Resistance brings about a subsequent choice for the Challenger at Node 3: if the Challenger backs down, the outcome *Challenger Defeated* results. If the Challenger holds firm, there is *Conflict*. The four outcomes in this game correspond, roughly, to the four outcomes in Chicken.

As before, the ordered pair beneath each outcome represents the ordinal payoff to the players, from best (i.e., 4) to worst (i.e., 1). The first entry represents the Challenger's evaluation, and the second, the Defender's. For example, in this variant of the Unilateral Deterrence Game, *Conflict* is the Challenger's next-best outcome (i.e., 3) and the Defender's next-worst outcome (i.e., 2). Since the present assumption is that the Challenger prefers *Conflict* to the *Status Quo*, the Defender's threat to resist at Node 2 lacks capability, by definition.

---

[9] The Generalized Mutual Deterrence Game (which will be examined in Chapter 8), the Asymmetric Escalation Game (analyzed in Chapters 5 and 6), and the Tripartite Crisis Game (described in Chapters 2 and 3) are also components of the theory. The Unilateral Deterrence Game was used in Chapter 2 to analyze the Rhineland crisis of 1936.

The version of the Unilateral Deterrence Game given in Figure 7.2 is an extensive form game with complete information. As indicated in Chapter 1, in such a game, rational play is determined by applying backward induction, starting with the Challenger's choice at Node 3 and working backward up the game tree. As the arrows indicate, a rational Challenger will hold firm at Node 3 in order to avoid its worst outcome, *Challenger Defeated*. Since concession results in the Defender's worst outcome, it will rationally resist should it face a choice at Node 2. Anticipating both these choices, the Challenger will rationally issue a demand at Node 1. After all, by assumption, the Challenger prefers *Conflict*, the unique subgame perfect equilibrium in this game, and the outcome that is implied by the backward induction process, to the *Status Quo*. In consequence, deterrence fails—as it always will—whenever a Defender's threat lacks capability.

Notice that, in this example, the Defender's threat is credible, that is, it prefers *Conflict* to *Defender Concedes*. Yet, deterrence does not succeed. Freedman (1989: 96) claims that credibility is the "magic ingredient" of deterrence. While there is a modicum of truth to this claim, a credible threat does not constitute a sufficient condition for deterrence success, as the game shown in Figure 7.2 demonstrates.

Nor is credibility a necessary condition. (Table 7.1 summarizes perfect deterrence theory's conclusions about the causal implications of different types of deterrent threats.) To see this, consider now the version of the Unilateral Deterrence Game depicted in Figure 7.3. In this example, neither player's threat is credible. Since *Conflict* is a mutually worst outcome, the Challenger rationally backs down at Node 3. But because it expects a rational Challenger to accept defeat at Node 3, the Defender intends to resist (because it prefers *Challenger Defeated* to *Defender Concedes*) if and when it is faced with a choice at Node 2. Node 2 is never reached in rational play, however, because the Challenger, preferring to avoid certain humiliation (i.e., *Challenger Defeated*), will suddenly find the *Status Quo* acceptable.

In this case, then, deterrence succeeds even though the Defender's threat lacks credibility. All of which illustrates that it may be the characteristics of the Challenger's threat, and not of the Defender's, that ultimately determines rational play in a game. This is not a trivial point. For the most part, classical deterrence theorists have fixated on a defender's threat in order to understand the conditions under which deterrence breaks down.[10] In so doing, they fail to recognize the interactive impact that deterrent threats have. Empirical studies that focus attention on the characteristics of a defender's threat also miss this important

Table 7.1 Causal characteristics of deterrent threats.

	Necessary Condition	**Sufficient Condition**
Capable threats	Yes	No
Credible threats	No	No
Credible and capable threats	No	No

[10] Lebow's (1981: 85) conclusion is typical. For Lebow and most strategic analysts, "four conditions emerge as crucial to successful deterrence. Nations must (1) define their commitment clearly, (2) communicate its existence to possible adversaries, (3) develop the means to defend it, or to punish adversaries who challenge it, and (4) demonstrate their resolve to carry out the actions this entails."

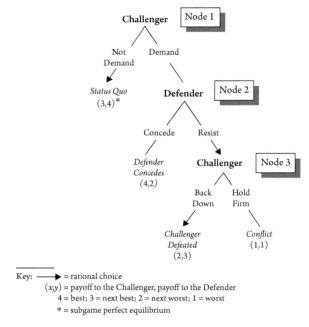

Key: ———▶ = rational choice
(*x,y*) = payoff to the Challenger, payoff to the Defender
4 = best; 3 = next best; 2 = next worst; 1 = worst
* = subgame perfect equilibrium

Figure 7.3 Unilateral Deterrence Game when neither player has a credible threat

dimension of not only direct deterrence relationships but also extended deterrence relationships and, in the process, introduce case-selection bias into their studies.[11] In a multi-level extended deterrence game where a challenger's strategic level (or endgame) threat lacks credibility, deterrence should prevail regardless of the configuration of a defender's tactical and strategic level threats (Zagare and Kilgour, 2000: ch. 9).

By now, it should be clear that the relationship between credibility and the operation of deterrence is anything but straightforward. More specifically, deterrence may fail even when all threats are credible, and it may succeed even when all threats lack credibility. But this is not the end of the story. Conditions also exist under which one player's possession of a credible threat will actually undermine the stability of the status quo. The game shown in Figure 7.4 is a case in point.

There is only one difference, and a relatively minor one at that, between the games shown in Figures 7.3 and 7.4. In the game shown in Figure 7.3, the Challenger's threat lacks credibility, yet deterrence succeeds. In the game shown in Figure 7.4, by contrast, the Challenger's threat is indeed credible, but as the arrows indicate, deterrence rationally fails. So once again, a credible threat (or the lack thereof) is central to the operation of deterrence game, but not in the way one might suspect. In the game shown in Figure 7.4, one player's (i.e., the Challenger's) possession of a threat that is rational to execute actually incentivizes a breakdown of deterrence.

---

[11] Not surprisingly, Lebow (1981) is a case in point (Most and Starr, 1989).

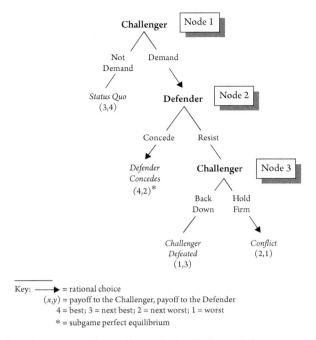

Key: ⟶ = rational choice
(x,y) = payoff to the Challenger, payoff to the Defender
4 = best; 3 = next best; 2 = next worst; 1 = worst
* = subgame perfect equilibrium

Figure 7.4 Unilateral Deterrence Game when only the Challenger's threat is credible

As noted in Section 7.2, classical deterrence theorists tend to focus on the defender's threat characteristics rather than the challenger's. One possible explanation is their mania for explaining deterrence failures, and their comparative lack of interest in explaining deterrence success. Their focus on deterrence breakdowns also helps to explain why they are similarly prone to overlook the role played by another important strategic variable, the value of the status quo. This is clearly a significant oversight. The extent to which a player is satisfied (or dissatisfied) with the existing order is not only a critical determinant of a threat's capability; it is also the only standard that can be used to establish a player's incentive (or willingness) to overturn it (Most and Starr, 1989). The failure to explore the consequences of this critical benchmark variable is yet another source of case-selection bias that plagues almost every empirical examination of classical deterrence theory's central propositions.[12]

7.3.2 Relationship Predictions

In some ways, the theoretical characteristics of classical deterrence theory and perfect deterrence theory are similar. For example, both frameworks assume egotistical or self-interested actors who make rational choices. But there are also some significant differences. As noted, perfect deterrence theory makes consistent use of the rationality postulate,

---

[12] See, for example, Mearsheimer (1983).

whereas classical deterrence theory does not. In contrast to the standard version of deterrence theory, perfect deterrence theory does not treat all actors as similarly motivated (i.e., as undifferentiated); nor does it take status quo evaluations and threat credibility to be fixed and constant. In perfect deterrence theory, these entities are important strategic variables. In consequence, the empirical expectations of these two competing specifications of deterrence theory are significantly at odds.

To be sure, it is not easy to summarize classical deterrence theory's core propositions. Since its sporadic application of the rationality postulate inevitably leads to logical inconsistencies, anything and its opposite can be deduced from its axiomatic base (Martin, 1999: 83). Nonetheless, although there are some exceptions, there also appears to be a general consensus among mainstream strategists. Table 7.2 summarizes both this confluence of opinion and the competing relationship propositions of perfect deterrence theory. Several of these differences have already been discussed.

One that has not concerns the overall nature of contentious deterrence relationships. As previously indicated, classical deterrence theorists see strategic (i.e., nuclear) deterrence relationships as more than robust, that is, as exceedingly stable. In their view, rational nuclear conflicts are extremely unlikely. Perfect deterrence theory, however, reaches a different conclusion. In the four incomplete information game models that collectively delineate the theory, a deterrence equilibrium under which no player contests the status quo always exists. But these equilibria are hardly ever unique, that is, more often than not they coexist with other equilibria, some of which admit the possibility of a complete deterrence breakdown.[13] In other words, the conditions that are most conducive to peace and stability are

Table 7.2 Classical deterrence theory and perfect deterrence theory: Empirical propositions.

Propositions	Classical Deterrence Theory	Perfect Deterrence Theory
Status quo	Unimportant/ignored	Significant
Strategic deterrence	Robust/all but certain	Fragile/contingent
Relationship between conflict costs and deterrence success	Strictly positive and monotonic	Non-monotonic
Asymmetric power relationships	Unstable	Potentially very stable
Parity relationships	Very stable	Potentially unstable
Capability	Sufficient for deterrence success	Necessary but not sufficient for deterrence success
Limited conflicts and escalation spirals	Unexplained	Placed in theoretical context

[13] For example, if and when they exist, the two plausible deterrence equilibria (Det_2 and Det_3) of the special case analysis of the Asymmetric Escalation Game (see Chapters 5 and 6) will always coexist with the No-Response Equilibrium. The same is true of the Sure-Thing Deterrence Equilibrium that sometimes exists in the Generalized Mutual Deterrence Game with incomplete information analyzed in Chapter 8. It too, is never unique.

also consistent with the possibility of war. Depending on leadership preferences and policy choices, either denouement is possible, that is, deterrence failures, limited conflicts, and all-out conflicts can almost never be ruled out as a rational strategic possibility. In perfect deterrence theory, then, deterrence success is highly contingent and anything but robust.

A related difference concerns the strategic characteristics of parity conditions. Classical deterrence theorists see them as generally stable and view asymmetric power relationships as potentially unstable. Once again, perfect deterrence theory reaches a different conclusion. In both theory and practice it is much easier to stabilize the status quo when only one player is a defender than it is when both play that role. Since the former category of games are more likely to exist when power is asymmetrically distributed, parity or balance of power conditions are judged, ceteris paribus, to be less stable than one-sided deterrence relationships, but especially those in which a dominant and highly satisfied great power clearly controls the agenda of the international system.

Finally, classical deterrence theorists see a monotonic and strictly positive relationship between the cost of conflict and the probability of deterrence success. Specifically, the higher war costs, the less likely there will be a deterrence breakdown, and conversely (e.g., Mearsheimer, 1990: 19). In perfect deterrence theory, by contrast, the relationship is neither. To be sure, there is a minimum cost threshold below which deterrence success is not possible. This is the point at which a non-capable threat becomes a threat that hurts. But there is also a maximum threshold beyond which additional increases in the cost of conflict are unrelated to the probability of successful deterrence. Moreover, in perfect deterrence theory, extended deterrence relationships are actually more likely to unravel, ceteris paribus, as war costs increase.

7.3.3 Empirical Support

There is considerable empirical support for both the relationship and, as will be seen in this section, for the equilibrium predictions of perfect deterrence theory (Morton, 1999). Perfect deterrence theory is, in other words, a more plausible theory of strategic interaction than is classical deterrence theory. In part, this advantage stems from classical deterrence theory's loose formulation and its inconsistent use of the rationality postulate. By contrast, perfect deterrence theory's straightforward empirical expectations are the result of an analytic framework that respects the requirements of strict logic.

Take, for instance, the empirics that support the stabilizing effect of satisfaction with the status quo.[14] For the most part, classical deterrence theorists have ignored this literature precisely because their conceptual lens has blinded them to its importance. This wide and largely straightforward literature, however, is more than consistent with the gestalt of perfect deterrence theory. It should be patently obvious that highly satisfied states are, ceteris paribus, less prone to contest the governing structure of either a regional or a systemic structure. Classical deterrence theorists, however, seem not to have noticed.

Much the same can be said about perfect deterrence theory's conclusion that a capable threat is a necessary (though not a sufficient) condition for deterrence success. Harvey's (1998: 691) empirical study "indirectly [supports perfect deterrence theory's] claim about

---

[14] For a detailed overview, see Geller and Singer (1998).

the crucial role of capabilities" in deterrence relationships. But so does Bueno de Mesquita's (1981: 154–6) finding that conflict initiators, in general, tend to be more capable than those they attack, regardless of the belligerents' alignment status or the level of initiated conflict.

There is also impressive empirical support for perfect deterrence theory's claim about the central role that credible threats play in deterrence relationships. For instance, in an important study of extended deterrence, Danilovic (2002) finds that a state's inherent credibility—as reflected in large part by its regional interests—is a far more important predictor of deterrence outcomes than are the coercive bargaining tactics recommended by decision-theoretic deterrence theorists. Huth's (1999) earlier review of the literature reaches a similar conclusion and, at the same time, undermines the empirical foundation of many of Schelling's most well-known policy prescriptions including, but not limited to, the "threat that leaves something to chance."[15] More recently Johnson, Leeds, and Wu's (2015: 309) comprehensive study of strategic alliances clearly shows that "defense pacts with more capability and more credibility reduce the probability that a member state will be the target of a militarized dispute." Their results reinforce both Danilovic's conclusions and perfect deterrence theory's propositions about the pacifying role played by both these variables in extended deterrence relationships. In this context, it should also be mentioned that the vast literatures supporting action–reaction models, reciprocity and tit-for-tat behavior[16] are also consistent with perfect deterrence theory's key finding that conditionally cooperative strategies (or policies) are associated with deterrence success and that unconditionally hard-line policies are generally prone to fail.

In another significant examination of the dynamics of deterrence, Senese and Quackenbush (2003) studied the long-term consequences of different types of conflict settlements on recurring conflicts. Their hypothesis that imposed settlements, which they contend create one-sided (or unilateral) deterrence relationships, should last longer than negotiated agreements, which they associate with bilateral (or mutual) deterrence situations, was independently deduced from perfect deterrence theory's axiomatic base. In an exhaustive analysis of 2,536 interstate conflict settlements between 1816 and 1992, this now clear implication of perfect deterrence theory was "strongly supported." Senese and Quackenbush's result held even when they controlled for regime type, relative capability, contiguity, decisive outcomes, power shifts, and war, leading them to observe that in their test of one of perfect deterrence theory's relationship predictions, the theory performed "quite well." Follow-up studies by Quackenbush and Venteicher (2008) and Quackenbush (2010b) reach the same conclusion.

Space and other considerations preclude a full discussion of empirical studies that support the relationship predictions of perfect deterrence theory.[17] But mention should be made of one study that attempted to test perfect deterrence theory's equilibrium predictions and also of those analytic narratives that have been constructed to illustrate the "theory in action."

[15] See, also, the discussion in Chapter 8 of his "commitment tactics."
[16] For overviews, see Cashman (1993: ch. 6) and Sullivan (2002: ch. 9). On reciprocity, see Crescenzi, Best, and Kwon (2010).
[17] For a detailed discussion, see Zagare (2011 184–6) or Zagare and Kilgour (2000: ch. 10).

In the literature of deterrence, attempts to test the equilibrium predictions of a game-theoretic model are virtually nonexistent, in large part because of the numerous methodological hurdles that stand in their way. Quackenbush (2011) is the rare exception in more than one way. Not only is it a direct test of a game model's equilibrium predictions, but it is also one of the few attempts to test a theory of general deterrence. The empirical literature of deterrence is dominated by tests of extended immediate deterrence (Morgan, 1977) in which general deterrence has already failed. Huth (1988) and Danilovic (2002) are representative of the best of this literature.

In order to test perfect deterrence theory's equilibrium predictions and, not coincidentally, minimize case-selection bias, Quackenbush had to construct a relevant data set. For this, he first developed a category of relationships he termed "politically active" (Quackenbush, 2006). He also had to develop measures that reflected the players' utility functions, perhaps the most daunting precondition for a meaningful test of any rational choice theory. Finally, using binary and multinomial logit methods to examine the predictions of various game outcomes, he found that the predictions of perfect deterrence theory were "generally supported by the empirical record." He also found that his results provide much stronger support for perfect deterrence theory than the support that Bennett and Stam (2000) found for Bueno de Mesquita and Lalman's (1992) international interaction game.

Because it is a direct test, Quackenbush's study is the most persuasive evidence to date that perfect deterrence theory stands on firm empirical grounds. Less systematic, but significant nonetheless, is the accumulated collection of case studies that have been constructed using the constituent models of perfect deterrence theory as an organizing mechanism, including each of the analytic narratives developed in Part II of this book. These case-specific applications should be considered as additional, albeit limited, evidence of perfect deterrence theory's empirical robustness (see also Zagare and Kilgour, 2006, and Zagare 2011a).

7.3.4 Policy Implications

Classical deterrence theory and perfect deterrence theory stand in sharp opposition to one another. The classical formulation is plagued by logical inconsistencies and empirical anomalies. By contrast, perfect deterrence theory respects the laws of logic and is not contradicted by the empirical record. As well, it has performed quite well when tested directly, better in fact than other theories, formal or not. The distinctions, axiomatic, theoretical and empirical, are simply not arcane academic nitpicks. The policy implications of these two competing formulations of deterrence theory are also widely divergent, as Table 7.3 shows. In other words, there are significant real world consequences for which variant is used, not only for understanding how interstate relationships operate, but also for determining how best to navigate them.

One important difference can be traced to the conclusions the two theoretical frameworks reach about the implications of manipulating the cost of conflict. As already noted in Section 7.2, classical deterrence theorists, but especially decision-theoretic deterrence theorists, assume fixed preferences. Specifically, players are assumed to always prefer an advantage to the status

Table 7.3 Classical deterrence theory and perfect deterrence theory: Policy prescriptions.

Policies	Classical Deterrence Theory	Perfect Deterrence Theory
Overkill capability	Supports	Opposes
Minimum deterrence	Opposes	Supports
"Significant" arms reductions	Opposes	Supports
Proliferation	Supports	Opposes
Negotiating stances	Coercive, based on increasing war costs and inflexible bargaining tactics	Conditionally cooperative, based on reciprocity

quo, and to always prefer backing off rather than enduring the costs associated with conflict. In consequence, any change in the cost of conflict will have the same impact relative to these other outcomes. Since they assume that the probability of successful deterrence increases monotonically as these costs rise, classical deterrence theorists see no obvious limit to the pacifying impact that expanded expenditures for weapons systems that are capable of imposing pain and suffering on an opponent can have. For this reason, most classical deterrence theorists favor an "overkill capability," oppose "significant" arms control agreements, and stand in opposition to minimum deterrence policies.

In perfect deterrence theory, however, the cost of conflict relative to other variables is not fixed. Rather it is gauged against two other important strategic variables. The first, of course, is the value of the status quo. The second reference point is the value of concession. In perfect deterrence theory, the players may, or may not, prefer to concede rather than execute a deterrent threat. In consequence, there are two strategically significant cost thresholds, one minimum and the other maximum. The minimum cost threshold is the point below which deterrence cannot succeed. This is the point separating threats that are capable from those that are not. The two theoretical frameworks are not at odds over the strategic impact of increasing conflict costs beyond this minimum cost threshold. Both agree that deterrence success can only be achieved once this threshold is crossed.

In perfect deterrence theory, however, the maximum cost threshold is also strategically significant. Once this point is reached, further increases in the cost of conflict do not contribute to the probability of deterrence success. Rather than an overkill capability, then, the logic of perfect deterrence theory is consistent with a policy of minimum deterrence, which rests on a threat that is costly enough to deter an opponent but is not so costly that the threat itself is rendered incredible.

An equally important difference stems from the fact that, in perfect deterrence theory, there is no simple monotonic relationship between the cost of conflict and the stability of the status quo, as there is in classical deterrence theory. Since extended deterrence becomes more and more difficult to maintain as conflict costs rise, an increased military capability may actually have a destabilizing implications. In other words, in perfect deterrence theory, increased conflict costs can, under some circumstances, be stabilizing but, under others, may have the opposite consequence. For this and other reasons, perfect deterrence theory stands in opposition to unconstrained armaments programs and an overkill capability and supports defense postures that are sufficiently capable but not unnecessarily redundant.

Perhaps the most significant policy difference between classical and perfect deterrence theories concerns proliferation policies. Those classical deterrence theorists who fully understand the logical implications of their paradigm favor them. In the past, Mearsheimer (1990, 1993), Posen (1993), Van Evera (1990/1991), and Waltz (1981), inter alia, have argued that nuclear weapons can help stabilize contentious interstate rivalries and that, therefore, nuclear weapons should be disseminated selectively. More recently, Waltz (2012) made the case that if Iran acquired nuclear weapons, the possibility of an all-out conflict between Iran and Israel would be all but eliminated.[18]

Of course, those deterrence theorists who favor proliferation argue that nuclear technology should not be distributed indiscriminately.[19] An exception, they assert, should be made for so-called rogue states that operate on the fringes of the interstate system and whose behavior is borderline "rational." Of course, there is no such category in realism, whether it be classical or neo, or in the standard formulations of deterrence theory, where all states are taken as like-units (Waltz, 1979; Mearsheimer, 1990). Nonetheless, this qualifier is frequently called upon to help those classical deterrence theorists who oppose proliferation avoid a conclusion they do not wish to reach. But the very fact that this anomalous category of states exists is further evidence that realism in general, and classical deterrence theory in particular, is a degenerative research program (Vasquez, 1997, 1998).

The proliferation of nuclear and other weapons of mass destruction is not a policy prescription that can be derived from the axioms of perfect deterrence theory. To be sure, the increased costs of conflict that are associated with these weapons imply, ceteris paribus, an increased probability of deterrence success. But all things are hardly ever equal in interstate relations. For one thing, the minimum cost threshold that renders a threat capable can also be achieved with more conventional weapons. For another, the increased probability of success that nuclear weapons imply must be balanced against the risks associated with the possibility of a massive breakdown of deterrence. Since, in perfect deterrence theory, these risks are seen as considerable, it opposes proliferation policies, selective or otherwise.

There is one additional significant policy dispute between classical deterrence theory and perfect deterrence theory that remains to be discussed: how best to manage a crisis, that is, how best to negotiate or bargain with an opponent. This is a question that has received considerable attention from decision-theoretic deterrence theorists from Schelling (1960, 1966) on down. In general, decision-theoretic deterrence theorists favor coercive bargaining tactics that involve either increasing the cost of conflict or making an irrevocably commitment to a hard-line policy so that the opponent is forced to make the difficult choice between war and peace. The standard example involves a player forfeiting control in a game of Chicken by pulling the steering wheel off the steering column. Snyder (1972) provides a useful summary of less colorful but equally counterintuitive tacit negotiating tactics. Empirically speaking, however, the stratagems that decision-theoretic deterrence theorists

---

[18] For a trenchant rebuttal, see Kugler (2012).
[19] On this point, Waltz (2012) is an outlier.

recommend have no basis in fact (Snyder and Diesing, 1977: 489–90; Betts, 1987: 30; Huth, 1999: 74; Danilovic, 2002).

By way of contrast, perfect deterrence theory recommends conditionally cooperative bargaining stances rooted in reciprocity. The large majority of the deterrence equilibria in its constituent models are supported by strategies that offer cooperation if the other player cooperates but threaten noncooperation if the other player does not reciprocate. Hard-line (unconditionally noncooperative) and soft-line (unconditionally cooperative) bargaining stances generally do not lead to either stable relationships or political compromises.

It is noteworthy that the efficacy of conditionally cooperative negotiating stances is well established not only in (perfect deterrence) theory, but also in customary diplomatic practice. For example, in a large-*n* statistical analysis of extended deterrence relationships, Huth (1988) found that firm-but-flexible negotiating styles and tit-for-tat military deployments are highly correlated with extended deterrence success. Each of these more or less standard diplomatic or military postures involves reciprocity, a behavioral norm that is pervasive in interstate relationships, contentious or not (Cashman, 1993: ch. 6). It is also a norm that decision-theoretic deterrence theorists have a hard time explaining. It is, in a word, inconsistent with their underlying conceptual framework (Zagare, 2011a: 55).

7.3.5 Some Specifics

What, then, are some of the specific implications of perfect deterrence theory for current foreign policy debates and/or problematic interstate relationships? Assuming that the United States has no future revisionist aspirations, perfect deterrence theory suggests that, in the short term at least, the national missile defense system currently under development would not seriously undermine strategic stability (Quackenbush, 2006). In perfect deterrence theory, the status quo is most likely to survive when a satisfied preponderant power exists.

The longer-term consequences of a missile shield, however, could be less benign. If other states, like China or Russia, respond to a fully operational (i.e., long-range) American ABM system by further expanding their offensive capability in order to blunt its effectiveness, then any purported benefits of a national missile defense may prove to be ephemeral. Worst still, if the abrogation of the ABM Treaty by the second Bush administration eventually leads to widespread proliferation of Russian or Chinese weapons technology, the interstate system will only become more dangerous and more likely to break down. Similarly, the recent decision by South Korea to deploy the Terminal High Altitude Area Defense system only increases the probability that the United States might be tempted proactively to overturn the status quo on the Korean peninsula, even if its initial strategic rationale was primarily defensive.

As mentioned in Section 7.3.4, in terms of nuclear weapons policy, perfect deterrence theory is consistent with a minimal deterrent combined with a "no first use" deployment. While neither approach, of course, can guarantee peace, combined they offer the best chance for avoiding a catastrophe. More generally, the United States should continue to build down (but not eliminate) its nuclear arsenal. It is far better for the United States to meet China, or other potential rivals, on its way down than on a rival's way up. Ceteris paribus, the more costly nuclear war, the less likely it is that the status quo will survive over time. All things considered, the Joint Comprehensive Plan of Action that the P5+1 (China, France, Germany, Russia, the United Kingdom, and the United States), the European Union, and Iran entered into was a positive development, although the fact that Israel continues to maintain a nuclear capability, and other states in the region might soon obtain one, is worrisome, Waltz (2012) notwithstanding.

The broad foreign policy orientation suggested by perfect deterrence theory is rooted in reciprocity (or conditional cooperation). In practice, this means avoiding inflexible hard-line policies that are not only likely to decrease the satisfaction of other states but are also likely to lead to a negative response that leaves all concerned worse off. For example, a firm pledge not to attack in return for demonstrable disarmament may have averted a costly US war with Iraq in 2003; a similar guarantee might help the United States peacefully resolve its current dispute with North Korea. However, reciprocity also means avoiding uncondition-ally cooperative stances (i.e., appeasement policies) that are patently one-sided. Unilateral concessions are generally invitations for exploitation.

Perhaps the most difficult foreign policy problem for the United States and its allies in the years ahead will be managing its relationship with China. Although some might disagree (e.g., Brooks and Wohlforth, 2015/2016), the strong consensus in the strategic community is that China will reach parity and likely overtake the United States sometime before mid century. In both perfect deterrence theory and its structural analog, power transition the-ory, parity constitutes a necessary, although not a sufficient, condition for major power conflict. Clearly, a range of issues separating the two leading states in the system, including disagreements over fiscal and monetary policies, trade, patents, and human rights, will need to be addressed. China's island-building activity in the South China Sea is a dangerous flash point. Another centers on Taiwan. A creative compromise on both of these territorial dis-putes will make a peaceful transition much more likely (Senese and Vasquez, 2008, chs. 3–4; Quackenbush, 2015: ch. 4). From the point of view of the United States, a rigid con-frontational foreign policy toward China will be counterproductive. Additional steps toward deployment of a either a national missile defense system or a localized system will continue to antagonize the Chinese, increasing their dissatisfaction, and likely prompt China to expand the size of its nuclear arsenal, and to possibly finance that growth by exporting dangerous technologies (Tammen et al., 2000). At the same time, the United States should maintain a strong presence in the Pacific, lest it be left with the unpalatable choices associated with all-or-nothing deployments. Obviously, instigating a trade war or clumsily pushing for Chinese internal reforms also will not help.

Much the same could be said about Russia. While the creation of the NATO–Russia Council in 2002 was seen as a good step at that time, needlessly provoking Russia by expanding NATO did not serve the cause of peace. Whether Russia's absorption of South Ossetia and Abkhazia after its war with Georgia in 2008, its 2014 annexation of Crimea, its cyber warfare in the Baltic region, and its continuing attempt to undermine Ukraine's sov-ereignty, not to mention its meddling in American and European elections, were direct consequences of its exclusion from NATO is largely irrelevant. But Russia's continuing iso-lation does not help, especially since it can only lead to a further intensification of its align-ment with the Chinese. Moving forward, finding common ground with Russia on a range of economic issues, coupled with military assurances, would make the world a safer place, ceteris paribus. But here too, however, reciprocity is the key. American, European, or Russian unilateralism—in either direction—would not be constructive. In the long run, neither appeasement nor coercive diplomacy works.

Beyond these specifics, the best hope for peace over the long haul is to promote an international environment in which grievances are addressed and not allowed to fester. To be sure, a peace that rests on credible and capable threats can be a seductive short-term fix. But such a peace is bound to unravel, eventually, as dissatisfied rational agents interact in an imperfect world.

7.4 Coda

In this chapter, I have drawn a sharp contrast between classical deterrence theory and perfect deterrence theory. Perfect deterrence theory is a completely general theory of conflict initiation, escalation, and resolution. It is applicable across time and space. Indeed, its empirical domain is not restricted to contentious nuclear relationships. Rather, its analytic framework can be used to understand the full range of situations wherein at least one actor's goal is to preserve the existing distribution of value, which is Arnold Wolfers's (1951) neologism for the status quo. Unlike classical deterrence theory, perfect deterrence theory is logically consistent and in accord with the empirical record. In a direct test, perfect deterrence theory even outperformed what is, according to Bennett and Stam (2000: 451), "one of the most important theories of international conflict." Observing all this, Quackenbush (2011: 74) recommends that the academic and policy communities "take note."

In perfect deterrence theory, a capable threat is a necessary condition for deterrence success. But it is not sufficient, as it is in classical deterrence theory. Threat credibility plays an important role in the operation of both direct and extended deterrence relationships, but it is neither necessary nor sufficient for deterrence to prevail; under certain conditions, the presence of a credible threat may actually precipitate a deterrence failure. In perfect deterrence theory, the cost of conflict and status quo evaluations are also important strategic variables in so far as their values determine the characteristics of the players' threats.

Significantly different policy recommendations also distinguish perfect deterrence theory from its classical rival. Perfect deterrence theory is consistent with a policy of minimum deterrence, it recommends a conditionally cooperative diplomatic approach to resolve disputes, opposes even the selective proliferation of nuclear and other weapons of mass destruction, and supports arms control agreements and other limitations on redundant military expenditures.

All this said, there are still some who dispute the utility of any theory of strategic interaction to describe, explain, or predict interstate behavior. In Chapter 8, I take issue with this general line of reasoning.

8

. . • . .

Explaining the Long Peace

8.1 Introduction

Game theory is best thought of as a logical system. Technically speaking, it is a branch of mathematics. As noted in Chapter 1, game theory entered the academic world as a distinct field of study in 1944 when John von Neumann and Oskar Morgenstern published their magisterial *Theory of Games and Economic Behavior* with Princeton University Press.[1] Its broad acceptance as a legitimate methodological tool and its application to tactical military affairs was almost immediate (see, inter alia, McDonald and Tukey, 1949; McDonald, 1950; and Williams, 1954). There were few objections. But when, shortly thereafter, game-theoretic models found wide currency among *strategic* analysts and international relations theorists, there was a backlash, both from leading game theorists, including Morgenstern (1961) himself and Anatol Rapoport (1964), one of the most prominent game theorists of his time, and from more traditional scholars, who objected to treating questions of war and peace as a mere "game" (Zuckerman, 1956).[2]

After the publication of Thomas Schelling's influential book, *The Strategy of Conflict*, in 1960, applications of game theory to national security affairs all but disappeared in the literature of international relations. There were, of course, a few exceptions. But, for the most part, strategic analysts concluded that Schelling had said all that could be said by a game theorist about deterrence in general, and coercive bargaining in particular (Martin 1999).

Toward the end of the 1970s and the beginning of the 1980s, however, applications of game theory began to reappear in some of the more specialized political science journals. At first, it was a trickle. But, by the end of the century, game models came to be accepted as part of the theoretical mainstream.[3] Predictably, there was another backlash (e.g., Johnson, 1997). One critic even accused formal modelers in general, but game theorists in particular, of hegemonic aspirations in the field of security studies, and warned that the increasing dominance of game-theoretic studies threatened to calcify the field by privileging some questions

---

[1] For the political, intellectual, historical, and psychological context in which game theory was developed, see Leonard (2010).
[2] This chapter is based on Zagare (2018a).
[3] Zagare and Slantchev (2012) trace the development of game-theoretic applications in international relations.

Game Theory, Diplomatic History and Security Studies. Frank C. Zagare. Oxford University Press (2019).
© Frank C. Zagare 2019. DOI: 10.1093/oso/9780198831587.001.0001

over others (Walt 1999). Not surprisingly, game theorists disagreed (Bueno de Mesquita and Morrow, 1999; Martin, 1999; Niou and Ordeshook, 1999; Powell, 1999; Zagare, 1999).

More recently, the utility of game-theoretic models for understanding interstate behavior has been challenged by some behavioral economists (aka cognitive psychologists) who contend that real world decision makers, who suffer from a number of cognitive and motivated biases, are not "rational" in a game-theoretic sense (Carlson and Dacey, 2014; Levy, 1997; Lewis, 2017; McDermott 2004).[4] The argument has significant implications. If one accepts it, as does Alexander J. Field (2014: 54), it follows that traditional game theory "offers little guidance, normatively or predictively, in thinking about behavior in a world of potential conflict."

My purpose in this chapter is to dispute this line of argumentation by demonstrating that, properly understood, game theory remains a powerful tool for understanding interstate conflict behavior. My purpose, however, is not to contest the significant insights behavioral economists have uncovered about human decision making. It is my contention that their insights and empirical observations are not necessarily inconsistent with the gestalt of game theory[5] (Camerer, 2003; Mercer, 2005: 16; Carlson and Dacey, 2006). As already mentioned, game theory should be understood as nothing more and nothing less than a potentially useful *methodology* for understanding human choice in an interactive decision-making environment (Morton, 1999). It is, therefore, one thing to find fault with an *application* of game theory or an *argument* that a game theorist might make. It is, however, quite a different thing to conclude that the methodology itself is at fault. For example, if a bridge fails or if a building collapses because of improper design, it does not follow that engineering as a field has nothing to contribute to the design of, say, physical objects. So it is with game theory. If a game-theoretic study is either logically inconsistent or empirically inaccurate, it is clear that it should be rejected. But it should also be obvious that the same is true of any theoretical argument, regardless of its microfoundation.

To develop this point, I begin by exploring Field's attempt to explain why the United States and the Soviet Union were able to avoid an all-out thermonuclear exchange during the Cold War era. His main conclusion is that deterrence worked "not because we are entirely rational" but "because we are human" (Field, 2014: 86). In so concluding, Field explicitly accepts a game-theoretic interpretation of the superpower relationship that he attributes to John von Neumann, and argues that inconsistencies in the work of Thomas Schelling reinforce his argument that a rationalist explanation of what Gaddis (1986) calls the "long peace" of the Cold War years is not possible.

Field's argument, however, cannot be sustained. Briefly, I argue that his interpretation of von Neumann's conceptualization of the game played between two nuclear adversaries, coupled with his own description of the dynamic nature of that game, leads to the exact opposite conclusion. Moreover, I show that Schelling's explanation of the absence of a superpower conflict during the Cold War period also falls apart. In consequence, so does Field's.

---

[4] The argument is not unique to behavioral economics. Indeed, it is common among foreign policy experts and security studies specialists. See, for example, Morgan (2003).

[5] Or expected utility theory (Bueno de Mesquita, McDermott and Cope, 2001: 165, fn. 6; McDermott and Kugler, 2001: 85).

It should be noted at the outset that Field's (2014: 55–6, fn. 8) argument begins with an overly rigid definition of what constitutes rationality. In addition to the standard axioms that are associated with a von Neumann–Morgenstern (1944) cardinal utility function, Field holds that rationality requires agents who are not only self-interested but also self-regarding.[6] As Field recognizes, neither altruists nor suicide bombers are self-regarding (by his definition) and, therefore, cannot be considered rational if this restriction on preferences is accepted. Thus, his nonstandard definition[7] unnecessarily places a great deal of human behavior outside the realm of scientific exploration within a rational choice framework. But even if this were not the case, to claim, as Field (2014: 57) does, that "the only defensible policy" for a self-regarding agent in a contentious bilateral nuclear relationship "was an immediate attack" that would bring about the destruction of tens and hundreds of millions of innocents is to offer a caricature of the very notion of rational choice and self-regarding behavior. As Field reports, during the Eisenhower years, there were indeed some rational, perhaps even self-regarding, individuals who pushed for a preemptive attack on the Soviet Union. Fortunately, there were also some rational self-regarding individuals, including Eisenhower himself, who saw things differently.

The point here, however, is not that Field's definition of rationality is wrong; rather, it is that this definition is overly demanding and, therefore, unnecessarily restrictive.[8] Nonetheless, Field's argument does not fail because of it, but in spite of it, as I next demonstrate. In other words, his conceptualization of rationality is not the reason his argument is less than persuasive.[9]

---

[6] For Field, "self-regarding" behavior requires that individuals "prefer more material goods to less, and life over death" *whatever the consequences or collateral implications.* Field claims that game theory can be tested only when preferences are so restricted. This would be the case if game theory per se is taken to be a descriptive as opposed to a normative framework. But game theory can also be considered a tool for theory construction. In other words, once preferences are postulated and strategic choices stipulated, a game-theoretic model can lead to testable propositions. Downs (1957), for example, assumes that political candidates and/or parties are rational vote maximizers. This assumption leads to specific behavioral expectations that are subject to empirical validation. Behavior inconsistent with these expectations undermines his theory of electoral competition.

[7] A von Neumann–Morgenstern utility function is a measure of an actor's *subjective* preference over outcomes, given uncertainty. Therefore, individuals with different preferences and/or risk propensities may have distinct utility functions. In other words, utility is defined by each individual. As Freedman (2013: 153) points out, game "theory assumes rationality, but on the basis of preferences and values that the players brought with them to the game." Field (2014, 74) misstates the facts, then, when he claims that game theorists commonly assume "that players are logical, rational, *and self-regarding*" (emphasis added) as he defines it. Self-regarding behavior, like beauty, is in the eye of the beholder (Kreps, 1988). So even if von Neumann considered a preemptive nuclear attack on the Soviet Union to be self-regarding (see Section 8.2), another (also) rational agent might think otherwise.

[8] Definitions, by their very nature, are arbitrary. In the deterrence literature, two definitions of rationality figure prominently. The concept of *procedural rationality* underlies the work of those who approach strategic behavior from the vantage point of individual psychology (Simon, 1976). Most rational choice deterrence theorists who study deterrence define rationality instrumentally. For a further discussion, see Zagare (1990a) and Quackenbush (2004). Schelling (1960, 1966) is inconsistent; he uses both. His "rationality of irrationality" stratagem assumes that instrumentally rational agents feign procedural irrationality to gain a bargaining advantage.

[9] It should also be noted that *once preferences are stipulated,* the distinction between rational and self-regarding behavior disappears. Behavior that is "rational" and behavior that is "self-regarding" are one and the same, that is, behavior that is and must be consistent with a player's utility function. As will be seen, Field seems to understand this when analyzing's Schelling's explanation of the absence of a nuclear war throughout the Cold War period but not when he discusses what he believes to be von Neumann's understanding of the game played by the United States and the Soviet Union.

8.2 John von Neumann and Prisoners' Dilemma

Field builds his case against game theory, rationality, and von Neumann's obviously incorrect fear that a nuclear exchange between the superpowers was all but certain by attributing to von Neumann the view that nuclear confrontations are essentially Prisoners' Dilemma games. Field (2014: 54, fn. 4) claims that his attribution is "indisputable." But it is far from it. In fact, the primary source that Field uses to make this case, William Poundstone (1992: 144), argues otherwise: "It's unlikely that von Neumann—or anyone else, circa 1950—explicitly thought of the U.S.–Soviet conflict as a prisoner's dilemma. If von Neumann did picture U.S.–Soviet relations as a game, it is more plausible that he saw it as a zero-sum game."[10]

For the sake of argument, however, let us accept Field's less-than-convincing contention that von Neumann saw the superpower relationship as a Prisoners' Dilemma game. The standard depiction of this well-known game is given in Figure 8.1. Prisoners' Dilemma is a two-person noncooperative game. In this representation, the players are nuclear adversaries, here called State A and State B. Each state has two strategies: Cooperate (C) or Defect (D). There are four possible outcomes: if both players choose to cooperate and do not attack one another, the outcome *Status Quo* results; if both defect and attack, *Conflict* (read nuclear war) takes place. But if one defects and attacks while the other cooperates (by waiting to attack), the state that attacks *Wins* and the state that waits *Loses*.

Prisoners' Dilemma is a 2 × 2 normal form game. There are seventy-eight such games (Rapoport and Guyer, 1966), but the preference assumptions that define this game are unique: both players are assumed to most prefer to *Win*, second most prefer the *Status Quo*,

	State B	
	Cooperate (C) (Wait)	Defect (D) (Attack)
State A Cooperate (C) (Wait)	*Status Quo* (3,3)	*B Wins* (1,4)
Defect (D) (Attack)	*A Wins* (4,1)	*Conflict* (2,2)*

Key: (x,y) = payoff to State A, payoff to State B
4 = best; 3 = next best; 2 = next worse; 1 = worst
* = Nash equilibrium

Figure 8.1 Prisoners' Dilemma

[10] If Poundstone is correct, Field must be wrong. Prisoners' Dilemma is not a zero-sum game.

third most prefer *Conflict*, and least prefer to *Lose*. As is well known and as discussed in Chapter 1 in the context of an arms race (see Chapter 1, Figure 1.1), both players in a Prisoners' Dilemma game have a strictly dominant strategy, that is, a strategy that is best regardless of the strategy selected by the other player. For instance, State A's Attack strategy (D) is A's best response should State B attack. But it is also State A's best response should State B wait to attack. Similarly, State B's Attack strategy is the best response to either of State A's two strategy choices.

Strictly dominant strategies, in other words, are unconditionally best. Thus, "rational" players, self-regarding or otherwise, will always select them. And when the players do, these strategies lead to a unique, *non-Pareto optimal* Nash equilibrium, *Conflict*.[11] Game theorists are in almost unanimous agreement that, in a one-shot game, which a thermonuclear war would surely be, rational players should select their dominant strategies and that *Conflict* is the outcome that is implied under rational play. Of course, there is a dilemma. In this game, two rational players are individually and collectively worse off than two irrational players who choose to cooperate. Unfortunately, in a one-shot Prisoners' Dilemma game, mutual cooperation cannot be supported under rational play.[12]

Although it is difficult to reconstruct his logic, Field's (2014: 54) argument that game theory "offers little guidance, normatively or predictively, in thinking about behavior or strategy in a world of potential conflict" rests on several indisputable facts: (1) von Neumann believed that an all-out nuclear war between the United States and the Soviet Union was inevitable and that, therefore, it was to the advantage of the United States to attack first; (2) rational players in a Prisoners' Dilemma game should select their dominant strategy and, if and when they do, conflict is implied; (3) a thermonuclear exchange between the United States and the Soviet Union did not occur; (4) there is a disconnect between von Neumann's belief, his policy recommendation, the strategic imperatives of the Prisoners' Dilemma game, and "the event that didn't occur."

But his argument also rests on some facts that are more than disputable: (1) von Neumann believed that the game played by the superpowers during the earliest years of the Cold War was indeed a Prisoners' Dilemma game; (2) "von Neumann was right to characterize … [the superpower relationship as] … a Prisoners' Dilemma" (Field, 2014: 55), and (3) that von Neumann's preferred policy (i.e., a *preemptive* strike) was supported by his belief.

As already noted, Field's assertion that von Neumann believed that a Prisoners' Dilemma game captured the underlying dynamic of the post-war relationship of the United States and the Soviet Union is speculative at best and factually incorrect at worst. On the other hand, von Neumann's beliefs are simply beside the point.[13] So even if Field's claim is accepted, the pertinent questions are (1) whether the Cold War competition between the superpowers was, in fact, a Prisoners' Dilemma game, and (2) assuming, for the sake of argument, that it

[11] An outcome is Pareto optimal if and only if at least one player prefers that outcome to any other outcome (see also Chapter 4, fn. 21).

[12] In a repeated game, there are conditions under which mutual cooperation is part of a Nash equilibrium. For a discussion, see Morrow (1994, ch. 9). See also Myerson (2007).

[13] They are beside the point because even if Field is correct about von Neumann's beliefs, Field's conclusions about the usefulness of game theory as either a normative theory or a tool for theory construction do not hold, as demonstrated in Section 8.3.

was, whether von Neumann's preference for a *preemptive* attack is consistent with the conventional wisdom of game theory.

It is clear that the answer to the second question is "no." As noted, Prisoners' Dilemma is a 2 × 2 normal form game in which each player has two strategies. In a normal form representation, the assumption is that the players make their strategy choice before the game begins. There are two logically equivalent interpretations of a 2 × 2 game. One is that the players choose their strategies simultaneously; the second is that they make their choices sequentially but without knowledge of each other's choice. In a 2 × 2 normal form game, therefore, there can be no first-mover advantage since there is, technically speaking, no first mover. And if there is no first mover, there is also no second mover and, hence, no possibility of retaliation. Thus, either von Neumann was not a very good game theorist or he had a different game in mind.[14]

It is important to point out that there also is no first-mover advantage in a Prisoners' Dilemma game, even if it is played sequentially. A normal form representation of such a game is depicted in Figure 8.2. In this representation, the assumption is that State A chooses its strategy first and then, after observing this choice, State B makes its choice. In the sequential version of this game, State A's strategy set is the same, but now State B has four strategies:

1. C Regardless (C/C): choose (C) whether State A chooses (C) or (D);
2. D Regardless (D/D): choose (D) whether State A chooses (C) or (D);
3. Tit for Tat (C/D): choose (C) if State A chooses (C), and choose (D) if State A chooses (D); and
4. Tat for Tit (D/C): choose (D) if State A chooses (C), and choose (C) if State A chooses (D).

In the sequential version of Prisoners' Dilemma, State B's D Regardless strategy weakly dominates all of its other strategies. And since State A's best response to B's weakly dominant

		State B			
		C Regardless (C/C)	D Regardless (D/D)	Tit for Tat (C/D)	Tat for Tit (D/C)
State A	C	(3,3)	(1,4)	(3,3)	(1,4)
	D	(4,1)	(2,2)*	(2,2)	(4,1)

Key: (x,y) = payoff to State A, payoff to State B
4 = best; 3 = next best; 2 = next worst; 1 = worst
* = Nash equilibrium

Figure 8.2 Sequential Prisoners' Dilemma

[14] He would not be a good game theorist because his argument for a first strike cannot be deduced from the strategic implications of a Prisoners' Dilemma game. He might have had a different game in mind because there are in fact some games (e.g., Chicken; see Section 8.4) in which there is a first-mover advantage.

strategy[15] is to choose (D) initially, the strategy pair (D,D/D) gives rise to the unique, Pareto-inferior Nash equilibrium (2,2), or *Conflict*. All of which is to say that there is no discernable strategic difference between a Prisoners' Dilemma game played simultaneously or played sequentially. There is no first-mover advantage in either version of the game.

It should be clear, then, that von Neumann's policy preference for a preemptive strike against the Soviet Union cannot be logically derived from a one-shot Prisoners' Dilemma game, however it is played out. In either the simultaneous or the sequential case, the fact that the United States did not preempt the Soviet Union at the dawn of the nuclear era does not support Field's (2014, 54) observation that "game theory leads to behavioral predictions which are simply not borne out...in the real world."[16]

8.3 Rational Deterrence

All of which strongly suggests that the standard version of Prisoners' Dilemma is not a very good model of the superpower relationship at the time that von Neumann recommended a preemptive attack. Assuming, then, that von Neumann was, in fact, a good game theorist, it follows that he must have had another game in mind. Which one, however, is unclear. The extensive form game shown in Figure 8.3 is, at minimum, a plausible possibility. As with the standard version of Prisoners' Dilemma, this game, which Zagare and Kilgour (2000) call the *Generalized Mutual Deterrence Game*, is symmetric. As well, both players are presented with the same *initial* choice that the players in a Prisoners' Dilemma game have, to either Cooperate (C) or Defect (D). Both players are also assumed to make their initial choices simultaneously or, what is an equivalent assumption, without knowledge of the initial choice of the other player. The Generalized Mutual Deterrence Game, however, allows for the possibility of a retaliatory strike if initially only one player chooses to defect by attacking. The possibility of retaliation is the *only* difference between the rules that govern play in the Generalized Mutual Deterrence Game and those that are associated with the standard version of Prisoners' Dilemma. The claim here is that the dynamic structure of the extensive form game of Figure 8.3 better captures the strategic situation envisioned by both von Neumann and Field.[17]

---

[15] One strategy weakly dominates another if it is as least as good, and sometimes better, than the other. Compare this definition to the definition of a strictly dominant strategy given in Chapter 1, Section 1.3.

[16] Much the same could be said about the experimental literature that suggests that players in Prisoners' Dilemma games do not always defect (Schecter and Gintis 2016, 13). If the standard version of Prisoners' Dilemma is an inappropriate model of the Cold War interaction of the superpowers, then the laboratory experiments that explore behavioral tendencies in an artificial environment are simply beside the point, and the conclusion that Field (2014, 86) draws from that literature, that "formal game theory has not been useful for understanding how people behave or how they necessarily should behave," is not germane to his analysis of the predictive or explanatory power of game theory in the field of security studies.

[17] The line connecting State B's choices at Nodes 2a and 2b is called an *information set*. An information sets is a graphical device that is used to indicate a player's knowledge about its place on a game tree. When two or more decision nodes are in the same information set, as they are here, the interpretation is that a player is unable to tell which node it is at and, by extension, what prior choice has been made by other players at previous nodes. An information set that contains only one decision node is called a *singleton*. As drawn here, the information set reveals that when State B makes its choice, it is unaware of the choice made by State A, either because the two players make

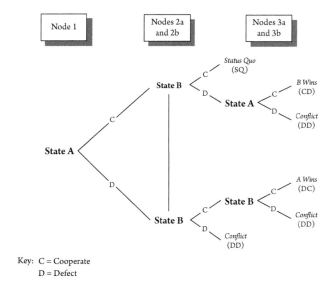

Figure 8.3 Generalized Mutual Deterrence Game

There are four distinct outcomes in the Generalized Mutual Deterrence Game. They are identical to the four outcomes in Prisoners' Dilemma. If both players cooperate, the outcome *Status Quo* is obtained. If both defect, either initially or subsequently, *Conflict* results. But if one defects initially and the other, after initially choosing to cooperate, does not retaliate, the defecting player wins and the cooperating player loses.

To retain as much of the structure of the game that Field claims that von Neumann had in mind when he pushed for a preemptive strike of the Soviet Union, I now analyze the Generalized Mutual Deterrence Game, using the same ordinal preference assumptions over the four distinct outcomes that define Prisoners' Dilemma. Specifically, the assumption here is that both players most prefer to win, next most prefer mutual cooperation, third most prefer *Conflict*, and least prefer to lose. The strategic form of this version of the Generalized Mutual Deterrence Game is given in Figure 8.4.

Notice that there are five Nash equilibria in the normal form game of Figure 8.4. But only two of them are subgame perfect. Interestingly, one of the subgame perfect equilibria results when both players intend to choose (D) (i.e., attack) whenever it is their turn to make a move. A D Regardless (D/D) strategy in the Generalized Mutual Deterrence Game is, in essence, the same as a D strategy in the standard 2 × 2 version of Prisoners' Dilemma and, not surprisingly, results in the same outcome, *Conflict*, which is the next-worse outcome for both players, that is, (2,2).

Significantly, however, the second subgame perfect equilibrium occurs when both players cooperate initially but intend to defect (retaliate) subsequently should the other player defect initially. Thus, a Tit for Tat (C/D) strategy of conditional cooperation that justifies a

their initial move at the same time or because State B has no knowledge of the move made by State A when it makes its initial move.

		State B			
		C/C	C/D	D/C	D/D
	C/C	(3,3)	(3,3)	(1,4)	(1,4)
	C/D	(3,3)	(3,3)**	(2,2)	(2,2)
State A	D/C	(4,1)	(2,2)	(2,2)*	(2,2)*
	D/D	(4,1)	(2,2)	(2,2)*	(2,2)**

Key: * = Nash equilibrium
     ** = subgame perfect Nash equilibrium

Figure 8.4 Ordinal strategic form representation of the Generalized Mutual Deterrence Game, given Prisoners' Dilemma-like preferences

non-preemptive (and, not incidentally, a no-first-use) strategy is also fully consistent with the requirements of rational choice. As well, unlike the *Conflict* subgame perfect equilibrium, the subgame perfect equilibrium that supports the outcome *Status Quo* is Pareto optimal. Thus, both self-interested and self-regarding players have a very good reason to select the strategies that bring it about.[18]

All of which is to say that even a slight modification of the game that Field contends drove von Neumann's thinking produces another rational strategic possibility that one could plausibly argue characterized the Cold War period. It is noteworthy that the modification that produced the Generalized Mutual Deterrence Game is clearly closer to Field's (2014: 76) and von Neumann's description of a contentious nuclear relationship. Recall that the condition that Field (2014: 57) used to justify von Neumann's view of the game included the possibility of retaliation. The Generalized Mutual Deterrence Game analyzed in Figures 8.3 and 8.4 incorporates not only von Neumann's presumed understanding of the game's payoff structure but the possibility of a retaliatory choice as well. In a 2 × 2 normal form representation of Prisoners' Dilemma, where the assumption is that the players make their strategy choices simultaneously (or in ignorance of each other's choice) before the play of the game, there is no possibility of a retaliatory strike.

Still, one might object. As Field (2014: 55) notes, his acceptance of von Neumann's characterization of contentious nuclear relationships required that there be "some possibility of destroying an adversary's offensive capability and/or its will to retaliate." There is no such possibility in the complete information version of the Generalized Mutual Deterrence Game. Given Prisoners' Dilemma-like preferences, rational players *always* retaliate.

---

[18] Notice that both players have capable and credible threats. Yet, deterrence is not assured. The fact that there are two rational strategic possibilities in this game, one of which is associated with conflict, renders deterrence less than certain. It is for this reason that mutually capable and credible threats do not constitute a sufficient condition for deterrence success. See Chapter 7, Table 7.1.

One way to model the *possibility* that one or both of the players might not be able to retaliate is to consider the equilibrium structure of the Generalized Mutual Deterrence Game *with incomplete information.*[19] When it is, the conclusions that can be drawn from an examination of the game under complete information are not disturbed. As Zagare and Kilgour (2000, ch. 4) show, there are four distinct perfect Bayesian equilibria in the incomplete information version of the Generalized Mutual Deterrence Game that are rationally consistent with maintenance of the status quo and an initially cooperative strategic stance (see Table 8.1). One of them, which they call the Sure-Thing Deterrence Equilibrium, *always* leads to the status quo, regardless of the players' actual preference between retaliating or not, that is, the players' types.

All of which is to say that it is difficult to agree with Field's (2014: 54) argument that game theory's "canonical behavioral assumptions predicted devastating conflict between nuclear adversaries," despite the fact that von Neumann did. When the preference assumptions associated with the standard version of the Prisoners' Dilemma game are maintained and used to analyze an extensive form game that more fully reflects von Neumann's understanding of the dynamics of deterrence, a devastating conflict remains but one of two rational strategic possibilities, and one that self-regarding players would reject at that. Thus, the predictive failure that Field attributes to game theory does not exist.

Table 8.1 Action choices of perfect Bayesian equilibria for the Generalized Mutual Deterrence Game with incomplete information.

	State A		State B	
	x_H	x_S	y_H	y_S
Sure-Thing Deterrence	0	0	0	0
Separating	1	0	1	0
Hybrid	u	0	v	0
Attack$_{1A}$	1	1	0	0
Attack$_{2A}$	1	1	v	0
Attack$_{3A}$	1	1	1	0
Attack$_{1B}$	0	0	1	1
Attack$_{2B}$	u	0	1	1
Attack$_{3B}$	1	0	1	1
Bluff	1	u	1	v

Key x_H = probability that State A chooses (D), given that it is Hard
  x_S = probability that State A chooses (D), given that it is Soft
  y_H = probability that State B chooses (D), given that it is Hard
  y_S = probability that State B chooses (D), given that it is Soft
  u = fixed value between 0 and 1
  v = fixed value between 0 and 1

[19] Under incomplete information, the players do not know each other's preferences. Thus, the assumption of incomplete information is logically equivalent to the assumption that one player believes that there is a distinct *possibility* that the other player will choose not to retaliate or, equivalently, will be incapable of retaliating effectively.

8.4 Thomas Schelling and Chicken

According to Field, von Neumann's policy prescription was opposed most forcefully by Thomas Schelling. But if Schelling wanted to develop a strategic rationale for a policy of deterrence, he chose a strange game form to do so. His starting point, and the starting point of the vast majority of those strategic thinkers who used game theory to study deterrence, was the game of Chicken (see Figure 8.5) (Schelling, 1966: 116–25).[20]

In some respects, Chicken resembles game theory's other canonical game and the purported object of von Neumann's attention, Prisoners' Dilemma. Chicken, like Prisoners' Dilemma, is a 2 × 2 normal form game in which each of the two players has two strategies: either Cooperate (C) or Defect (D) from cooperation. As is the case in Prisoners' Dilemma, each player is assumed to most prefer to win and second most prefer to cooperate when the other player also cooperates. The two games diverge, however, with respect to the relative ranking of the two remaining outcomes. In Chicken, *Conflict* is a mutually worst outcome, and losing is the second-worst outcome. In Prisoners' Dilemma, these preferences are reversed.

The difference is not trivial. In Prisoners' Dilemma, *Conflict* is the unique Nash equilibrium. By contrast there are two (subgame perfect) Nash equilibria in Chicken. As indicated by the asterisks in Figure 8.5, one is associated with a victory for the row player (State A) and the other with a victory for column (State B).[21] Significantly, the outcome *Status*

		State B	
		Cooperate (C) (Wait)	Defect (D) (Attack)
State A	Cooperate (C) (Wait)	*Status Quo* (3,3)	*B Wins* (2,4)*
	Defect (D) (Attack)	*A Wins* (4,2)*	*Conflict* (1,1)

Key: (x,y) = payoff to State A, payoff to State B
 4 = best; 3 = next best; 2 = next worst; 1 = worst
 * = Nash equilibrium

Figure 8.5 Chicken

---

[20] In Chapter 7, I label those strategic thinkers who accept the Chicken analogy "decision-theoretic deterrence theorists."

[21] There is also a mixed strategy equilibrium that, for the sake of brevity, will be ignored. Under the mixed strategy equilibrium, deterrence succeeds sometimes, but not necessarily often. The mixed strategy equilibrium in Chicken fails as a normative device. As O'Neill (1992: 471–2) shows, it prescribes behavior that is "just the opposite of what one would expect." Under the mixed strategy equilibrium in Chicken, the worse the *Conflict* outcome

Quo (or mutual cooperation) is not a Nash equilibrium in Chicken, an observation which, on its face, presents a challenge to those who want to explain why two rational agents might choose to cooperate. On the other hand, rational *Conflict* is also ruled out once the assumption is that it is each actor's least preferred outcome. A mutually worst outcome can never be part of a (pure) strategy Nash equilibrium in a strictly ordinal normal form game.

Making Schelling's task of justifying a non-preemptive policy even more difficult is the fact that there is, indeed, a first-move advantage in Chicken, as Figure 8.6 shows. There are three Nash equilibria in sequential Chicken. As indicated by the asterisks, two correspond to equilibria in the original (simultaneous choice) game while a third—(D,D/C)—is strictly a product of a sequential structure. But this additional equilibrium has special properties that distinguishes it from the other two and, therefore, gives it a singular status: it alone is subgame perfect; as well, it is the product of State A's best response to State B's weakly dominant strategy (i.e., D/C or Tat for Tit). It is more than significant that, under this equilibrium, State A, the player assumed to move first, wins.

Schelling was clearly aware of the fact that, in sequential Chicken, the player who moves first always wins. In one of the most cited chapters of his book *Arms and Influence*, entitled "The Art of Commitment," Schelling (1966: ch. 2) prescribed a variety of tactics for statesmen (and stateswomen) who wish to seize the initiative and prevail in a nuclear crisis by forcing a rational opponent to "chicken out." He was not the only defense intellectual to do so. Herman Kahn (1962), Robert Jervis (1972), Glenn Snyder (1971a), and several others also trafficked in this dangerous drug.

Schelling's musings with respect to what George (1993) calls "forceful persuasion" and what Field (2014: 56, fn. 10) refers to as "aggressive" deterrence have little basis in fact. For instance, the multiple bargaining tactics that Schelling proposed in *Arms and Influence* have been shown in more than one large-*n* study to be empirically dubious (e.g., Huth, 1999; Danilovic, 2002). And, as Jervis (1988b: 80) has pointed out, in the specific case of the

		State B			
		C Regardless (C/C)	D Regardless (D/D)	Tit for Tat (C/D)	Tat for Tit (D/C)
State A	C	(3,3)	(2,4)*	(3,3)	(2,4)
	D	(4,2)*	(1,1)	(1,1)	(4,2)**

Key: (*x,y*) = payoff to State A, payoff to State B
   4 = best; 3 = next best; 2 = next worst; 1 = worst
   * = Nash equilibrium
   ** = subgame perfect Nash equilibrium

Figure 8.6 Sequential Chicken

is for one player, the *more* likely it is that the other player will concede. Like the unique pure strategy Nash equilibrium in Prisoners' Dilemma, the mixed strategy Nash equilibrium in Chicken is not Pareto optimal.

Cold War interaction of the superpowers, the reality is that "the United States and the USSR have not behaved like reckless teenagers" in a game of Chicken. Given the discrepancy between theory and fact, it is no small wonder then that Jervis (1979: 292) claimed that many of Schelling's policy prescriptions were "contrary to common sense," that Rapoport (1992: 482) found them to be just plain "bizarre," and that Morgenstern (1961: 105) concluded that they would be "dangerous should they have an influence on policy."

Field (2014: 61, fn. 21), who was well aware of the empirical shortcomings of Schelling's work on coercive bargaining, thought it significant that Schelling "had a strong influence on academic thinking about strategic policy," but "limited influence" on actual policy makers in either Washington or Moscow. For Field (2014: 54), this was contributing evidence that that formal game theory itself "offers little guidance, normatively or predictively, in thinking about behavior or strategy in a world of potential conflict."

But it was actually Schelling's work on what he termed "traditional" deterrence that Field (2014: 56, fn. 10) believed conclusively demonstrated game theory's inability to explain adequately the long peace. If Chicken were, in fact, the proper game form to use to represent the strategic relationship of two nuclear adversaries, as the majority of strategic analysts thought, and if preemption was rational in Chicken, as Schelling and others had properly inferred, how might mutual deterrence evolve (or be explained)?

To answer this question, Schelling abandoned both logical consistency and the rationality postulate. His response was that nuclear deterrence would only work if an aggressor was convinced that its opponent would retaliate—*irrationally*. As he put it so succinctly: "another paradox of deterrence is that it does not always help to be, or to be believed to be, fully rational, cool headed, and in control of one's country" (Schelling, 1966, 37). In other words, it would be (instrumentally) rational to be thought of as (procedurally) irrational. Schelling, of course, was not the first strategic thinker to play fast and loose with the rationality postulate. Bernard Brodie (1959a: 293), considered by many to be the seminal deterrence theorist, put it this way: "For the sake of deterrence before hostilities, the enemy must expect us to be vindictive and irrational if he attacks us."

If, for Schelling, each superpower was able to deter the other only by threatening to retaliate, irrationally, how can one explain the fact that neither superpower followed von Neumann's advice and preempted the other? The only defensible answer for Schelling was that both were *rationally* deterred. In other words, Schelling's explanation assumes, simultaneously, that decision makers in the United States and the Soviet Union were rational when they were being deterred, and irrational when they were deterring one another.

Schelling's inconsistent application of the rationality postulate is more than problematic. If players can be either rational or irrational, *any* course of action can be justified or explained away.[22] Walt (1999) notwithstanding, logical consistency is clearly the sine qua non of sound

---

[22] For a contemporary policy relevant example, consider the strategic rationale offered by some US officials who, in 2018, were advocating a preventive limited strike against North Korea, based on the assumption that, otherwise, the North Korean dictator Kim Jong-un would remain undeterred. As Victor Cha (2018) wrote at the time: Their hope was "that a military strike would shock Pyongyang into appreciating U.S. strength, after years of inaction, and force the regime to the denuclearization negotiating table. . . . Yet there is a point at which hope must give in to logic. If we believe that Kim is undeterrable without such a strike, how can we also believe that a strike will deter him from responding in kind? And if Kim is unpredictable, impulsive and bordering on irrational, how can

theory. Since any proposition and its opposite can be derived from a logically inconsistent theoretical framework, empirical validation is foreclosed. And, as Field (2014: 56, fn. 8; 61) correctly argues, empirical validation is a cornerstone of scientific inquiry.

Field (2014: 71, fn. 43) saw the logical problem with Schelling's explanation and dismissed it because, as he saw it, Schelling assumed players who were not self-regarding. Field's (2014: 55, fns. 6 and 7) argument was that a self-regarding actor would/should rationally attack its opponent since a self-regarding opponent would *not* retaliate ex post. Schelling's explanation assumed the opposite. What is both remarkable and significant is that Field is absolutely correct—if the game is Chicken and not Prisoners' Dilemma—because, in Chicken, a self-regarding player would not retaliate (because retaliation would lead to its worst outcome and not retaliating would lead to its second-worst outcome), which is why a self-regarding player would also have an incentive to strike first (see Figure 8.6).

Notice that when Field dismissed Schelling's logic, he simply reversed it. Schelling argued that each superpower was rationally deterred by the threat of irrational retaliation. Field argued that each superpower was irrationally deterred by a threat of rational non-retaliation. His argument, in other words, was that it was irrational for either superpower *not* to attack the other, since it would have been rational for the other not to retaliate. In the end, Schelling and Field (2014: 79) both conclude that "nonaggressive deterrence (peaceful coexistence) requires a combination of irrational and rational behavior on the part of both parties." All of which helps to explain Field's (2014: 55, fn. 7) claim that his "paper argues (and Schelling suggests the same), that deterrence worked and von Neumann's predictions failed because humans are not entirely self-regarding."

It is no inconsequential fact, however, that the actual incentive structure of Chicken is entirely consistent with von Neumann's policy prescription. So if von Neumann was a good game theorist and if he had any game in mind when he campaigned for a preemptive strike of the Soviet Union, it could have been Chicken, but not Prisoners' Dilemma. In Prisoners' Dilemma, as we have seen, a self-regarding player will, in fact, retaliate, because retaliation will result in its next-worst outcome, but non-retaliation will bring about its worst outcome (see Figure 8.2). Moreover, when both players are afforded an equal opportunity to retaliate, mutual deterrence is entirely consistent with the rationality assumption (see the discussion of the Generalized Mutual Deterrence Game in Section 8.3)[23] and, not insignificantly, self-regarding behavior, as clearly implied by the payoff assumptions. To argue otherwise, as does Field, is to ignore the incentive structure implicit in any game with preferences that mirror those of a standard version of a Prisoners' Dilemma game.

All of which is to say that Field's assumption that the strategic relationship of the superpowers during the Cold War period is captured by the payoff structure of a Prisoners' Dilemma game, and that neither player possessed a first-strike capability that eliminated the possibility of a retaliatory strike, leads to a conclusion that is exactly the opposite of his. More

we control the escalation ladder, which is premised on an adversary's rational understanding of signals and deterrence?"

[23] To say that there are conditions under which mutual deterrence is consistent with rational choice is not the same thing as saying that mutual deterrence is either robust or all but certain. For the argument that mutual deterrence is both rational and fragile, see Chapter 7.

specifically, *if* the superpowers were in fact involved in a mutual deterrence relationship,[24] then deterrence worked not only because we are human but also because we were rational (Zagare, 2004). It also follows, contrary to the view held by Field and some other behavioral economists, that game theory indeed has much to offer, both normatively and descriptively, about behavior and strategy in a world of intense interstate conflict, even though von Neumann, Schelling (and Field) got it wrong.

8.5 Coda

In a far-ranging article, Alexander J. Field makes an argument common among some strategic theorists that, during the Cold War period, peace was preserved not because the superpowers were rational but because they were human. He claims, without convincing evidence, that John von Neumann, one of the cofounders of game theory, saw the superpower relationship during that period as a Prisoners' Dilemma game and contends that von Neumann's policy preference for a preemptive nuclear attack of the Soviet Union by the United States, while game-theoretically sound, was empirically inaccurate. By contrast, he argued that Thomas Schelling's explanation of the long peace was game-theoretically faulty but empirically correct. Field (2014, 54) concludes that "game theory leads to behavioral predictions which are simply not born out in the laboratory or . . . in the real world."

By contrast, in this chapter, I showed that Field's inference about the rational basis for a *preemptive* attack in a Prisoners' Dilemma game is not logically supported. I also showed that a slight modification of that game's strategic structure that more closely mirrors the dynamic decision-making context that both von Neumann and Field have in mind leads to exactly the opposite conclusion, namely, that self-regarding players with Prisoners' Dilemma-like preferences might rationally be deterred. Finally, I showed that Field's analysis of Schelling's explanation of the long peace, which is game-theoretically sound, is at odds with his analysis of the game he attributes to von Neumann and that he accepts as the proper representation of a contentious nuclear relationship.

The larger point, however, is that Field's outright rejection of game theory as an explanatory or predictive tool and, by extension, that of many other behavioral economists and security studies experts, cannot be sustained. Game theory constitutes a powerful methodological tool for theory construction. Even if one accepts Field's argument that von Neumann's policy preference for a preemptive attack on the Soviet Union was game-theoretically correct and that his prediction based on it was empirically inaccurate, it does not follow that formal game theory is of limited utility "for thinking about behavior or strategy in a world of potential conflict" (Field, 2014: 54).

---

[24] It is also possible that one side or the other was not incentivized to upset the status quo. In this case, the relationship would be one of unilateral deterrence. (For the argument that, during the Cold War period, the United States was a satisfied status quo power, see Organski and Kugler (1980)). Under specific conditions, unilateral deterrence relationships are also consistent with rational choice (see Chapter 7). Of course, if neither side preferred to upset the existing order, deterrence was not germane. It remains an important empirical puzzle whether the long peace was an instance of unilateral or mutual deterrence, or whether deterrence was even relevant.

Game-theoretic models should be thought of as empty vessels that can be filled in, or even shaped by, conflict theorists. As Morrow (1994, 70) points out, the specification of a model is the single most important step in the development of theory. What is sometimes unappreciated, however, is the fact that model design is not a game-theoretic exercise. Rather, it more properly falls under the purview of the conflict theorist qua conflict theorist. All of which is to say that game theory itself is silent on how the rules of a game are interpreted and represented, and on what particular preference and information assumptions are appropriate (Jervis, 1988a: 319). Different interpretations of the rules and conflicting preference and information assumptions lead to distinct theories, all of which are subject to formal (i.e., game-theoretic) analysis. In this regard, neither von Neumann nor Schelling has any special expertise. The models they relied on and the conclusions they drew remain subject to both logical and empirical scrutiny. And the same is true of any theory of interstate conflict, game theoretic or otherwise.

Postscript

To some mathematicians, game theory's appeal is largely aesthetic. But, for many social scientists, it is game theory's ability to generate testable propositions that accounts for its widespread use in economics, sociology, psychology, and political science. To be sure, game theory can, and has been, construed as a strictly normative theory that prescribes optimal behavior to rational actors (or players) involved in an interactive decision-making situation (i.e., a game) wherein the outcome depends on the choice made by at least one other actor (Rapoport, 1958). But descriptive (or positive) theory development lies at the heart of the social science enterprise, and it is here that game theory shines. Rather than simply asserting this, however, it has been my intention in this book to demonstrate game theory's vitality by observing it in action. Counterintuitively, perhaps, the aesthetic appeal of game theory is no more apparent than when it is actually *at work*. As David Berlinski (1995: xii) has insightfully observed, "it is in the world of things and places, times and troubles and dense turbid processes, that mathematics is not so much applied as *illustrated*." While Berlinski clearly had the calculus in mind when he wrote this, his observation applies equally to the mathematical theory of games.

Although game theory is gainfully employed in Chapters 1 and 2, it is still only working part time. In these two chapters, the broad parameters and basic concepts of game theory are described and illustrated. By far, the central concept is that of an equilibrium outcome (or strategy set). Four distinct equilibrium concepts are placed in theoretical context: Nash, subgame perfect, Bayesian, and perfect Bayesian equilibria. Not mentioned or discussed are a number of refinements and extensions of these standard definitions. Several theoretical competitors, including meta- and nonmyopic equilibria, also exist.

But, however an equilibrium is defined, the underlying concept is central, because game-theoretic explanations and game-theoretic predictions derive from them. When players in a real world game make choices that can plausibly be associated with an equilibrium outcome, an explanation is at hand. Similarly, a game-theoretic prediction about future play presumes rational choice, that is, the assumption is that an equilibrium choice will be made by all of the players.

Of course, problems arise when there is more than one rational strategic possibility in a game. These and related difficulties are discussed, and proper analytic responses offered. In the end, the acknowledged limitations of the rational choice paradigm are offset by the

Game Theory, Diplomatic History and Security Studies. Frank C. Zagare. Oxford University Press (2019).
© Frank C. Zagare 2019. DOI: 10.1093/oso/9780198831587.001.0001

many advantages of using game theory to construct an analytic narrative or to develop a general theory. Game models, by their very nature, facilitate the assessment of logical consistency, minimize the probability of its absence, and encourage counterfactual or "off-the-equilibrium-path" reasoning, allowing for contingent theorizing, inter alia. It should go without saying that game models also explicitly take account of the interactive dynamic typically present in intense interstate disputes. And that is no minor achievement.

The real work, however, begins in Part II, when game theory is asked to till the (sub-)field of diplomatic history. The most important point to be taken from this collection of analytic narratives, besides the specific explanations of the first Moroccan crisis, the Cuban missile crisis, and the July Crisis of 1914, is game theory's ability to deliver an explicit causal mechanism upon which explanations can be constructed, and to do this in a transparent and logically consistent way. Hindmoor (2006) claims that this is, in fact, its primary function. Causal mechanisms are especially valuable because they provide the element of necessity absent in standard (historical) narratives. This does not mean that an explanation has been definitively established simply because a game model has been used to explain a particular event, as the various competing explanations of the Cuban crisis surveyed in Chapter 4 demonstrate. Reasonable objections to the model used to organize a case study, the assumptions made about the players and their preferences and beliefs, the plausibility of ad hoc adjustments, if any are made, and any and all other elements of a model are always possible. But the transparent characteristic of game models makes it easier not only to distinguish competing explanations from one another but also to adjudicate them.

Game theory is still at work in Part III, where the heavy lifting continues. In Chapter 7, it is used to describe a theory of interstate conflict initiation, limitation, escalation, and resolution: perfect deterrence theory. Lisa Martin (1999: 76) defines theory as "integrated complexes of assumptions, insights, and testable propositions." Such complexes do not occur naturally in nature. They must be constructed. Perfect deterrence theory is one example. Others abound in the theoretical literature.

Perfect deterrence theory is composed of a set of four logically interconnected game models, two of which were used to construct the analytic narratives in Part II. The argument, however, is not that the theory is perfect. No theory is. Nor do I contend that perfect deterrence theory is the only way to describe and explain interstate conflict. There are competing models, not all of which are game theoretic. The differences in both the empirical and the policy implications of contending models, however, can sometimes be substantial. For example, Bueno de Mesquita and Riker's (1982) deterrence model supports selected proliferation policies, while perfect deterrence theory opposes them. Both models are explicit about their underlying assumptions. Unlike loosely stated arguments that favor or oppose this or any other policy, then, the logical structure of the competing formal models can easily be counterpoised, revealing, in the process, the assumptions that give rise to the differences. From this, it follows that if a model, game theoretic or otherwise, either suffers from a logical deficiency or leads to empirically inaccurate expectations, it should be cast aside and replaced by its theoretical superior. When models and the theories that are developed from them compete, winners and losers can sometimes be identified. But the fact that there are losers does not mean that the modeling enterprise itself is flawed, as the discussion in Chapter 8 should make clear. The same is, of course, true of competing analytic narratives.

One final thought: no modeling effort or theoretical construct is ever quite finished. No explanation of a complex event or important political, social, or economic process is ever definitive. Since its inception in 1944, formal game theory has made tremendous strides. Over time, its focus has moved away from zero-sum games to a concentration on nonzero-sum games, away from cooperative game theory to noncooperative game theory, away from static strategic form games to dynamic extensive form games, away from analyses of games of complete information to games of incomplete information, and away from minimax and maximin strategies and Nash equilibria to subgame perfect and perfect Bayesian equilibria and the latter's multitude of refinements. All of which strongly suggests that game-theoretic knowledge, like all knowledge, is always provisional and that an attitude of intellectual modesty, and not arrogance, is the proper one. Or as Dani Rodrik (2015) might put it, always remember that game theory is *a* methodology, not *the* methodology.

GLOSSARY OF BASIC CONCEPTS

Bayes's Rule Standard method that rational players are assumed to use to update their beliefs about the others' types in an extensive form game with incomplete information as new information is received.

Bayesian equilibrium A Nash equilibrium in a strategic form game with incomplete information that maximizes each player's expected utility given its beliefs about the other players' types.

Cardinal utility (von Neumann–Morgenstern utility) A measurement of utility on an interval scale that reflects both the rank and the intensity of a player's preferences.

Common conjecture The assumption that, in equilibrium, each of the players is able to correctly anticipate the strategy choice of all the other players.

Common information Information that is shared by all the players in a game.

Complete information A game in which every player knows the preferences of all the other players.

Cooperative game A game in which commitments to a particular course of action are possible.

Dominant strategy A strategy which is always better than all other strategies. Sometimes referred to as a "strictly dominant strategy." (See *weakly dominant strategy*.)

Dominated strategy A strategy that is always worse than another. Sometimes referred to as a "strictly dominated strategy." (See *weakly dominated strategy*.)

Extensive form game A game represented on a game tree in which the players make moves, sequentially, at nodes of the tree.

Game Any situation in which the outcome depends upon the choices of two or more players.

Information set A graphical device that indicates a player's knowledge of his or her place on a game tree.

Incomplete information A game in which at least one player does not know for sure the preferences of another player.

Instrumental rationality Purposeful behavior.

Matrix form game See *strategic form game*.

Metaequilibrium A Nash equilibrium of a metagame.

Metagame A game played in the heads of the players before an actual game is played.

Metastrategy A strategy in a metagame.

Mixed strategy A probability distribution over a set of a player's pure strategies.

Nash equilibrium A strategy combination in a strategic or normal form game with complete information such that no player has an incentive to switch to another strategy.

Noncooperative game A game in which the players are unable to commit themselves to a particular course of action.

Nonequivalent equilibria Two or more equilibria with different payoffs.

Noninterchangeable equilibria Two equilibria are noninterchangeable if every combination of strategies associated with them is not in equilibrium.

Nonmyopic equilibrium A state of the world in a two-person game that neither player, looking ahead indefinitely, has an incentive to move from.

Non-Pareto optimal (or Pareto deficient) An outcome is non-Pareto optimal if another outcome exists wherein at least one player is better off and no other player is worse off.

Non-zero-sum game A game in which the players have some interests in common.

Normal form game See *strategic form game*.

Ordinal utility A measurement of utility on an ordinal scale that reflects the rank but not the intensity of a player's preference.

Pareto inferior See non-*Pareto optimal*.

Pareto optimal (or Pareto superior) An outcome is Pareto optimal if no other outcome exists such that at least one player is better off and all other players are no worse off.

Perfect Bayesian equilibrium A strategy combination in an extensive form game with incomplete information in which each player updates its beliefs rationally (according to Bayes's rule) about the other players' types in the light of new information obtained as the game is played out.

Perfect information A game in which all the players are fully aware of all of the choices of the other player or players as the game is played out.

Pooling equilibrium A perfect Bayesian equilibrium in which all types of the same player play the same strategy.

Posited preferences Preferences that are assumed.

Private information Information that a player alone possesses.

Procedural rationality A demanding definition of rationality that assumes, inter alia, that players have an accurate perception of the implications of all conceivable alternatives and a well-defined set of preferences over the entire outcome set.

Pure strategy The certain selection of a particular course of action.

Revealed preference Preferences that are inferred from an observation of a player's action choice.

Separating equilibrium A perfect Bayesian equilibrium in which all types of the same player play different strategies.

Singleton An information set in an extensive form game that contains one decision node.

Strategic form game A game in which players select strategies.

Strategy A complete course of action.

Subgame That part of an extensive form game that can be considered a game unto itself.

Subgame perfect equilibrium A Nash equilibrium in an extensive form game with complete information that is consistent with a rational choice by every player at every decision node on a game tree, including those nodes that are never reached.

von Neumann Morgenstern utility See *cardinal utility*.

Weakly dominant strategy A strategy that is at least as good and sometimes better than another. See *dominant strategy*.

Weakly dominated strategy A strategy that is sometime worse, but never better, than another.

Zero-sum game A game in which the interests of the players are diametrically opposed.

BIBLIOGRAPHY

Achen, Christopher H. (1987). "A Darwinian View of Deterrence." In Jacek Kugler and Frank C. Zagare (eds.), *Exploring the Stability of Deterrence*. Denver: Lynne Rienner Publishers, pp. 148–50.

Achen, Christopher H., and Duncan Snidal (1989). "Rational Deterrence Theory and Comparative Case Studies." *World Politics*, 41(2): 143–69.

Albertini, Luigi (1952). *The Origins of the War of 1914*. Translated and edited by Isabella M. Massie. 3 vols. Oxford: Oxford University Press.

Allison, Graham (2017). "Thinking the Unthinkable with North Korea." *New York Times*, May 31.

Allison, Graham, and Philip Zelikow (1999). *Essence of Decision: Explaining the Cuban Missile Crisis*. 2nd ed. New York: Longman.

Anderson, Eugene N. (1966). *The First Moroccan Crisis 1904–1906*. Hamden, CT: Archon Books.

Axelrod, Robert (1984). *The Evolution of Cooperation*. New York: Basic Books.

Baldwin, David A. (ed.) (1993). *Neorealism and Neoliberalism: The Contemporary Debate*. New York: Columbia University Press.

Banks, Jeffrey S. (1990). "Equilibrium Behavior in Crisis Bargaining Games." *American Journal of Political Science*, 34(3): 599–614.

Bapat, Navin A., Luis De La Calle, Kaisa H. Hinkkainen, and Elena V. McLean (2016). "Economic Sanctions, Transnational Terrorism, and the Incentive to Misrepresent." *American Journal of Political Science*, 78(1): 249–64.

Bates, Robert H., Avner Greif, Margaret Levi, Jean-Laurent Rosenthal, and Barry R. Weingast (1998). *Analytic Narratives*. Princeton: Princeton University Press.

Bates, Robert H., Avner Greif, Margaret Levi, Jean-Laurent Rosenthal and Barry R. Weingast (2000a). "*Analytic Narratives* Revisited." *Social Science History*, 24(4): 685–96.

Bates, Robert H., Avner Greif, Margaret Levi, Jean-Laurent Rosenthal, and Barry R. Weingast (2000b). "The Analytic Narrative Project." *American Political Science Review*, 94(3): 696–702.

Bennett, D. Scott, and Allan C. Stam (2000). "A Universal Test of an Expected Utility Theory of War." *International Studies Quarterly*, 44(3): 451–80.

Benson, Brett V. (2012). *Constructing International Security: Alliances, Deterrence, and Moral Hazard*. Cambridge: Cambridge University Press.

Berlinski, David (1995). *A Tour of the Calculus*. New York: Random House.

Berghahn, Volker R. (1993). *Germany and the Approach of War in 1914*. 2nd ed. New York: St. Martin's.

Bethmann-Hollweg, Theobald von (1920). *Reflections on the World War*. London: Butterworth.

Betts, Richard K. (1987). *Nuclear Blackmail and Nuclear Balance*. Washington, DC: Brookings.

Brams, Steven J. (1985). *Superpower Games*. New Haven, CT: Yale University Press.

Brams, Steven J. (1994). *Theory of Moves*. Cambridge: Cambridge University Press.

Brams, Steven J. (2001). "Game Theory and the Cuban Missile Crisis." *Plus Magazine*, January 1, http://plus.maths.org/content/os/issue13/features/brams/index, accessed August 22, 2018.

Brams, Steven J. (2002). "Game Theory in Practice: Problems and Prospects in Applying It to International Relations." In Frank P. Harvey and Michael Brecher (eds.), *Evaluating Methodology in International Studies*. Ann Arbor: University of Michigan Press, pp. 81–96.

Brams, Steven J. (2011). *Game Theory and the Humanities*. Cambridge, MA: MIT Press.

Brams, Steven J., and D. Marc Kilgour (1988). *Game Theory and National Security*. New York: Basil Blackwell.

Brodie, Bernard (ed.) (1946). *The Absolute Weapon: Atomic Power and World Order*. New York: Harcourt Brace.

Brodie, Bernard (1959a). *Strategy in the Missile Age*. Princeton: Princeton University Press.

Brodie, Bernard (1959b). "The Anatomy of Deterrence." *World Politics*, 11(2): 173–9.

Brooks, Stephen G., and William Wohlforth (2015/2016). "The Rise and Fall of Great Powers in the Twenty-First Century: China's Rise and the Fate of America's Global Position." *International Security*, 40(3): 7–53.

Bueno de Mesquita, Bruce (1981). *The War Trap*. New Haven, CT: Yale University Press.

Bueno de Mesquita, Bruce (2002). "Accomplishments and Limitations of a Game-Theoretic Approach to International Relations." In Frank P. Harvey and Michael Brecher (eds.), *Evaluating Methodology in International Studies*. Ann Arbor: University of Michigan Press, pp. 59–80.

Bueno de Mesquita, Bruce (2009). *The Predictioneer's Game: Using the Logic of Brazen Self-Interest to See and Shape the Future*. New York: Random House.

Bueno de Mesquita, Bruce, and David Lalman (1992). *War and Reason: A Confrontation Between Domestic and International Imperatives*. New Haven, CT: Yale University Press.

Bueno de Mesquita, Bruce, Rose McDermott, and Emily Cope (2001). "The Expected Prospects for Peace in Northern Ireland." *International Interactions*, 27(2): 129–67.

Bueno de Mesquita, Bruce, and James D. Morrow (1999). "Sorting Through the Wealth of Notions." *International Security*, 24(2): 56–73.

Bueno de Mesquita, Bruce, and William H. Riker (1982). "An Assessment of the Merits of Selective Nuclear Proliferation." *Journal of Conflict Resolution*, 26(2): 283–306.

Bueno de Mesquita, Ethan (2005). "The Terrorist Endgame: A Model with Moral Hazard and Learning." *Journal of Conflict Resolution*, 49(2): 237–58.

Bundy, McGeorge (1983). "The Bishops and the Bomb." *New York Review of Books*, June 16: 3–8.

Butterfield, Herbert (1965). "Sir Edward Grey in July 1914." *Historical Studies*, 5: 1–25.

Büthe, Tim (2002). "Taking Temporality Seriously: Modeling History and the Use of Narratives as Evidence." *American Political Science Review*, 96(3): 481–93.

Cairns, John C. (1965). "March 7, 1936 Again: The View from Paris." *International Journal*, 20(2): 230–46.

Camerer, Colin F. (2003). *Behavioral Game Theory: Experiments in Strategic Interaction*. Princeton: Princeton University Press.

Carlson, Lisa J. (1995). "A Theory of Escalation and International Conflict." *Journal of Conflict Resolution*, 39(3): 511–34.

Carlson, Lisa J., and Raymond Dacey (2006). "Sequential Analysis of Deterrence Games with a Declining Status Quo." *Conflict Management and Peace Science*, 23(2): 181–98.

Carlson, Lisa J., and Raymond Dacey (2014). "The Use of Fear and Anger to Alter Crisis Initiation." *Conflict Management and Peace Science*, 31(2): 168–92.

Carr, Edward Hallett (1939). *The Twenty Years' Crisis: 1919–1939*. New York: Harper & Row.

Cashman, Greg (1993). *What Causes War? An Introduction to Theories of International Conflict*. New York: Lexington Books.

Cha, Victor (2018). "Giving North Korea a 'Bloody Nose' Carries a Huge Risk to Americans." *Washington Post*, January 30.

Christensen, Thomas J., and Jack Snyder (1990). "Chain Gangs and Passed Bucks: Predicting Alliance Patterns in Multipolarity." *International Organization*, 44(2): 137–68.

Clark, Christopher M. (2000). *Kaiser Wilhelm II*. Harlow, England: Longman.

Clark, Christopher M. (2012). *The Sleepwalkers.* New York: Harper.

Cohen, Raymond (1987). *Theatre of Power: The Art of Diplomatic Signalling.* London: Longman.

Copeland, Dale C. (2000). *The Origins of Major War.* Ithaca, NY: Cornell University Press.

Crawford, Timothy W. (2003). *Pivotal Deterrence: Third-Party Statecraft and the Pursuit of Peace.* Ithaca, NY: Cornell University Press.

Crawford, Timothy W. (2014). "World War I Alignment Counterfactuals: Why Italy and Turkey Mattered." Paper delivered at the Annual Meeting of the American Political Science Association, September 28–31.

Crescenzi, Mark J. C., Rebecca H. Best, and Bo Ram Kwon (2010). "Reciprocity in International Studies." In Robert A. Denemark et al. (eds.), *The International Studies Encyclopedia.* Oxford: Wiley-Blackwell, http://www.oxfordreference.com/view/10.1093/acref/9780191842665.001.0001/acref-9780191842665-e-0322, accessed August 20, 2018.

Danilovic, Vesna (2001). "The Sources of Threat Credibility in Extended Deterrence." *Journal of Conflict Resolution,* 45(3): 34–69.

Danilovic, Vesna (2002). *When the Stakes Are High: Deterrence and Conflict among Major Powers.* Ann Arbor: University of Michigan Press.

Dixit, Avinash, and Susan Skeath (2004). *Games of Strategy.* 2nd ed. New York: Norton.

Dobbs, Michael (2008). *One Minute to Midnight: Kennedy, Khrushchev, and Castro on the Brink of Nuclear War.* New York: Knopf.

Dobbs, Michael (2012). "The Price of a 50-Year Myth." *New York Times,* October 16.

Dobrynin, Anatoly (1995). *In Confidence.* New York: Random House.

Dodge, Robert V. (2012). *Schelling's Game Theory.* Oxford: Oxford University Press.

Downs, Anthony (1957). *An Economic Theory of Democracy.* New York: Harper & Row.

Ellsberg, Daniel (1959). "The Theory and Practice of Blackmail." Lecture at the Lowell Institute, Boston, MA, March 10. Reprinted in: Oran R. Young (ed.) (1975). *Bargaining: Formal Theories of Negotiation.* Urbana, IL: University of Illinois Press.

Erlanger, Steven (1996). "'Ambiguity' on Taiwan? Will the U.S. Fight a Chinese Attack?" *New York Times,* March 12.

Farrar, L.L., Jr. (1972). "The Limits of Choice: July 1914 Reconsidered." *Journal of Conflict Resolution,* 16(1): 1–23.

Fay, Sidney Bradshaw (1930). *The Origins of the World War.* 2nd ed., rev. New York: Macmillan.

Fearon, James D. (1994). "Domestic Political Audiences and the Escalation of International Disputes." *American Political Science Review,* 88(3): 577–92.

Fearon, James D. (1997). "Signaling Foreign Policy Interests." *Journal of Conflict Resolution,* 41(1): 68–90.

Ferguson, Niall (1998). *The Pity of War: Explaining World War I.* New York: Basic Books.

Field, Alexander J. (2014). "Schelling, von Neumann, and the Event that Didn't Occur." *Games,* 5(1): 53–89.

Fischer, David Hackett (1970). *Historians' Fallacies: Toward a Logic of Historical Thought.* New York: Harper & Row.

Fischer, Fritz (1967). *Germany's Aims in the First World War.* New York: Norton.

Fischer, Fritz (1975). *War of Illusions: German Policies from 1911 to 1914.* New York: Norton.

Förster, Stig (1999). "Dreams and Nightmares: German Military Leadership and the Images of Future Warfare, 1871–1914." In Manfred F. Boemeke, Roger Chickering, and Stig Förster (eds.), *Anticipating Total War: The German and American Experiences, 1871–1914.* Washington, DC: German Historical Institute, pp. 343–76.

Fraser, Niall M., and Keith W. Hipel (1982–3). "Dynamic Modelling of the Cuban Missile Crisis." *Conflict Management and Peace Science,* 6(2): 1–18.

Freedman, Lawrence (1989). *The Evolution of Nuclear Strategy*. 2nd ed. New York: St. Martin's.

Freedman, Lawrence (2013). *Strategy*. Oxford: Oxford University Press.

Fromkin, David (2004). *Europe's Last Summer: Who Started the Great War in 1914?* New York: Knopf.

Fursenko, Aleksandr, and Timothy Naftali (1997). *"One Hell of a Gamble:" Khrushchev, Castro, and Kennedy, 1958–1964*. New York: Norton.

Fursenko, Aleksandr, and Timothy Naftali (2006). *Khrushchev's Cold War: The Inside Story of an American Adventure*. New York: Norton.

Gaddis, John Lewis (1986). "The Long Peace: Elements of Stability in the Postwar International System." *International Security*, 10(4): 99–142.

Gaddis, John Lewis (1987). *The Long Peace: Inquiries into the History of the Cold War*. New York: Oxford University Press.

Gaddis, John Lewis (1997). *We Now Know. Rethinking Cold Wear History*. Oxford: Clarendon Press.

Geiss, Imanuel (ed.) (1967). *July 1914. The Outbreak of the First World War: Selected Documents*. New York: Scribner's.

Geller, Daniel S., and J. David Singer (1998). *Nations at War: A Scientific Study of International Conflict*. Cambridge: Cambridge University Press.

George, Alexander L. (1993). *Forceful Persuasion: Coercive Diplomacy as an Alternative to War*. Washington, DC: United States Institute of Peace Press.

George, Alexander L., and Andrew Bennett (2005). *Case Studies and Theory Development in the Social Sciences*. Cambridge, MA: MIT Press.

Gibbons, Robert (1992). *Game Theory for Applied Economists*. Princeton, NJ: Princeton University Press.

Gilboa, Itzhak (2010). *Rational Choice*. Cambridge, MA: MIT Press.

Goemans, Hein, and William Spaniel (2016). "Multimethod Research: A Case for Formal Theory." *Security Studies*, 25(1): 25–33.

Gooch, G. P. (1936). *Before the War: Studies in Diplomacy*. Vol. 1. New York: Longmans, Green and Co.

Gooch, G. P., and Harold Temperley (eds.) (1926). *British Documents on the Origins of the War: 1898–1914*. Vol. 11. London: His Majesty's Stationery Office.

Grey, Sir Edward, Viscount of Fallodon (1925). *Twenty-Five Years 1892–1916*. 2 vols. New York: Stokes.

Hamilton, K. A. (1977). "Great Britain and France, 1905–1911." In F. H. Hinsley (ed.), *British Foreign Policy under Sir Edward Grey*. Cambridge: Cambridge University Press, pp. 113–32.

Hanson, Norwood Russell (1958). *Patterns of Discovery: An Inquiry into the Conceptual Foundations of Science*. Cambridge: Cambridge University Press.

Harsanyi, John C. (1967). "Games with Incomplete Information Played by 'Bayesian' Players, Part 1." *Management Science*, 14(3): 159–82.

Harsanyi, John C. (1968a). "Games with Incomplete Information Played by 'Bayesian' Players, Part 2." *Management Science*, 14(5): 320–34.

Harsanyi, John C. (1968b). "Games with Incomplete Information Played by 'Bayesian' Players, Part 3." *Management Science*, 14(7): 486–502.

Harsanyi, John C. (1973). "Review of *Paradoxes of Rationality*." *American Political Science Review*, 67(2): 599–600.

Harsanyi, John C. (1974a). "Communications." *American Political Science Review*, 68(2): 731–2.

Harsanyi, John C. (1974b). "Communications." *American Political Science Review*, 68(4): 1694–5.

Harsanyi, John C. (1977a). "Advances in Understanding Rational Behavior." In Robert E. Butts and Jaakko Hintikka (eds.), *Foundational Problems in the Special Sciences*. Dordrecht: D. Reidel, pp. 315–43.

Harsanyi, John C. (1977b). *Rational Behavior and Bargaining Equilibrium in Games and Social Situations.* Cambridge: Cambridge University Press.

Harvey, Frank P. (1998). "Rigor Mortis, or Rigor, More Tests: Necessity, Sufficiency, and Deterrence Logic." *International Studies Quarterly*, 42(4): 675–707.

Hausman, Daniel H. (2012). *Preference, Value, Choice and Welfare.* Cambridge: Cambridge University Press.

Haywood, O. J., Jr. (1954). "Military Decision and Game Theory." *Operations Research*, 2(4): 365–85.

Hempel, Carl G. (1965). *Aspects of Scientific Explanation: And Other Essays in the Philosophy of Science.* New York: Free Press.

Hesse, Andrei (2010). *Game Theory and the Cuban Missile Crisis: Using Schelling's Strategy of Conflict to Analyze the Cuban Missile Crisis.* München: Martin Meidenbauer Verlagsbuchhandlung.

Hindmoor, Andrew (2006). *Rational Choice.* New York: Palgrave MacMillan.

Holsti, Ole R., Robert C. North, and Richard A. Brody (1968). "Perception and Action in the 1914 Crisis." In J. David Singer (ed.), *Quantitative International Politics: Insights and Evidence.* New York: Free Press, pp. 123–58.

Howard, Nigel (1971). *Paradoxes of Rationality: Theory of Metagames and Political Behavior.* Cambridge, MA: MIT Press.

Howard, Nigel (1973). "Comment on a Mathematical Error in a Review by Harsanyi." *International Journal of Game Theory*, 2(1): 251–2.

Howard, Nigel (1974a). "Communications." *American Political Science Review*, 68(2): 729–30.

Howard, Nigel (1974b). "Communications." *American Political Science Review*, 68(4): 1692–3.

Huth, Paul. K. (1988). *Extended Deterrence and the Prevention of War.* New Haven, CT: Yale University Press.

Huth, Paul K. (1999). "Deterrence and International Conflict: Empirical Findings and Theoretical Debates." *Annual Review of Political Science*, 2(1): 61–84.

Hastings, Max (2013). *Catastrophe 1914: Europe Goes to War.* New York: Vintage Books.

Intriligator, Michael D., and Dagobert L. Brito (1981). "Nuclear Proliferation and the Probability of Nuclear War." *Public Choice*, 37(2): 247–60.

Intriligator, Michael D., and Dagobert L. Brito (1984). "Can Arms Races Lead to the Outbreak of War?" *Journal of Conflict Resolution*, 28(1): 63–84.

Jannen, William, Jr. (1983). "The Austro-Hungarian Decision for War in July 1914." In Samuel R. Williamson and Peter Pastor, (eds.), *Essays on World War I: Origins and Prisoners of War.* New York: Columbia University Press, pp. 55–83.

Jannen, William, Jr. (1996). *The Lions of July: Prelude to War.* Novato, CA: Presidio Press.

Jarausch, Konrad H. (1969). "The Illusion of Limited War: Chancellor Bethmann Hollweg's Calculated Risk, July 1914." *Central European History*, 2(1): 48–76.

Jervis, Robert (1972). "Bargaining and Bargaining Tactics." In J. Roland Pennock and John W. Chapman (eds.), *Coercion. Nomos XIV: Yearbook of the American Society for Political and Legal Philosophy.* Chicago: Aldine, pp. 272–88.

Jervis, Robert (1979). "Deterrence Theory Revisited." *World Politics*, 31(2): 289–324.

Jervis, Robert (1985). "Introduction: Approach and Assumptions." In Robert Jervis, Richard Ned Lebow, and Janice Gross Stein (eds.), *Psychology and Deterrence.* Baltimore: The Johns Hopkins University Press, pp. 1–12.

Jervis, Robert (1988a). "Realism, Game Theory, and Cooperation." *World Politics*, 40(3): 317–49.

Jervis, Robert. (1988b). "The Political Effects of Nuclear Weapons: A Comment." *International Security*, 13(2): 80–90.

Johnson, Chalmers (1997). "Preconception vs. Observation, or the Contributions of Rational Choice Theory and Area Studies to Contemporary Political Science." *PS: Political Science and Politics*, 30(2): 170–4.

Johnson, Jesse C, Brett Ashley Leeds, and Ahra Wu (2015). "Capability, Credibility, and Extended Deterrence." *International Interactions*, 41(2): 309–36.

Kagan, Donald (1995). *On the Origins of War and the Preservation of Peace*. New York: Doubleday.

Kahn, Herman (1962). *Thinking About the Unthinkable*. New York: Horizon Press.

Kaplan, Abraham (1964). *The Conduct of Inquiry*. San Francisco: Chandler Publishing Company.

Kaufmann, William (1956). "The Requirements of Deterrence." In William Kaufmann (ed.), *Military Policy and National Security*. Princeton: Princeton University Press, pp. 12–38.

Kautsky, Karl (1924). *Outbreak of the World War: German Documents Collected by Karl Kautsky*, Max Montgelas and Walter Schücking (eds.). New York: Oxford University Press.

Kennedy, Robert F. (1969). *Thirteen Days: A Memoir of the Cuban Missile Crisis*. New York: Norton.

Khrushchev, Nikita S. (1970). *Khrushchev Remembers*. Boston: Little, Brown.

Khrushchev, Nikita S. (1990). *Khrushchev Remembers: The Glasnost Tapes*. Boston: Little, Brown.

Kilgour, D. Marc, and Frank C. Zagare (2007). "Explaining Limited Conflicts." *Conflict Management and Peace Science*, 24(1): 65–82.

King, Gary, Robert O. Keohane, and Sidney Verba (1994). *Designing Social Inquiry: Scientific Inference in Qualitative Research*. Princeton: Princeton University Press.

Kraig, Michael R. (1999). "Nuclear Deterrence in the Developing World: A Game-Theoretic Treatment." *Journal of Peace Research*, 36(2): 141–67.

Kreps, David M. (1988). *Notes on the Theory of Choice*. Boulder, CO: Westview.

Kreps, David M., and Robert Wilson (1982). "Sequential Equilibria." *Econometrica*, 50(4): 863–57.

Kugler, Jacek (2012). "A World beyond Waltz: Neither Iran nor Israel Should Have the Bomb." *Tehran Bureau*, September 25, http://www.pbs.org/wgbh/pages/frontline/tehranbureau/2012/09/opinion-a-world-beyond-waltz-neither-iran-nor-israel-should-have-the-bomb.html#ixzz28I tL7mq1, accessed August 22, 2018.

Lakatos, Imre (1970). "Falsification and the Methodology of Scientific Research Programs." In Imre Lakatos and Alan Musgrave (eds.), *Criticism and the Growth of Knowledge*. Cambridge: Cambridge University Press, pp. 91–196.

Langdon, John W. (1991). *July 1914: The Long Debate, 1918–1990*. New York: Berg.

Lebow, Richard Ned (1981). *Between Peace and War: The Nature of International Crisis*. Baltimore: Johns Hopkins University Press.

Lebow, Richard Ned (1984). "Windows of Opportunity: Do States Jump Through Them?" *International Security*, 9(1): 147–86.

Leeds, Brett Ashley, Andrew G. Long, and Sara McLaughlin Mitchell (2000). "Reevaluating Alliance Reliability: Specific Threats, Specific Promises." *Journal of Conflict Resolution*, 44(5): 686–99.

Legro, Jeffrey W., and Andrew Moravcsik (1999). "Is Anybody Still a Realist?" *International Security*, 24(2): 5–55.

Leonard, Robert (2010). *Von Neumann, Morgenstern, and the Creation of Game Theory: From Chess to Social Science, 1900–1960*. Cambridge: Cambridge University Press.

Levy, Jack S. (1988). "When Do Deterrent Threats Work?" *British Journal of Political Science*, 18(4): 485–512.

Levy, Jack S. (1990/1991). "Preferences, Constraints, and Choices in July 1914." *International Security*, 15(3): 151–86.

Levy, Jack S. (1997). "Prospect Theory, Rational Choice, and International Relations." *International Studies Quarterly*, 41(1): 87–112.

Levy, Jack S., and John A. Vasquez (eds.) (2014). *The Outbreak of the First World War: Structure, Politics, and Decision-Making.* Cambridge: Cambridge University Press.

Lewis, Michael (2017). *The Undoing Project: A Friendship that Changed Our Minds.* New York: Norton.

Luce, R. Duncan, and Howard Raiffa (1957). *Games and Decisions: Introduction and Critical Survey.* New York: Wiley.

MacMillan, Margaret (2013). *The War that Ended Peace: The Road to 1914.* New York: Random House.

Malin, V. N., and Khrushchev, Nikita Sergeevich. (1962). "Central Committee of the Communist Party of the Soviet Union Presidium Protocol 61" October 25, 1962, History and Public Policy Program Digital Archive, Russian State Archive of Contemporary History (RGANI), F. 3, Op. 16, D. 165, L. 170–173. Translated and edited by Mark Kramer, with assistance from Timothy Naftali, http://digitalarchive.wilsoncenter.org/document/115136, accessed August 22, 2018.

Martin, Lisa (1999). "The Contributions of Rational Choice: A Defense of Pluralism." *International Security*, 24(2): 74–83.

Massie, Robert K. (1991). *Dreadnought: Britain, Germany, and the Coming of the Great War.* New York: Ballantine Books.

May, Ernest R., and Philip D. Zelikow (1997). *The Kennedy Tapes: Inside the White House During the Cuban Missile Crisis.* Cambridge, MA: Belknap Press.

McDermott, Rose (2004). "Prospect Theory in Political Science: Gains and Losses from the First Decade." *Political Psychology*, 25(2): 289–312.

McDermott, Rose, and Jacek Kugler (2001). "Comparing Rational Choice and Prospect Theory Analyses: The US Decision to Launch Operation 'Desert Storm', January 1991." *The Journal of Strategic Studies*, 24(3): 49–85.

McDonald, John (1950). *Strategy in Poker, Business and War.* New York: Norton.

McDonald, John, and John W. Tukey (1949). "Colonel Blotto: A Problem of Military Strategy." *Fortune*, June: 102.

McGrayne, Sharon Bertsch (2011). *The Theory That Would Not Die: How Bayes' Rule Cracked the Enigma Code, Hunted Down Russian Submarines, and Emerged Triumphant from Two Centuries of Controversy.* New Haven, CT: Yale University Press.

McMeekin, Sean (2011). *Russian Origins of the First World War.* Cambridge, MA: Harvard University Press.

McMeekin, Sean (2013). *July 1914.* New York: Basic Books.

Mearsheimer, John J. (1983). *Conventional Deterrence.* Ithaca, NY: Cornell University Press.

Mearsheimer, John J. (1990). "Back to the Future: Instability in Europe After the Cold War." *International Security*, 15(1): 5–56.

Mearsheimer, John J. (1993). "The Case for a Ukrainian Nuclear Deterrent." *Foreign Affairs*, 72(3): 50–66.

Mercer, Jonathan (1996). *Reputation and International Politics.* Ithaca, NY: Cornell University Press.

Mercer, Jonathan (2005). "Prospect Theory and Political Science." *Annual Review of Political Science*, 8: 1–21.

Mikoyan, Sergo (2012). *The Soviet Cuban Missile Crisis: Castro, Mikoyan, Kennedy, Khrushchev, and the Missiles of November.* Washington, DC: Woodrow Wilson Center Press.

Miller, Gregory Daniel (2012). *The Shadow of the Past. Reputation and Military Alliances Before the First World War.* Ithaca, NY: Cornell University Press.

Mombauer, Annika (2001). *Helmuth von Moltke and the Origins of the First World War.* Cambridge: Cambridge University Press.

Mombauer, Annika (2002). *The Origins of the First World War: Controversies and Consensus.* London: Longman.

Mombauer, Annika (2013). "The Fischer Controversy, Documents and the 'Truth' about the Origins of the First World War." *Journal of Contemporary History*, 48(2): 290–314.

Mongin, Philippe (2018). "A Game-Theoretic Analysis of the Waterloo Campaign and Some Comments on the Analytic Narrative Project." *Cliometrica*, 12: 451–80.

Morgan, Patrick M. (1977). *Deterrence: A Conceptual Analysis*. Beverly Hills: Sage.

Morgan, Patrick M. (2003). *Deterrence Now*. Cambridge: Cambridge University Press.

Morgenstern, Oskar (1961). "Reviewed Work: Fights, Games and Debates by Anatol Rapoport." *Southern Economic Journal*, 28(1): 103–5.

Morrow, James D. (1994). *Game Theory for Political Scientists*. Princeton: Princeton University Press.

Morrow, James D. (1997). "A Rational Choice Approach to International Conflict." In Nehemia Geva and Alex Mintz (eds.), *Decisionmaking on War and Peace: The Cognitive–Rational Debate*. Boulder: Lynne Rienner, pp. 11–31.

Morton, Rebecca B. (1999). *Methods and Models: A Guide to the Empirical Analysis of Formal Models in Political Science*. Cambridge: Cambridge University Press.

Most, Benjamin A., and Harvey Starr (1989). *Inquiry, Logic and International Politics*. Columbia: University of South Carolina Press.

Mulligan, William (2010). *The Origins of the First World War*. Cambridge: Cambridge University Press.

Myerson, Roger B. (2007). "Force and Restraint in Strategic Deterrence: A Game Theorist's Perspective," http://permanent.access.gpo.gov/websites/ssi.armywarcollege.edu/pdffiles/PUB823.pdf, accessed August 22, 2018.

Myerson, Roger B. (2009). "Learning from Schelling's *Strategy of Conflict*." *Journal of Economic Literature*, 47(4): 1109–25.

Nash, John (1951). "Non-Cooperative Games." *Annals of Mathematics*, 54(2): 286–95.

Niou, Emerson M. S., and Peter C. Ordeshook (1999). "The Return of the Luddites." *International Security*, 24(2): 84–96.

Niou, Emerson M.S., Peter C. Ordeshook, and Gregory F. Rose (1989). *The Balance of Power: Stability in International Systems*. Cambridge: Cambridge University Press.

Nomikos, Eugenia V., and Robert C. North (1976). *International Crisis: The Outbreak of World War I*. Montreal: McGill-Queen's University Press.

O'Neill, Barry (1992). "Are Game Models of Deterrence Biased towards Arms-Building? Wagner on Rationality and Misperception." *Journal of Theoretical Politics*, 4(4): 459–77.

O'Neill, Barry (1994a). "Game Theory Models of Peace and War." In Robert J. Aumann and Sergiu Hart (eds.). *Handbook of Game Theory*. Vol. 2. Amsterdam: Elsevier, pp. 995–1090.

O'Neill, Barry (1994b). "Sources in Game Theory for International Relations Specialists." In Michael D. Intriligator and Urs Luterbacher (eds.), *Cooperative Models in International Relations Research*. Boston: Kluwer, pp. 9–30.

O'Neill, Barry (2007). "Game Models of Peace and War: Some Recent Themes." In Rudolf Avenhaus and I. William Zartman (eds.), *Diplomacy Games: Formal Models and International Negotiations*. Berlin: Springer, pp. 25–44.

Organski, A. F. K. (1958). *World Politics*. New York: Knopf.

Organski, A. F. K., and Jacek Kugler (1980). *The War Ledger*. Chicago: University of Chicago Press.

Otte, T. G. (2014a). "A 'Formidable Factor in European Politics': Views of Russia in 1914." In Jack S. Levy and John A. Vasquez (eds.), *The Outbreak of the First World War: Structure, Politics, and Decision-Making*. Cambridge: Cambridge University Press, pp. 87–112.

Otte, T. G. (2014b). *July Crisis. The World's Descent into War, Summer 1914*. Cambridge: Cambridge University Press.

Oye, Kenneth A. (1986). *Cooperation Under Anarchy*. Princeton: Princeton University Press.

Palmer, Glenn, and T. Clifton Morgan (2006). *A Theory of Foreign Policy*. Princeton: Princeton University Press.

Pogge von Strandmann, Hartmut (1988). "Germany and the Coming of War." In R. J. W. Evans and Hartmut Pogge von Strandmann (eds.), *The Coming of the First World War*. Oxford: Clarendon Press, pp. 87–124.

Posen, Barry R. (1993). "The Security Dilemma and Ethnic Conflict." *Survival*, 35(1): 27–47.

Poundstone, William (1992). *Prisoner's Dilemma*. New York: Doubleday.

Powell, Robert (1987). "Crisis Bargaining, Escalation, and MAD." *American Political Science Review*, 81(3): 717–35.

Powell, Robert (1990). *Nuclear Deterrence Theory: The Search for Credibility*. New York: Cambridge University Press.

Powell, Robert (1999). "The Modeling Enterprise and Security Studies." *International Security*, 24(2): 97–106.

Powell, Robert (2002). "Bargaining Theory and International Conflict." *Annual Review of Political Science*, 5(1): 1–30.

Powell, Robert (2003). "Nuclear Deterrence Theory, Nuclear Proliferation, and National Missile Defense." *International Security*, 27(4): 86–118.

Quackenbush, Stephen L. (2001). "Deterrence Theory: Where Do We Stand?" *Review of International Studies*, 37(2): 741–62.

Quackenbush, Stephen L. (2004). "The Rationality of Rational Choice Theory." *International Interactions*, 30(2): 87–107.

Quackenbush, Stephen L. (2006). "National Missile Defense and Deterrence." *Political Research Quarterly*, 59(4): 533–41.

Quackenbush, Stephen L. (2010a). "General Deterrence and International Conflict: Testing Perfect Deterrence Theory." *International Interactions*, 36(1): 60–85.

Quackenbush, Stephen L. (2010b). "Territorial Issues and Recurrent Conflict." *Conflict Management and Peace Science*, 27(3): 239–52.

Quackenbush, Stephen L. (2011). *Understanding General Deterrence: Theory and Application*. New York: Palgrave Macmillan.

Quackenbush, Stephen L. (2015). *International Conflict: Logic and Evidence*. Washington, DC: CQ Press.

Quackenbush, Stephen L., and Jerome F. Venteicher II (2008). "Settlements, Outcomes, and the Recurrence of Conflict." *Journal of Peace Research*, 45(6): 723–42.

Quackenbush, Stephen L., and Frank C. Zagare (2006). "A Game-Theoretic Analysis of the War in Kosovo." In Jennifer Sterling-Folker (ed.), *Making Sense of IR Theory*. Boulder: Lynne Rienner Publishers, pp. 98–114.

Rapoport, Anatol (1958). "Various Meanings of 'Theory.'" *American Political Science Review*, 52(4): 972–88.

Rapoport, Anatol (1964). *Strategy and Conscience*. New York: Harper & Row.

Rapoport, Anatol (1992). "Comments on 'Rationality and Misperceptions in Deterrence Theory.'" *Journal of Theoretical Politics*, 4(4): 479–84.

Rapoport, Anatol, and Melvin J. Guyer (1966). "A Taxonomy of 2 × 2 Games." *General Systems: Yearbook of the Society for General Systems Research*, 11: 203–14.

Reed, William, and Katherine Sawyer (2013). "Bargaining Theory of War." *Oxford Bibliographies Online*. DOI: 10.1093/obo/9780199743292-0040.

Reiter, Dan (2003). "Exploring the Bargaining Model of War." *Perspectives on Politics*, 1(1): 27–43.

Remak, Joachim (1971). "1914—The Third Balkan War: Origins Reconsidered." *Journal of Modern History*, 43(3): 353–66.

Rich, Motoko, and David E. Sanger (2017). "Motives of a Young Dictator Baffle Americans and Allies." *New York Times*, September 4.

Riker, William H. (1990). "Political Science and Rational Choice." In James E. Alt and Kenneth A. Shepsle (eds.), *Perspectives on Positive Political Economy*. Cambridge: Cambridge University Press, pp. 163–81.

Riker, William H. (1992). "The Entry of Game Theory into Political Science." In E. Roy Weintraub (ed.), *Toward a History of Game Theory*. Durham, NC: Duke University Press, pp. 207–23.

Riker, William H. (1995). "The Political Psychology of Rational Choice Theory." *Political Psychology*, 16(1): 23–44.

Riker, William H., and Peter C. Ordeshook (1973). *An Introduction to Positive Political Theory*. Englewood Cliffs, NJ: Prentice-Hall.

Rodrik, Dani (2015). *Economics Rules: The Rights and Wrongs of the Dismal Science*. New York: Norton.

Rosenau, James N., and Mary Durfee (2000). *Thinking Theory Thoroughly: Coherent Approaches to an Incoherent World*. Boulder, CO: Westview Press.

Röhl, John C. G. (2014). *Wilhelm II: Into the Abyss of War and Exile, 1900–1941*. Translated by Sheila de Bellaigue and Roy Bridge. Cambridge: Cambridge University Press.

Sabrosky, Alan Ned (1980). "Interstate Alliances: Their Reliability and the Expansion of War." In J. David Singer (ed.), *The Correlates of War II: Testing Some Realpolitik Models*. New York: Free Press, pp. 161–98.

Sartori, Anne E. (2005). *Deterrence by Diplomacy*. Princeton: Princeton University Press.

Sazonov, Serge (1928). *The Fateful Years, 1909–1916: The Reminiscences of Serge Sazonov*. New York: Frederick A. Stokes.

Schecter, Stephen, and Herbert Gintis (2016). *Game Theory in Action: An Introduction to Classical and Evolutionary Models*. Princeton: Princeton University Press.

Schelling, Thomas C. (1960). *The Strategy of Conflict*. Cambridge, MA: Harvard University Press.

Schelling, Thomas C. (1966). *Arms and Influence*. New Haven, CT: Yale University Press.

Schilling, M. F. (1925). *How the War Began in 1914*. London: George Allen.

Schlesinger, Arthur M., Jr. (1965). *A Thousand Days*. Boston: Houghton-Mifflin.

Schmitt, Bernadotte E. (1930). *The Coming of the War 1914*. 2 vols. New York: Scribner's.

Schmitt, Bernadotte E. (1934). *Triple Alliance and Triple Entente*. New York: Henry Holt.

Schmitt, Bernadotte E. (1944). "July 1914: Thirty Years After." *Journal of Modern History*, 16(3): 169–204.

Schroeder, Paul W. (1972). "World War I as Galloping Gertie: A Reply to Joachim Remak." *Journal of Modern History*, 44(3): 319–45.

Schroeder, Paul W. (2007). "Necessary Conditions and World War I as an Unavoidable War." In Gary Goertz and Jack S. Levy (eds.), *Explaining War and Peace: Case Studies and Necessary Condition Counterfactuals*. New York: Routledge, pp. 163–210.

Schultz, Kenneth A. (2001). *Democracy and Coercive Diplomacy*. Cambridge: Cambridge University Press.

Selten, Reinhard (1975). "Reexamination of the Perfectness Concept for Equilibrium Points in Extensive Games." *International Journal of Game Theory*, 4(1): 25–55.

Senese, Paul D., and Stephen L. Quackenbush (2003). "Sowing the Seeds of Conflict: The Effect of Dispute Settlements on Durations of Peace." *Journal of Politics*, 65(3): 696–717.

Senese, Paul D., and John A. Vasquez (2008). *The Steps to War: An Empirical Study*. Princeton: Princeton University Press.

Shirer, William L. (1962). *The Rise and Fall of the Third Reich*. New York: Crest.

Signorino, Curt S., and Ahmer Tarar (2006). "A Unified Theory and Test of Extended Immediate Deterrence." *American Journal of Political Science*, 50(3): 586–605.

Simon, Herbert A. (1976). "From Substantive to Procedural Rationality." In S. J. Latsis (ed.), *Method and Appraisal in Economics*. Cambridge: Cambridge University Press, 65–86.

Slantchev, Branislav L. (2011). *Military Threats: The Costs of Coercion and the Price of Peace*. Cambridge: Cambridge University Press.

Smith, Alastair (1995). "Alliance Formation and War." *International Studies Quarterly*, 39(4): 405–25.

Snidal, Duncan (2002). "Rational Choice and International Relations." In Walter Carlsnaes, Thomas Risse and Beth A. Simmons (eds.), *Handbook of International Relations*. Thousand Oaks: Sage, pp. 73–94.

Snyder, Glenn H. (1971a). "'Prisoner's Dilemma' and 'Chicken' Models in International Politics." *International Studies Quarterly*, 15(1): 66–103.

Snyder, Glenn H. (1971b). "The Moroccan Crisis of 1905–1906." Crisis Bargaining Project, Center for International Conflict Studies, State University of New York at Buffalo.

Snyder, Glenn H. (1972). "Crisis Bargaining." In Charles F. Hermann (ed.), *International Crises: Insights from Behavioral Research*. New York: Free Press, pp. 217–66.

Snyder, Glenn H. (1984). "The Security Dilemma in Alliance Politics." *World Politics*, 36(4): 461–95.

Snyder, Glenn H. (1997). *Alliance Politics*. Ithaca, NY: Cornell University Press.

Snyder, Glenn H., and Paul Diesing (1977). *Conflict among Nations: Bargaining, Decision Making and System Structure in International Crises*. Princeton: Princeton University Press.

Sorensen, Theodore C. (1965). *Kennedy*. New York: Harper & Row.

Sörenson, Karl (2017). "Comparable Deterrence: Target, Criteria and Purpose." *Defence Studies*, 17(2): 198–213.

Spring, D. W. (1988). "Russia and the Coming of War." In R. J. W. Evans and Hartmut Pogge von Strandmann (eds.), *The Coming of the First World War*. Oxford: Clarendon Press, pp. 57–86.

Steiner, Zara S. (1977). *Britain and the Origins of the First World War*. New York: St. Martin's.

Stern, Sheldon M. (2012). *The Cuban Missile Crisis in American Memory: Myths versus Reality*. Stanford, CA: Stanford University Press.

Stokes, Gale (1976). "The Serbian Documents from 1914: A Preview." *Journal of Modern History*, 48(S3): 69–83.

Stone, Norman (2009). *World War One*. New York: Basic Books.

Sullivan, Michael P. (2002). *Theories of International Relations: Transition vs. Persistence*. New York: Palgrave MacMillan.

Tammen, Ronald L., et al. (2000). *Power Transitions: Strategies for the 21st Century*. New York: Chatham House.

Taylor, A. J. P. (1954). *The Struggle for Mastery in Europe: 1848–1918*. Oxford: Oxford University Press.

Thompson, William R. (2003). "A Streetcar Named Sarajevo: Catalysts, Multiple Causation Chains, and Rivalry Structures." *International Studies Quarterly*, 47(3): 453–74.

Trachtenberg, Marc (1985). "The Influence of Nuclear Weapons in the Cuban Missile Crisis." *International Security*, 10(1): 137–63.

Trachtenberg, Marc (1990/1991). "The Meaning of Mobilization in 1914." *International Security*, 15(3): 120–50.

Trachtenberg, Marc (1991). *History and Strategy*. Princeton: Princeton University Press.

Trachtenberg, Marc (2006). *The Craft of International History: A Guide to Method*. Princeton: Princeton University Press.

Tuchman, Barbara (1962). *The Guns of August*. New York: Dell.

Turner, L. C. F. (1968). "The Russian Mobilization in 1914." *Journal of Contemporary History*, 3(1): 65–88.

Van Evera, Stephen (1990/1991). "Primed for Peace: Europe After the Cold War." *International Security*, 15(3): 7–57.

Van Evera, Stephen (1999). *Causes of War*. Ithaca, NY: Cornell University Press.

Vasquez, John A. (1997). "The Realist Paradigm and Degenerative versus Progressive Research Programs: An Appraisal of Neotraditional Research on Waltz's Balancing Proposition." *American Political Science Review*, 91(4): 899–912.

Vasquez, John A. (1998). *The Power of Power Politics: From Classical Realism to Neotraditionalism*. Cambridge: Cambridge University Press.

Von Neumann, John, and Oskar Morgenstern (1944). *Theory of Games and Economic Behavior*. Princeton: Princeton University Press.

Wagner, R. Harrison (1989). "Uncertainty, Rational Learning, and Bargaining in the Cuban Missile Crisis." In Peter C. Ordeshook (ed.), *Models of Strategic Choice in Politics*. Ann Arbor, University of Michigan Press, pp. 177–205.

Walt, Stephen M. (1999). "Rigor or Rigor Mortis? Rational Choice and Security Studies." *International Security*, 23(4): 5–48.

Waltz, Kenneth N. (1964). "The Stability of a Bipolar World." *Daedalus*, 93(3): 881–909.

Waltz, Kenneth N. (1979). *Theory of International Politics*. Reading, MA: Addison-Wesley.

Waltz, Kenneth N. (1981). "The Spread of Nuclear Weapons: More May Be Better." Adelphi Paper No. 171. London: International Institute for Strategic Studies.

Waltz, Kenneth N. (1990). "Nuclear Myths and Political Realities." *The American Political Science Review*, 84(3): 731–45.

Waltz, Kenneth N. (1993). "The Emerging Structure of International Politics." *International Security*, 18(2): 44–79.

Waltz, Kenneth N. (2012). "Why Iran Should Get the Bomb: Nuclear Balancing Would Mean Stability." *Foreign Affairs*, 91(4): 2–5.

Williams, J. D. (1954). *The Compleat Strateyst*. Santa Monica, CA: Rand.

Williamson, Samuel R., Jr. (1969). *The Politics of Grand Strategy: Britain and France Prepare for War, 1904–1914*. Cambridge, MA: Harvard University Press.

Williamson, Samuel R., Jr. (1983). "Vienna and July 1914: The Origins of the Great War Once More." In Samuel R. Williamson and Peter Pastor (eds.), *Essays on World War I: Origins and Prisoners of War*. New York: Columbia University Press, pp. 9–36.

Williamson, Samuel R., Jr. (1991). *Austria-Hungary and the Origins of the First World War*. London: MacMillan.

Williamson, Samuel R., Jr., and Ernest R. May (2007). "An Identity of Opinion: Historians and 1914." *Journal of Modern History*, 79(2): 335–87.

Wolfers, Arnold (1951). "The Pole of Power and the Pole of Indifference." *World Politics*, 4(1): 39–63.

Young, Oran R. (ed.) (1975). *Bargaining: Formal Theories of Negotiation*. Urbana, IL: University of Illinois Press.

Zagare, Frank C. (1984). *Game Theory: Concepts and Applications*. Sage University Paper Series on Quantitative Applications in the Social Sciences. Beverly Hills: Sage Publications.

Zagare, Frank C. (1987). *The Dynamics of Deterrence*. Chicago: University of Chicago Press.

Zagare, Frank C. (1990a). "Rationality and Deterrence." *World Politics*, 42(2): 238–60.

Zagare, Frank C. (1990b). "The Dynamics of Escalation," *Information and Decision Technologies*, 16(3): 249–61.

Zagare Frank C. (1996a). "Classical Deterrence Theory: A Critical Assessment." *International Interactions*, 21(4): 365–87.

Zagare, Frank C. (1996b). "The Rites of Passage: Parity, Nuclear Deterrence and Power Transitions." In Jacek Kugler and Douglas Lemke (eds.), *Parity and War: Evaluations and Extensions of "The War Ledger."* Ann Arbor: University of Michigan Press, pp. 249–68.

Zagare, Frank C. (1999). "All Mortis, No Rigor." *International Security*, 24(2): 107–14.

Zagare Frank C. (2004). "Reconciling Rationality with Deterrence: A Re-examination of the Logical Foundations of Deterrence Theory." *Journal of Theoretical Politics*, 16(2): 107–41.

Zagare, Frank C. (2007). "Toward a Unified Theory of Interstate Conflict." *International Interactions*, 33(3): 305–27.

Zagare Frank C. (2008). "Game Theory and Security Studies." In Paul D. Williams (ed.), *Security Studies: An Introduction.* London: Routledge, pp. 44–58.

Zagare, Frank C. (2009a). "After Sarajevo: Explaining the Blank Check." *International Interactions*, 35(1): 106–27.

Zagare, Frank C. (2009b). "Explaining the 1914 War in Europe: An Analytic Narrative." *Journal of Theoretical Politics*, 21(1): 63–95.

Zagare, Frank C. (2011a). "Analytic Narratives, Game Theory, and Peace Science." In Manas Chatterji (ed.), *Frontiers of Peace Economics and Peace Science, Contributions to Conflict Management, Peace Economics and Development.* Vol. 16. Bingley: Emerald Group, 2011, pp. 19–35.

Zagare, Frank C. (2011b). *The Games of July: Explaining the Great War.* Ann Arbor: University of Michigan Press.

Zagare, Frank C. (2014). "A Game-Theoretic History of the Cuban Missile Crisis." *Economies*, 2(1): 20–44.

Zagare, Frank C. (2015a). "Reflections on the Great War." *Review of History and Political Science*, 3(2): 1–5.

Zagare, Frank C. (2015b). "The Moroccan Crisis of 1905–1906: An Analytic Narrative." *Peace Economics, Peace Science and Public Policy*, 21(3): 1–24.

Zagare, Frank C. (2016). "A General Explanation of the Cuban Missile Crisis." *International Journal of Peace Economics and Peace Science*, 1(1): 91–118.

Zagare, Frank C. (2018a). "Explaining the Long Peace: Why von Neumann (and Schelling) Got it Wrong." *International Studies Review*, 20: 422–37.

Zagare, Frank C. (2018b). "Perfect Deterrence Theory." In William R. Thompson (ed.), *Oxford Encyclopedia of Empirical International Relations Theory.* 4 vols. New York: Oxford University Press, pp. 34–52.

Zagare, Frank C., and D. Marc Kilgour (1993). "Asymmetric Deterrence." *International Studies Quarterly*, 37(1): 1–27.

Zagare, Frank C., and D. Marc Kilgour (1998). "Deterrence Theory and the Spiral Model Revisited." *Journal of Theoretical Politics*, 10(1): 59–87.

Zagare, Frank C., and D. Marc Kilgour (2000). *Perfect Deterrence.* Cambridge: Cambridge University Press.

Zagare, Frank C., and D. Marc Kilgour (2003). "Alignment Patterns, Crisis Bargaining, and Extended Deterrence: A Game-Theoretic Analysis." *International Studies Quarterly*, 47(4): 587–615.

Zagare, Frank C., and D. Marc Kilgour (2006). "The Deterrence-versus-Restraint Dilemma in Extended Deterrence: Explaining British Policy in 1914." *International Studies Review*, 8(4): 623–41.

Zagare, Frank C., and Branislav L. Slantchev (2012). "Game Theory and Other Modeling Approaches." In Robert A. Denemark et al. (eds.), *The International Studies Encyclopedia.* Vol. 4. Oxford: Wiley-Blackwell, pp. 2591–2610.

Zakaria, Fareed (2001). "Don't Oversell Missile Defense: The Old Theory of Nuclear Deterrence still Makes Sense. Just Ask the Man Who Invented It." *Newsweek*, 14 May.

Zuckerman, Solly (1956). *Scientists and War: The Impact of Science on Military and Civil Affairs.* London: Hamish Hamilton.

INDEX

OPL

OXFORD PSYCHIATRY LIBRARY

Cognition in Major Depressive Disorder

OXFORD PSYCHIATRY LIBRARY

Cognition in Major Depressive Disorder

Roger S. McIntyre

Mood Disorders Psychopharmacology Unit, University Health Network (UHN)
Institute of Medical Science (IMS); Departments of Psychiatry
and Pharmacology
University of Toronto
Toronto, Ontario
Canada

Danielle S. Cha

Mood Disorders Psychopharmacology Unit, University Health Network (UHN)
Institute of Medical Science (IMS)
University of Toronto
Toronto, Ontario
Canada

Joanna K. Soczynska

Mood Disorders Psychopharmacology Unit, University Health Network (UHN)
Institute of Medical Science (IMS)
University of Toronto
Toronto, Ontario
Canada

OXFORD
UNIVERSITY PRESS

UNIVERSITY PRESS

Great Clarendon Street, Oxford, OX2 6DP,
United Kingdom

Oxford University Press is a department of the University of Oxford.
It furthers the University's objective of excellence in research, scholarship,
and education by publishing worldwide. Oxford is a registered trade mark of
Oxford University Press in the UK and in certain other countries

Published in the United States of America by Oxford University Press
198 Madison Avenue, New York, NY 10016, United States of America

British Library Cataloguing in Publication Data
Data available

Library of Congress Control Number: 2013947891

ISBN 978–0–19–968880–7

Printed in Great Britain by
Ashford Colour Press Ltd, Gosport, Hampshire

Table of Contents

Preface

During the past decade, major depressive disorder (MDD) has gone through a significant transformation in its conceptualization from a population health perspective. Rather than the anachronistic and inaccurate description of MDD as a relatively rare, mild, and transient condition, MDD occupies the ignominious position as the leading cause of disability amongst all psychiatric disorders in both developed and developing nations.

The human capital costs attributable to MDD are substantial and increasing. The global economic landscape has gone through a tectonic plate shift wherein the human knowledge economy is the principal determinant of economic competitiveness. Towards the aim of realizing the human potential of individuals (and populations) in this human knowledge economy, optimal psychological and physical health is *sine qua non*.

Emanating and surrounding extant evidence is the observation that cognitive deficits are a core disturbance in MDD, risk factor for illness onset, recurrence, chronicity, and importantly, a principal cause of functional impairment. The latter observation provides the rationale for prioritizing cognitive deficits in the diagnosis, assessment, ongoing evaluation, and treatment of individuals with MDD. Moreover, the identification of cognitive deficits in individuals with MDD provides the impetus for parsing out neurobiological substrates subserving this critical domain of psychopathology.

The overarching aim of this Oxford Psychiatry Library pocketbook is to encourage a pivot towards cognitive dysfunction in MDD. Our hope is to encourage practitioners to probe and evaluate cognitive dysfunction in the clinical ecosystem, and to introduce vistas for future research. Towards the aim of full recovery of MDD (and in the future, prevention of MDD), as well as optimization of individual and population human capital potential, the identification, characterization, evaluation, treatment, and prevention of cognitive dysfunction are warranted.

Roger S. McIntyre

Symbols and Abbreviations

AD	Alzheimer's disease
AMPA	α-Amino-3-hydroxy-5-methyl-4-isoxazoleproprionic acid
ANCOVA	analysis of covariance
APTD	acute phenylalanine and tyrosine depletion
BC-CCI	British Columbia Cognitive Complaints Inventory
BD	bipolar disorder
BDI	Beck depression inventory
BDNF	brain-derived neurotrophic factor
BLC	big/little circle
CalCAP	California computerised assessment package
CANTAB	Cambridge neuropsychological test automated battery
CGI-I	clinical global impression-improvement
CGI-S	clinical global impression-severity
CGT	Cambridge gambling task
CNS	central nervous system
COWAT	controlled oral word association test
CPFQ	Massachusetts General Hospital cognitive and physical functioning questionnaire
CPT	continuous performance task
CRT	choice reaction time
CSF	cerebral spinal fluid
CST	concept shifting task
CVLT	California verbal learning test
D-KEFS	Delis–Kaplan executive function system
DA	dopamine
DBS	deep brain stimulation
DMN	default mode network
DMS	delayed matching to sample
DRST	delayed recognition span test
DSM-5	*Diagnostic and Statistical Manual of Mental Disorders (5th edition)*

DSST	digit symbol substitution test
DTMS	deep transcranial magnetic stimulation
ECT	electroconvulsive therapy
FAS	full analysis set
fMRI	functional magnetic resonance imaging
GABA	gamma-aminobutyric acid
GLP-1	glucagon-like peptide-1
GNAT	go/no-go association task
GR	glucocorticoid receptor
HAM-A	Hamilton rating scale for anxiety
HPA	hypothalamic-pituitary-adrenal
HVLT-R	Hopkins verbal learning test revised
HRSD	Hamilton rating scale for depression
HRSD-21	Hamilton rating scale for depression (21 items)
HRSD-24	Hamilton rating scale for depression (24 items)
IED	intra-extra dimensional set shift
IL	interleukin
IFN	interferon
IST	information sampling task
JOLO	judgement of line orientation
LOCF	last observation carried forward
LNS	letter-number sequencing
LTP	long-term potentiation
LVLT	Luria's verbal learning test
MADRS	Montgomery-Åsberg depression rating scale
MCI	mild cognitive impairment
MDD	major depressive disorder
MDE	major depressive episode
MFFT-20	matching familiar figures test 20
mGlu	metabotropic glutamate
MMRM	mixed model for repeated measures
MMSE	mini-mental state exam
MOCA	Montreal cognitive assessment
MR	mineralocorticoid receptor

MST	magnetic seizure therapy
NA	noradrenaline
nAChR	nicotinic acetylcholine receptors
NART	national adult reading test
NEAR	neuropsychological educational approach remediation
NMDA	N-methyl-D-aspartate
NREM	non-rapid eye movement
PAL	paired associates learning
PASAT	paced auditory serial addition test
PCP	phencyclidine
PDQ-D	perceived deficits questionnaire
PRM	pattern recognition memory
RAVLT	rey auditory verbal learning test
RBANS	repeatable battery for the assessment of neuropsychological status
RBMT	Rivermead behavioural memory test
REM	rapid eye movement
RFFT	Ruff figural fluency test
RFT	Kimura's recurring figures test
ROCF	Rey-Osterrieth complex figure test
RTI	reaction time
rTMS	repetitive transcranial magnetic stimulation
SAMe	S-adenosylmethionine
SCWT	Stroop colour-word interference test
SD	standard deviation
SE	standard error
SNRI	serotonin-norepinephrine reuptake inhibitor
SOC	stockings of cambridge
SRM	spatial recognition memory
SRT	Buschke selective reminding test
SRT	simple reaction time
SSP	spatial span
SSRI	selective serotonin reuptake inhibitor
SSST	serial sevens subtraction test

SWM	spatial working memory
SWS	slow-wave sleep
TCAs	tricyclic antidepressants
tDCS	transcranial direct current stimulation
TMT A	trail making test part A
TMT B	trail making test part B
TNF	tumour necrosis factor
TOL	tower of London
TONI-3	test of nonverbal intelligence-3
TRP	tryptophan
UHR-MDD	unaffected healthy relatives of MDD patients
VFD	Benton visual form discrimination test
VNS	vagus nerve stimulation
VPA	verbal paired associates
VRM	verbal recognition memory test
VRT	Benton visual retention test
VSVT	Victoria symptom validity test
VVLT	visual verbal learning test
WAIS-R	Wechsler adult intelligence test-revised
WCST	Wisconsin card sorting test
WMS-R	Wechsler memory scale revised
WTAR	Wechsler test of adult reading

Chapter 1

Introduction: the relevance of cognitive dysfunction in major depressive disorder

Key Points

- Cognitive dysfunction is a core psychopathological dimension of MDD.
- Cognitive dysfunction is documented in both acute and maintenance phases of MDD.
- Cognitive deficits are a principal mediator of psychosocial impairment in many individuals with MDD.

Major depressive disorder (MDD) is a multidimensional mental disorder that affects approximately one in seven individuals at some time in their life (Kessler et al 2012). Major depressive disorder is associated with a high rate of non-recovery and recurrence, with chronicity rates estimated at approximately 20% (van Randenborgh et al 2012). The estimated annual costs attributable to MDD in the USA are approximately $83 billion, with indirect costs due to decreased psychosocial function (notably workforce performance) being a major contributor (Greenberg et al 2003). For example, it is estimated that MDD is associated with an annual loss of 27.2 workdays per ill worker (Kessler et al 2006). The human capital cost attributable to MDD, and the significant reduction in the quality of life, provides the impetus for identifying determinants of functional impairment. Available evidence indicates that cognitive dysfunction is a critical mediator of adverse psychosocial outcomes in this population (Jaeger et al 2006; Conradi et al 2010; Buist-Bouwman et al 2008). For example, the impact of a major depressive episode (MDE) on role functioning and psychosocial outcomes in MDD has been reported to be mediated by impairments in cognition in a substantial proportion of individuals (Buist-Bouwman et al 2008).

The *Diagnostic and Statistical Manual of Mental Disorders* (2013; DSM-5) identifies cognitive impairment (e.g. indecisiveness or lack of concentration) as a criterion for a MDE. Cognitive complaints are commonly reported by individuals with MDD during the symptomatic and 'remitted' phases (Greenberg et al 2003; Kessler et al 2006; Conradi et al 2010; Reppermund 2007). When compared to other mental disorders (e.g. schizophrenia, bipolar disorder (BD)), relatively fewer studies have critically reviewed the affected cognitive domains and estimated the effect sizes of these deficits in younger adults with MDD. In addition, there have been fewer original reports that primarily aim to parse out the neurobiological substrate(s) of cognitive dysfunction, determine the contribution of cognitive dysfunction to psychosocial impairment, and evaluate the procognitive effects of treatment (regardless of modality) in the MDD population.

The mediational contribution of cognitive dysfunction to psychosocial impairment in MDD has become a recent topic of inquiry. Available evidence suggests that cognitive dysfunction is a critical determinant of functional outcomes in MDD (Buist-Bouwman et al 2008). It remains to be determined if the neurobiological substrates that subserve cognitive dysfunction in MDD are discrete, and/or if they overlap with substrates implicated in other mental disorders. Questions regarding which, if any, treatment modality is most effective in mitigating cognitive deficits, preventing their occurrence and/or enhancing cognitive function in MDD, remain unanswered. A general impression, albeit based on relatively few empirical studies, is that cognitive deficits are sub-optimally treated with conventional treatment approaches [e.g. selective serotonin reuptake inhibitors (SSRIs)]; (Herrara-Guzman et al 2010; Raskin et al 2007; Millan et al 2012).

The persistence of cognitive dysfunction beyond resolution of the acute episode indicates that in some individuals with MDD it may represent a trait and/or residual phenomenon in some individuals with MDD. Taken together, these results provide the impetus to mentally reconceptualize the definition of 'remission'. Moreover, the moderational influence of cognitive dysfunction on treatment efficacy as well as the influence of cognitive dysfunction on relapse/recurrence vulnerability of MDD has been relatively less studied (Conradi et al 2010). Studies assessing temporality of onset provide evidence that, in a subgroup of younger individuals, cognitive dysfunction may predate the onset of the first MDE (Airaksinen et al 2007). Cognitive dysfunction as an antecedent to MDD is more frequently reported in late onset depression. The preponderance of evidence indicates that cognitive deficits are a consequence as well as a residual phenomenon of MDD, and an antecedent in a smaller proportion of cases.

Cognitive dysfunction in MDD can be disaggregated into two overarching components: 1) General cognitive deficits in one or more non-emotionally valenced domains (e.g. learning and memory, attention, psychomotor speed, executive dysfunction) and 2) processing bias, representing attentional allocation towards negatively valenced stimuli and/or aberrant interpretation of social cues. It has been further hypothesized that cognitive bias represents a trait characteristic in individuals who are temperamentally predisposed to MDD (e.g. neuroticism) (Kendler and Myers 2010).

The effect of conventional antidepressants on cognitive performance has been studied more frequently in non-geriatric individual with MDD; nevertheless, available studies in non-geriatric samples demonstrate that SSRIs, serotonin norepinephrine inhibitors (SNRIs), dopamine (DA) modulators (e.g. bupropion), and norepinephrine inhibitors (e.g. reboxetine) improve cognitive performance in adults with MDD (Furtado et al 2012; Wagner et al 2012; McLennan and Mathias 2010; Herrera-Guzman et al 2010; Herrera-Guzman et al 2008; Ferguson et al 2003). There is also a paucity of evidence comparing the different classes of antidepressants on cognitive function (Herrera-Guzman et al 2010; Herrera-Guzman et al 2008; Ferguson et al 2003). Questions pertaining to the degree that improvement in cognitive deficits can be dissociated from the effect of antidepressants on other domains of MDD psychopathology (e.g. chronicity, subtype, number and duration of MDEs) and/or its co-morbidities have been insufficiently addressed.

This Oxford Psychiatry Library pocketbook aims to emphasize the relevance of cognitive deficits in MDD. Towards this aim we: 1) succinctly review cognitive deficits and their determinants in adults with MDD; 2) identify the underlying substrates that subserve cognitive deficits; and 3) discuss the effects of different treatment modalities on cognitive performance, proposing treatment approaches that primarily aim to mitigate, reverse, and prevent deficits in cognitive function.

References

Airaksinen E, Wahlin A, Forsell Y, Larsson M. Low episodic memory performance as a premorbid marker of depression: evidence from a 3-year follow-up. Acta Psychiatr Scand 2007 Jun;115(6):458–65.

American Psychiatric Association. Major depressive disorder, In Diagnostic and statistical manual of mental disorders (5th ed). Washington, DC: Author; 2013, p.160–188.

Buist-Bouwman MA, Ormel J, de Graaf R, de Jonge JP, van Sonderen E, Alonso J, et al. Mediators of the association between depression and role functioning. Acta Psychiatr Scand 2008 Dec;118(6):451–8.

Conradi HJ, Ormel J, de Jonge JP. Presence of individual (residual) symptoms during depressive episodes and periods of remission: a 3-year prospective study. Psychol Med 2010 Jun;41(6):1165–74.

Ferguson JM, Wesnes KA, Schwartz GE. Reboxetine versus paroxetine versus placebo: effects on cognitive functioning in depressed patients. Int Clin Psychopharmacol 2003 Jan;18(1):9–14.

Furtado CP, Hoy KE, Maller JJ, Savage G, Daskalakis ZJ, Fitzgerald PB. An investigation of medial temporal lobe changes and cognition following antidepressant response: A prospective rTMS study. Brain Stimul 2012 Jul; 6(3):346–54.

Greenberg PE, Kessler RC, Birnbaum HG, Leong SA, Lowe SW, Berglund PA, et al. The economic burden of depression in the United States: how did it change between 1990 and 2000? J Clin Psychiatry 2003 Dec;64(12):1465–75.

Herrera-Guzman I, Gudayol-Ferre E, Herrera-Abarca JE, Herrera-Guzman D, Montelongo-Pedraza P, Padros BF, et al. Major Depressive Disorder in recovery and neuropsychological functioning: effects of selective serotonin reuptake inhibitor and dual inhibitor depression treatments on residual cognitive deficits in patients with Major Depressive Disorder in recovery. J Affect Disord 2010 Jun;123(1-3):341–50.

Herrera-Guzman I, Gudayol-Ferre E, Lira-Mandujano J, Herrera-Abarca J, Herrera-Guzman D, Montoya-Perez K, et al. Cognitive predictors of treatment response to bupropion and cognitive effects of bupropion in patients with major depressive disorder. Psychiatry Res 2008 Jul;160(1):72–82.

Herrera-Guzman I, Herrera-Abarca JE, Gudayol-Ferre E, Herrera-Guzman D, Gomez-Carbajal L, Pena-Olvira M, et al. Effects of selective serotonin reuptake and dual serotonergic-noradrenergic reuptake treatments on attention and executive functions in patients with major depressive disorder. Psychiatry Res 2010 May;177(3):323–9.

Jaeger J, Berns S, Uzelac S, Davis-Conway S. Neurocognitive deficits and disability in major depressive disorder. Psychiatry Res 2006 Nov;145(1):39–48.

Kessler RC, Akiskal HS, Ames M, Birnbaum H, Greenberg P, Hirschfeld RM, et al. Prevalence and effects of mood disorders on work performance in a nationally representative sample of U.S. workers. Am J Psychiatry 2006 Sep;163(9):1561–8.

Kessler RC, Petukhova M, Sampson NA, Zaslavsky AM, Wittchen HU. Twelve-month and lifetime prevalence and lifetime morbid risk of anxiety and mood disorders in the United States. Int J Methods Psychiatr Res 2012 Aug;21(3):169–84.

McLennan SN, Mathias JL. The depression-executive dysfunction (DED) syndrome and response to antidepressants: a meta-analytic review. Int J Geriatr Psychiatry 2010 Oct;25(10):933–44.

Millan MJ, Agid Y, Brune M, Bullmore ET, Carter CS, Clayton NS, et al. Cognitive dysfunction in psychiatric disorders: characteristics, causes and the quest for improved therapy. Nat Rev Drug Discov 2012 Feb;11(2):141–68.

Raskin J, Wiltse CG, Siegal A, Sheikh J, Xu J, Dinkel JJ, et al. Efficacy of duloxetine on cognition, depression, and pain in elderly patients with major depressive disorder: an 8-week, double-blind, placebo-controlled trial. Am J Psychiatry 2007 Jun;164(6):900–9.

Reppermund S, Zihl J, Lucae S, Horstmann S, Kloiber S, Holsboer F, et al. Persistent cognitive impairment in depression: the role of psychopathology and altered hypothalamic-pituitary-adrenocortical (HPA) system regulation. Biol Psychiatry 2007 Sep;62(5):400–6.

van Randenborgh RA, Huffmeier J, Victor D, Klocke K, Borlinghaus J, Pawelzik M. Contrasting chronic with episodic depression: An analysis of distorted socio-emotional information processing in chronic depression. J Affect Disord 2012 Apr;141(2-3):177–84.

Wagner S, Doering B, Helmreich I, Lieb K, Tadic A. A meta-analysis of executive dysfunctions in unipolar major depressive disorder without psychotic symptoms and their changes during antidepressant treatment. Acta Psychiatr Scand 2012 Apr;125(4):281–92.

Chapter 2

Measurement and evaluation of cognitive function

Key Points

- No measures have been specifically developed to evaluate cognitive function in MDD.
- Existing screening tools for dementia are insufficiently sensitive for evaluating cognitive dysfunction in younger individuals with MDD.
- Notwithstanding the limitations, several existing general and specific measures of cognitive dysfunction can be applied to the MDD population.

The heterogeneity of cognitive batteries and measures, as well as the inclusion of mixed sample compositions and treatment regimens, have contributed to variable findings as it relates to the 'cognitive profile' of MDD. Furthermore, available tests and batteries that measure multiple cognitive domains (e.g., attention, executive function, verbal learning and memory), have not been specifically developed or validated for the detection of cognitive deficits in the MDD population (Table 2.1; McIntyre et al 2013). A separate and replicated observation in adults with MDD is the lack of correlation between subjective and objective measures of cognitive dysfunction (Naismith et al 2007).

Other methodological factors that may affect the interpretation of the available data include, but are not limited to, the comparison of group means without stratification for cognitive dysfunction (e.g. self-reported or observable deficits). For example, an operational definition of cognitive dysfunction could include clinically significant deviation (e.g. 1–2 standard deviations below the age-matched population mean) in one or more cognitive domains (Gualtieri and Morgan 2008). Non-stratified evaluation may inadvertently decrease 'assay sensitivity' and underestimate the magnitude of cognitive deficits in subgroups of individuals with MDD. However, a limitation to deficit stratification based on objective measures is that it potentially precludes individuals with clinically significant deficits (i.e. self-reported cognitive decline but normal objective cognitive performance) from receiving pro-cognitive interventions. The discrepancy between subjective and objective cognitive dysfunction may potentially be due to cognitive reserve (e.g. intelligence, education, occupation; McIntyre et al 2013). Identification of predisposing and resiliency factors for cognitive decline is a research

CHAPTER 2 Measurement and evaluation

6

Cognitive domain	Neurocognitive tests
Executive function (concept formation, abstraction, set shifting, set maintenance, planning, self-monitoring, & divided attention)	• Wisconsin Card Sorting Test (WCST) • Trail Making Test Part B (TMT B) • Stroop Colour-Word Interference Test (SCWT) • Categories Test • Block Design (Wechsler Adult Intelligence Test-Revised (WAIS-R)) • Picture Completion (WAIS-R) • Concept Shifting Task (CST) • Tower of London (TOL; Cambridge Neuropsychological Test Automated Battery (CANTAB)) • Stockings of Cambridge (SOC; CANTAB) • Intra-/extra Dimensional Set Shift (IED; CANTAB) • Spatial Span (SSP; CANTAB) • Ruff Figural Fluency Test (RFFT) • Verbal, Letter, and Category Fluencies • Controlled Oral Word Association Test (COWAT) • The Delis–Kaplan Executive Function System (D-KEFS)
Attention and processing speed	• Digit Symbol Substitution Test (DSST; WAIS-R) • Digit Span Forwards and Backwards (WAIS-R) • Continuous Performance Task (CPT) • Reaction Time (RTI; CANTAB) • Choice Reaction Time (CRT; CANTAB) • Simple Reaction Time (SRT; CANTAB) • Trail Making Test Part A (TMT A) • Paced Auditory Serial Addition Test (PASAT) • Serial sevens subtraction test (SSST)
Working memory	• Arithmetic (WAIS-R) • Digit Span Forwards and Backwards (WAIS-R) • Delayed Recognition Span Test (DRST) • Spatial Working Memory (SWM; CANTAB) • Letter-Number Sequencing (LNS; Wechsler Memory Scale Revised (WMS-R)) • Logical Memory (WMS-R) • n-Back Test
Verbal learning and memory	• California Verbal Learning Test (CVLT) • Rey Auditory Verbal Learning Test (RAVLT) • Rivermead Behavioral Memory Test (RBMT) • Hopkins Verbal Learning Test Revised (HVLT-R) • Logical Memory (WMS-R) • Verbal Paired Associates (VPA; WMS-R) • Visual Verbal Learning Test (VVLT) • Digit Span Forwards and Backwards (WAIS-R) • Luria's verbal learning test (LVLT) • SSST • Verbal Recognition Memory Test (VRM; CANTAB)
Visual learning and memory	• Visual Reproduction (WMS-R) • Benton Visual Retention Test (VRT) • Benton Visual Form Discrimination (VFD) • Rey-Osterrieth Complex Figure Test (ROCF) • Kimura's Recurring Figures Test (RFT) • Visual Verbal Learning Test (VVLT)

Table 2.1 (Continued)	
Cognitive domain	**Neurocognitive tests**
	• Pattern Recognition Memory (PRM; CANTAB) • Spatial Recognition Memory (SRM; CANTAB) • Delayed Matching to Sample (DMS; CANTAB) • Paired Associates Learning (PAL; CANTAB) • Matching Familiar Figures Test 20 (MFFT-20)
Language and verbal comprehension	• COWAT • Verbal, Category, and Letter Fluencies • Similarities (WAIS-R) • Vocabulary (WAIS-R) • Information (WAIS-R) • Comprehension (WAIS-R) • Token Test
Visuospatial/ perceptual processing	• Judgement of Line Orientation (JOLO) • Benton Visual Form Discrimination (VFD) • Block Design (WAIS-R) • Visuospatial Span Forwards and Backwards (WMS-R)
Brief mental status	• Mini-Mental State Exam (MMSE)
General intelligence	• Raven's Progressive Matrices • WAIS-R • National Adult Reading Test (NART) • Wechsler Test of Adult Reading (WTAR) • Test of Nonverbal Intelligence-3 (TONI-3)
Cognitive battery	• CANTAB • WMS-R • WAIS-R • Victoria Symptom Validity Test (VSVT) • California Computerised Assessment Package (CalCAP) • Central nervous system (CNS) Vital Signs • Massachusetts General Hospital Cognitive and Physical Functioning Questionnaire (CPFQ) • D-KEFS
Psychomotor performance	• Finger Tapping • Grooved Pegboard Test • Purdue Pegboard Test
Decision making and response control	• The Go/No-go Association Task (GNAT; CANTAB) • Information Sampling Task (IST; CANTAB) • Cambridge Gambling Task (CGT; CANTAB)
Induction	• Big/Little Circle (BLC; CANTAB)

priority. Taken together, the hazards posed by cognitive deficits on functional outcome in individuals with MDD provide the impetus for refining disease models and extrapolating from the results reported in individuals with schizophrenia and BD. A consensus-driven development of measurement tools that can effectively detect cognitive decline in this population and batteries assessing cognitive deficits for both descriptive and interventional research is warranted.

References

Gualtieri CT, Morgan DW. The frequency of cognitive impairment in patients with anxiety, depression, and bipolar disorder: an unaccounted source of variance in clinical trials. J Clin Psychiatry 2008 Jul;69(7):1122–30.

McIntyre RS, Cha DS, Soczynska JK, Woldeyohannes HO, Gallaugher LA, Kudlow P, et al. Cognitive deficits and functional outcomes in major depressive disorder: determinants, substrates, and treatment interventions. Depress Anxiety 2013 Mar;30(6):515–27.

Naismith SL, Longley WA, Scott EM, Hickie IB. Disability in major depression related to self-rated and objectively-measured cognitive deficits: a preliminary study. BMC Psychiatry 2007;7:32.

Chapter 3

Cognitive deficits in major depressive disorder

Key Points

- Individuals with MDD exhibit deficits across multiple cognitive domains, including attention, psychomotor speed, verbal learning and memory, as well as executive function.
- Deficits are evident early in the illness and persist during euthymic states.
- MDD is a risk factor for, and a prodrome to, mild cognitive impairment and dementing disorders.

3.1 Introduction

A compelling body of evidence indicates that disturbances in multiple domains of cognitive function, including learning and memory (verbal and nonverbal), attention, psychomotor speed, executive function, emotional processing and social cognition are core psychopathological features of MDD (See Table 3.2; Porter et al 2003). The domains of cognitive function affected in MDD are not specific to this diagnosis and the effect size across each domain are highly variable between studies. Separate lines of evidence also indicate that deficits in cognitive function can be identified as early as the first episode (Lee et al 2012), and in some cases may predate the onset of MDD (Mannie et al 2009). Cognitive deficits often persist in euthymic individuals and may worsen with age, number of depressive episodes, and illness duration (Bora et al 2012; Elgamal et al 2010; Hammar et al 2003; for definitions of different cognitive domains see Table 3.1, and see Figure 3.1 for the breakdown of the different components of executive function). For example, it has been documented that for every additional MDE, memory declines by 2 to 3% for the first several episodes (See Figure 3.5; Gorwood 2008).

3.2 Cognitive deficits in MDD

Disturbances in cognitive function are evident throughout adulthood and present early in the course of MDD. In a recent meta-analysis of individuals experiencing their first MDE, decreased performance was evident on measures of psychomotor speed, attention, visual learning, memory, and executive function (i.e. attentional switching, verbal fluency, cognitive flexibility), with effect sizes falling in the small to moderate range. Verbal learning and memory as well as working memory were unaffected,

Table 3.1 Definition of cognitive domains

Domain	Description
Attention	Ability to focus and sustain attention
Memory/learning	Episodic memory (relating to past or future events), verbal memory/learning, and visuo-spatial memory/learning
Executive functioning	Ability to monitor and regulate cognitive processes, employing attention, planning, working memory, mental flexibility, inhibition, task initiation and monitoring, multitasking, and decision making
Psychomotor speed	Speed at which the brain controls the body to perform

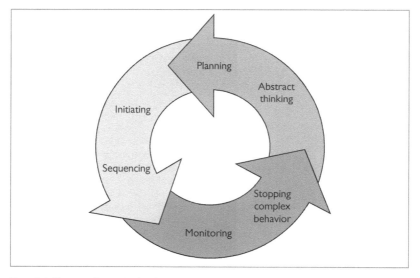

Figure 3.1 Elements of executive function

although there was a moderate to high heterogeneity of effect sizes between studies (Lee et al 2012). Executive function is a frequently reported abnormality in MDD with an average of 0.5 to 1.0 standard deviations below the population mean (Wagner et al 2012). Overgeneral autobiographical memory is also a phenomenon observed in some individuals with MDD (Sumner et al 2010). In addition, it has been reported that individuals with MDD who exhibit a more categorical/overgeneral memories have a greater susceptibility to more severe depressive symptomatology (See Figures 3.2–3.4 and Table 3.1 for review; Sumner et al 2010).

Study (Test)	Weight	Std. Mean difference IV, Random, 95% CI
Ilonen 2000 (2)	6.4%	0.66 [−0.16, 1.49]
Reischies 2000 (1)	12.4%	0.97 [0.61, 1.32]
Grant 2001 (1, 2)	12.2%	0.14 [−0.23, 0.51]
Neu 2005 (1)	9.3%	0.97 [0.40, 1.55]
Castaneda 2008a (1, 2)	12.2%	0.04 [−0.33, 0.41]
Preiss 2009 (1)	13.5%	0.54 [0.26, 0.83]
Reppermund 2009 (1)	8.8%	0.32 [−0.29, 0.93]
Kaymak 2010 (1)	7.2%	1.34 [0.59, 2.08]
van Wingen 2010 (curr) (1, 3)	9.0%	0.04 [−0.55, 0.63]
van Wingen 2010 (prev) (1, 3)	8.9%	0.05 [−0.55, 0.66]

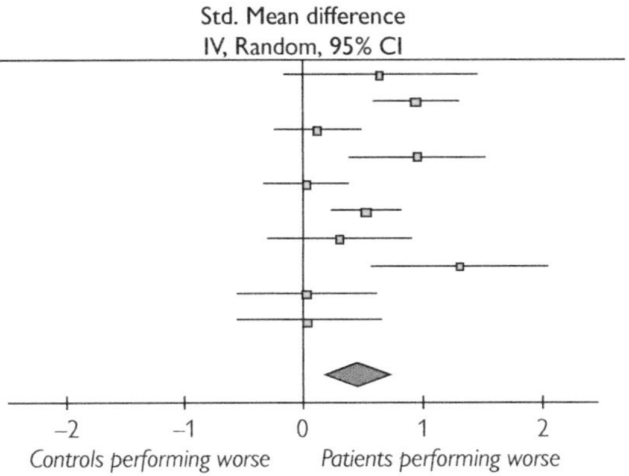

Heterogeneity: $Chi^2 = 28.37$, $df = 9$ ($P=0.0008$); $I^2 = 68\%$
Test for overall effect: $Z = 3.49$ ($P=0.0005$)

(1) Trail making test; (2) Digit symbol-coding; (3) Symbol digit modalities test

Figure 3.2 Psychomotor speed in first depressive episode

Reprinted from J Affect Disord, 140(2), Lee RSC, Lee R, Hermens D, Porter MA, Redoblado-Hodge MA. A meta-analysis of cognitive deficits in the first-episode Major Depressive Disorder, 113–24. Copyright (2012), with permission from Elsevier.

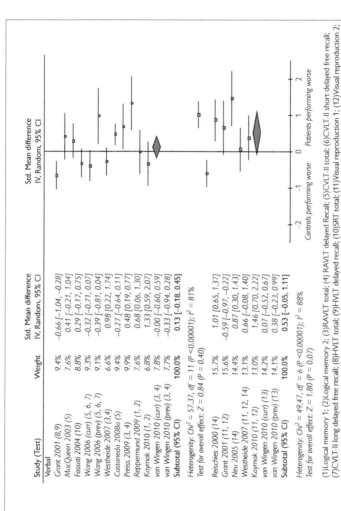

Study (Test)	Weight	Std. Mean difference IV, Random, 95% CI
Verbal		
Grant 2001 (8,9)	9.4%	−0.66 [−1.04, −0.28]
MacQueen 2003 (5)	7.6%	0.41 [−0.21, 1.04]
Fossati 2004 (10)	8.8%	0.29 [−0.17, 0.75]
Wang 2006 (curr) (5, 6, 7)	9.3%	−0.32 [−0.71, 0.07]
Wang 2006 (prev) (5, 6, 7)	9.1%	−0.39 [−0.81, 0.04]
Westheide 2007 (3,4)	6.6%	0.98 [0.22, 1.74]
Castaneda 2008a (5)	9.4%	−0.27 [−0.64, 0.11]
Preiss 2009 (3, 4)	9.9%	0.48 [0.19, 0.77]
Reppermund 2009 (1, 2)	7.6%	0.68 [0.06, 1.30]
Kaymak 2010 (1, 2)	6.8%	1.33 [0.59, 2.07]
van Wingen 2010 (curr) (3, 4)	7.8%	−0.00 [−0.60, 0.59]
van Wingen 2010 (prev) (3, 4)	7.7%	−0.33 [−0.94, 0.28]
Subtotal (95% CI)	**100.0%**	**0.13 [−0.18, 0.45]**

Heterogeneity: Chi^2 = 57.37, df = 11 (P <0.00001); I^2 = 81%
Test for overall effect: Z = 0.84 (P = 0.40)

Study (Test)	Weight	Std. Mean difference IV, Random, 95% CI
Reischies 2000 (14)	15.7%	1.01 [0.65, 1.37]
Grant 2001 (11, 12)	15.6%	−0.59 [−0.97, −0.22]
Neu 2005 (14)	14.4%	0.87 [0.30, 1.43]
Westheide 2007 (11, 12, 14)	13.1%	0.66 [−0.08, 1.40]
Kaymak 2010 (11, 12)	13.0%	1.46 [0.70, 2.22]
van Wingen 2010 (curr) (13)	14.2%	0.07 [−0.52, 0.67]
van Wingen 2010 (prev) (13)	14.1%	0.38 [−0.23, 0.99]
Subtotal (95% CI)	**100.0%**	**0.53 [−0.05, 1.11]**

Heterogeneity: Chi^2 = 49.47, df = 6 (P <0.00001); I^2 = 88%
Test for overall effect: Z = 1.80 (P = 0.07)

(1)Logical memory 1; (2)Logical memory 2; (3)RAVLT total; (4) RAVLT delayed Recall; (5)CVLT-II total; (6)CVLT-II short delayed free recall;
(7)CVLT-II long delayed free recall; (8)HVLT total; (9)HVLT delayed recall; (10)SRT total; (11)Visual reproduction 1; (12)Visual reproduction 2;
(13)RCFT delayed recall; (14) WMS visual memory index

Figure 3.3 Visual and verbal learning and memory in the first depressive episode
Reprinted from J Affect Disord, 140(2), Lee RSC, Lee R, Hermens D, Porter MA, Redoblado-Hodge MA. A meta-analysis of cognitive deficits in first-episode Major
Depressive Disorder, 113–24. Copyright (2012), with permission from Elsevier.

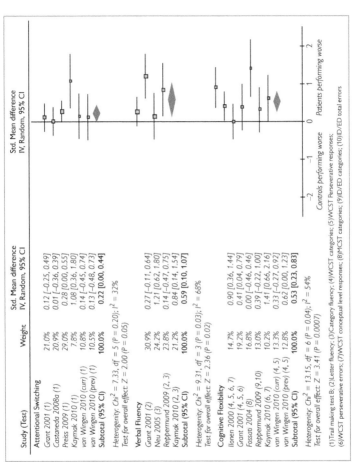

Study (Test)	Weight	Std. Mean difference IV, Random, 95% CI
Attentional Switching		
Grant 2001 (1)	21.0%	0.12 [−0.25, 0.49]
Castaneda 2008a (1)	20.9%	0.01 [−0.36, 0.39]
Preiss 2009 (1)	29.0%	0.28 [0.00, 0.55]
Kaymak 2010 (1)	7.8%	1.08 [0.36, 1.80]
van Wingen 2010 (curr) (1)	10.8%	0.14 [−0.45, 0.74]
van Wingen 2010 (prev) (1)	10.5%	0.13 [−0.48, 0.73]
Subtotal (95% CI)	**100.0%**	**0.22 [0.00, 0.44]**
Heterogenity: Chi² = 7.33, df = 5 (P = 0.20); I² = 32%		
Test for overall effect: Z = 2.00 (P = 0.05)		
Verbal Fluency		
Grant 2001 (2)	30.9%	0.27 [−0.11, 0.64]
Neu 2005 (3)	24.2%	1.21 [0.62, 1.80]
Reppermund 2009 (2, 3)	23.8%	0.14 [−0.47, 0.75]
Kaymak 2010 (2, 3)	21.2%	0.84 [0.14, 1.54]
Subtotal (95% CI)	**100.0%**	**0.59 [0.10, 1.07]**
Heterogenity: Chi² = 9.31, df = 3 (P = 0.03); I² = 68%		
Test for overall effect: Z = 2.36 (P = 0.02)		
Cognitive Flexibility		
Ilonen 2000 (4, 5, 6, 7)	14.7%	0.90 [0.36, 1.44]
Grant 2001 (4, 5, 6)	19.2%	0.41 [0.04, 0.79]
Fossati 2004 (8)	16.8%	0.00 [−0.46, 0.46]
Reppermund 2009 (9, 10)	13.0%	0.39 [−0.22, 1.00]
Kaymak 2010 (6, 7)	10.2%	1.41 [0.66, 2.16]
van Wingen 2010 (curr) (4, 5)	13.3%	0.33 [−0.27, 0.92]
van Wingen 2010 (prev) (4, 5)	12.8%	0.62 [0.00, 1.23]
Subtotal (95% CI)	**100.0%**	**0.53 [0.23, 0.83]**
Heterogenity: Chi² = 13.15, df = 6 (P = 0.04); I² = 54%		
Test for overall effect: Z = 3.41 (P = 0.0007)		

(1)Trail making test B; (2)Letter fluency; (3)Category fluency; (4)WCST categories; (5)WCST categories; (6)WCST perseverative errors; (7)WCST conceptual level responses; (8)MCST categories; (9)ID/ED categories; (10)ID/ED total errors

Controls performing worse   Patients performing worse

Figure 3.4 Executive function in the first depressive episode

Reprinted from J Affect Disord, 140(2), Lee RSC, Lee R, Hermens D, Porter MA, Redoblado-Hodge MA. A meta-analysis of cognitive deficits in first-episode Major Depressive Disorder, 113–24. Copyright (2012), with permission from Elsevier.

Table 3.2 Cognitive deficits reported in MDD[a]

Author	Experimental Subjects (n)	Control subjects (n)	Cognitive measures	Cognitive deficits reported
Austin et al (1992)	60	20	NART, RAVLT, DSST, Block Design, Digit Span F & B, TMT A & B, verbal fluency	• attention & processing speed
Smith et al (1994)	36	26	RAVLT, verbal fluency, signal detection	• verbal learning & memory • language & verbal comprehension
Ilsley et al (1995)	15	15	MMSE, NART Revised, Digit Span, RBMT, verbal fluency, implicit learning and subsequent recall, DSST, paper-and-pencil coding task	• attention & processing speed • verbal learning & memory
Brebion et al (1997)	26	26	VRM	• verbal learning & memory
Purcell et al (1997)	20	20	SSP, SWM, TOL, IED, DMS, SRM, PRM	• executive function
Del'Innocenti et al (1998)	17	17	WCST, SCWT, verbal fluency	• executive function • language & verbal comprehension
Austin et al (1999)	77	28	MMSE, NART, Digit Span F & B, CRT, SRT, TMT A & B, SCWT, verbal fluency, WCST, RAVLT, Visual Reproduction, DSST	• executive function • attention & processing speed
Grant et al (2001)	123	36	TMT A & B, Digit Span, CPT, HVLT-R, Visual Reproduction, Categories Test, COWAT, WCST, BLC, RTI, DMS, SOC, SWM, IED, SRM, PRM, PAL, SSP	• executive function
Landro et al (2001)	22	30	finger tapping, RTI, TMT A & B, DSST, PASAT, Digit Span, Randt Memory Test, RFT, COWAT, Block Design	• selective attention • verbal learning & memory
Fossati et al (2002)	49	70	VRM, WCST	• executive function • verbal learning & memory • psychomotor performance
Ravnkilde et al (2002)	40	49	WAIS-R, SSST, SCWT, DSST, TMT A & B, verbal fluency, token test, Brown–Peterson test, Visual Reproduction, Logical Memory, LVLT, WCST	• executive function • language & verbal comprehension • verbal learning & memory

Study			Tests	Cognitive domains
Watkins et al (2002)	14	14	random number generation task: performed after both a rumination induction and after a distraction induction, with order of inductions counterbalanced within each group	• executive function • emotional processing
Hammar et al (2003)	21	21	visual search paradigm with two testing sessions: at inclusion and at six months	• attention & processing speed
Naismith et al (2003)	55	22	TMT B, RAVLT, Raven's Matrices, Vocabulary, Block Design, VRT, category fluency, letter fluency, WCST, SRT, CRT, TOL	• attention & processing speed
Porter et al (2003)	44	44	DSST, RAVLT, PAL, PRM, SRM, DMS, SWM, TOL, COWAT, verbal fluency, CPT	• executive function • attention & processing speed • visuospatial/perceptual processing
Harvey et al (2004)	22	22	n-back test, Digit Span F & B, TMT A & B, WCST, SCWT, verbal fluency	• working memory
Sordal et al (2004)	45	50	COWAT, TOL, PASAT, Digit Span B, SCWT, WCST, VSVT, CalCAP	• language & verbal comprehension • executive function • attention & processing speed • psychomotor performance • visuospatial/perceptual processing
Langenecker (2005)	21	20	VFD, category fluency, RTI, visual processing cost, Purdue Pegboard, GNAT	• executive function • emotional processing
Rose et al (2006)	20	20	NART, two subtests of the Test of Everyday Attention (i.e. Elevator counting with distraction, Visual elevator with n-back task)	• attention & processing speed
Mondal et al (2007)	30	30	Cancellation test, TMT A & B, RFFT, Digit Span F & B	• attention & processing speed • verbal learning & memory • executive function

(Continued)

Table 3.2 (Continued)

Author	Experimental Subjects (n)	Control subjects (n)	Cognitive measures	Cognitive deficits reported
Sarosi et al (2007)	71	30	TMT A & B, SCWT, RAVLT, ROCF	• attention & processing speed • executive function • working memory • psychomotor performance
Sarosi et al (2008)	96	52	RAVLT, ROCF, TMT A & B, SCWT	• verbal learning & memory • executive function
Thomas et al (2009)	75	82	COWAT, CPT, RAVLT, CANTAB: SWM, SRM, PRM	• verbal learning & memory • psychomotor performance
Bhardwaj et al (2010)	20	20	WCST, Digit Span, MMSE, Vocabulary	• executive function
Hammar et al (2010)	19	19	SCWT	• attention & processing Speed • executive function
Zobel et al (2010)	30	30	Digit Span, Logic Memory, phasic alertness task, GNAT	• attention & processing speed • working memory • decision making & response control
Oral et al (2012)	39	40	TMT, SCWT, WCST, Test of Variables of Attention, Auditory consonant trigram test, Digit Span, RAVLT, COWAT	• executive function • attention & processing speed • verbal learning & memory

Abbreviations: BLC, Big/Little Circle; CalCAP, California Computerised Assessment Package; CANTAB, Cambridge Neuropsychological Test Automated Battery; COWAT, Controlled Oral Word Association Test; CPT, Continuous Performance Task; CRT, Choice Reaction Time; Digit Span B, Digit Span Backwards; Digit Span F & B, Digit Span Forwards and Backwards; DMS, Delayed Matching to Sample; DSST, Digit Symbol Substitution Test; GNAT, Go/No go Association Task; HVLT R, Hopkins Verbal Learning Test-Revised; IED, Intra Extra Dimensional Set Shift; LVLT, Luria's verbal learning test; MMSE, mini-mental state exam; NART, National Adult Reading Test; PAL, Paired Associates Learning; PASAT, Paced Auditory Serial Addition Test; PRM, Pattern Recognition Memory; RAVLT, Rey Auditory Verbal Learning Test; RBMT, Rivermead Behavioural Memory Test; ROCF, Rey Osternieth Complex Figure Test; RTI, Reaction Time; SCWT, Stroop Colour-Word Interference Test; SOC, Stockings of Cambridge; SRM, Spatial Recognition Memory; SRT, Simple Reaction Time; SSP, Spatial Span; SSST, Serial sevens subtraction test, SWM, Spatial Working Memory; TMT A & B, Trail Making Test Part A and Part B; TMT B, Trail Making Test Part B; TOL, Tower of London; VFD, Benton Visual Form Discrimination Test; VRM, Verbal Recognition Memory Test; VSVT, Victoria Symptom Validity Test; WAIS-R, Wechsler Adult Intelligence Test-Revised; WCST, Wisconsin Card Sorting Test.

ᵃ Studies listed in chronological order.

3.3 Cognitive deficits during euthymia

Persistence of cognitive deficits during states of euthymia is a replicated finding. Deficits have been observed in global cognition, psychomotor speed, working memory, verbal learning and memory, and semantic fluency. Less affected are phonemic fluency, perseveration, attention and list recognition. Moreover, euthymic individuals with late-onset depression have been reported to exhibit more severe deficits than individuals with early-onset MDD (Bora et al 2012). Although results are variable between studies, worse performance in at least one domain of sustained and selective attention, memory and executive function, or global cognitive function was identified in nine out of eleven studies included in a systematic review of remitted individuals with MDD (Table 3.3; Hasselbalch et al 2011).

3.4 MDD and dementia

In addition to being highly associated with clinically significant cognitive impairment, MDD is also a risk factor and prodrome for mild cognitive impairment (MCI) as well as dementing disorders (e.g. Alzheimer's disease (AD); Enache et al 2011). For example, in a retrospective cohort study of 280,000 veterans aged 55 years or older, individuals diagnosed with MDD or dysthymia were twice as likely to develop incident dementia when compared to those without MDD or dysthymia (Byers et al 2012).

3.5 Objectively vs subjectively measured deficits

A frequent finding in MDD is that objectively measured cognitive performance does not correlate well with subjective reporting of cognitive complaints. Subjective deficits represent more subtle deviations and may not be easily captured by standard neuropsychological tests, however, such deficits are clinically relevant and may be a warning sign for early intervention. The foregoing observation of subjective cognitive dysfunction predating measurable objective deficits has been reported in older populations with MCI. The converse phenomenon has also been observed, wherein patients reporting normal subjective cognitive function evince objective cognitive deficits (Cappa 2012).

Table 3.3 Subjective cognitive deficits in MDD		
Number of domains	**Below 2nd percentile**	
	MDD (%)	**Healthy subjects (%)**
0	63	93
1 or more	38	7
2 or more	22	2
3 or more	14	1
4 or more	8	0

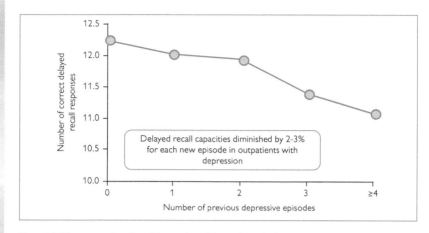

Figure 3.5 Memory as a function of the number of depressive episodes

Adapted from Gorwood P, Corruble E, Falissard B, Goodwin GM. Toxic effects of depression on brain function: impairment of delayed recall and the cumulative length of depressive disorder in a large sample of depressed outpatients. Am J Psychiatry 2008 Jun;165(6):731–9.

Future research will need to more clearly delineate MDD subgroups that are more likely to manifest cognitive deficits at an earlier age and/or be susceptible to progressive cognitive decline.

References

Bora E, Harrison B, Yücel M, Pantelis C. Cognitive impairment in euthymic major depressive disorder: a meta-analysis. Psychol Med. 2012 Oct 26;1–10. [Epub ahead of print]

Byers AL, Covinsky KE, Barnes DE, Yaffe K. Dysthymia and depression increase risk of dementia and mortality among older veterans. Am J Geriatr Psychiatry 2012 Aug;20(8):664–72.

Cappa SF. Subjective cognitive complaints: not to be dismissed. Eur J Neurol. 2012 May;19(5):665.

Elgamal S, Denburg S, Marriott M, MacQueen G. Clinical factors that predict cognitive function in patients with major depression. Can J Psychiatry 2010 Oct;55(10):653–61.

Enache D, Winblad B, Aarsland D. Depression in dementia: epidemiology, mechanisms, and treatment. Curr Opin. Psychiatry 2011 Nov;24(6):461–72.

Gorwood P, Corruble E, Falissard B, Goodwin GM. Toxic effects of depression on brain function: impairment of delayed recall and the cumulative length of depressive disorder in a large sample of depressed outpatients. Am J Psychiatry 2008 Jun;165(6):731–9.

Hammar A, Lund A, Hugdahl K. Long-lasting cognitive impairment in unipolar major depression: a 6-month follow-up study. Psychiatry Res 2003 May;118(2):189–96.

Hasselbalch BJ, Knorr U, Kessing LV. Cognitive impairment in the remitted state of unipolar depressive disorder: A systematic review. J Affect Disord 2011 Nov;134(1-3):20–31.

Lee RSC, Lee R, Hermens D, Porter MA, Redoblado-Hodge MA. A meta-analysis of cognitive deficits in first-episode Major Depressive Disorder. J Affect Disord 2012 Oct;140(2):113–24.

Mannie ZN, Barnes J, Bristow GC, Harmer CJ, Cowen PJ. Memory impairment in young women at increased risk of depression: influence of cortisol and 5-HTT genotype. Psychol Med 2009 May;39(05):757–62.

Porter RJ, Gallagher P, Thompson JM, Young AH. Neurocognitive impairment in drug-free patients with major depressive disorder. Br J Psychiatry 2003 Mar;182:214–20.

Sumner JA, Griffith JW, Mineka S. Overgeneral autobiographical memory as a predictor of the course of depression: a meta-analysis. Behav Res Ther 2010 Jul;48(7):614–25.

Wagner S, Doering B, Helmreich I, Lieb K, Tadić A. A meta-analysis of executive dysfunctions in unipolar major depressive disorder without psychotic symptoms and their changes during antidepressant treatment. Acta psychiatrica Scandinavica 2012 Apr;125(4):281–92.

Chapter 4

'Hot' versus 'cold' cognition

Key Points
- Individuals with MDD display abnormal patterns of emotional processing.
- Impairments at the cognitive-affective interface are a core feature of MDD.
- Negative bias represents a susceptibility factor for new onset and recurrent affective episodes.

21

4.1 Introduction

Major depressive disorder is characterized by altered emotional and cognitive function, commonly described as 'affective cognition' (Elliott et al 2011). Impairments at the cognitive-affective interface are increasingly recognized as a core feature of MDD. The cognitive-affective interface can be broadly described as either "cold" cognition (emotion-independent) or "hot" cognition (emotionally valenced). Executive cognitive processes (i.e. 'cold' cognition) facilitate the coordination and monitoring of thought(s) and action(s) to complete various complex tasks, some of which are influenced by emotional stimuli and/or the emotionality of the individual (i.e. 'hot' cognition) (Diener et al 2012; Murphy et al 2012; Roiser and Sahakian 2013; Foland-Ross and Gotlib 2012).

4.2 Affective cognition

Affective cognition describes the interaction between 'cold' (e.g. executive functions including problem solving, planning, and inhibition) and 'hot' (e.g. emotional processing in the perception, recognition, and modulation of emotionally valenced stimuli) cognitive processes (Figure 4.1; Diener et al 2012; Murphy et al 2012). Models of cognitive-affective interactions suggest that individuals with MDD display heightened 'bottom-up' responses (e.g. neurobiological processes) to emotional stimuli and diminished 'top-down' cognitive control (e.g. inhibiting prepotent responses, maintaining affective goals, and recruiting further resources) in response to emotionally valenced experiences (Roiser, JP and Sahakian 2013; Lu et al 2012). These foregoing models are instantiated by reports describing altered valence-specific patterns of

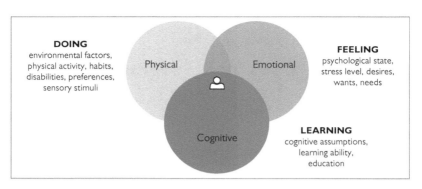

Figure 4.1 Domains affecting mental health
"Personal-Behavioral Context: The New User Persona." © Daniel Eizans, 2010. Modified from a diagram by Andrew Hinton.

response among individuals with MDD (Hu et al 2012; Vanderhasselt 2012; Demeyer et al 2012; Everaert et al 2012).

Healthy individuals differ from those with MDD in reactivity to emotionally valenced task-irrelevant stimuli. For example, although both positively and negatively valenced emotional stimuli rapidly capture attentional resources and influence executive functions among healthy individuals, positively valenced stimuli primarily engage healthy individuals' attention, exhibiting a 'positive bias' when compared to individuals with MDD (Everaert et al 2012). Abnormal emotional processes have also been reported to contribute to and/or moderate biases in causal attribution of events to internal versus external factors; internal attribution of positive events and external attribution of negative events represent a 'self-serving' bias (posited to serve a psychologically protective function) observed in healthy individuals but not in those with MDD (Everaert et al 2012; Seidel et al 2012).

4.3 Negative bias in MDD

Individuals with MDD exhibit deficits in positive emotional processing and preferentially process mood-congruent information commonly referred to as 'negative bias', wherein negative emotional stimuli act as potent distractors and impede executive cognitive processes, contributing to poor task performance (Hu et al 2008). For example, in a recent study evaluating attention and memory abnormalities among currently depressed and currently remitted individuals with MDD, results indicated that both groups displayed selective attention for negative information. Moreover, positive memory bias, which is characteristic of individuals with no current or past history of depression, was not observed (Gupta and Kar 2012). According to cognitive models of MDD (Figure 4.2), negative biases affect most cognitive processes, including perception (Sterzer et al 2011). Moreover, perceptual biases towards mood-congruent information have been reported to reinforce depressed mood (Demeyer et al 2012; Strand et al 2012; Sterzer et al 2011). Taken together,

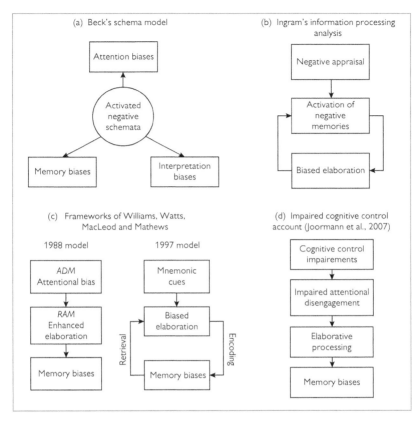

Figure 4.2 Schematics of cognitive bias according to cognitive models of MDD
Reproduced with permission from J. Everaert et al Clinical Psychology Review 32 (2012).

individuals with MDD demonstrate differential patterns of emotional processing, monitoring, and control when compared to unaffected individuals. The documented abnormal emotional processing may be moderated by 'cold' cognitive deficits observed in individuals with MDD; however, the mechanisms underlying the neural substrates, circuits, and networks involved, subserving emotional processing, require further refinement (McIntyre et al 2013).

Neuroimaging studies of emotion and cognition in MDD have instantiated differential brain structure, activity, and connectivity in cognitive-emotional information processing across broadly overlapping neural networks involved in both 'hot' and 'cold' cognitive processes (Diener et al 2012). Structural imaging studies in MDD consistently report differences in amygdalar and hippocampal brain volumes, suggesting that morphological changes may be relevant to negative bias; however, the causal and temporal relationship between these structural abnormalities and/or changes remains undetermined (Gerritsen et al 2011; Holmes et al 2012). Increased emotional distractibility and impaired concentration on task-relevant information commonly observed among individuals with MDD are reported to be a consequence of altered interactions between dorsal brain systems involved in 'cold' cognitive processes (e.g. prefrontal cortices) and ventral systems involved in 'hot' cognition (e.g. limbic structures; Roiser et al 2013). For example, several neuroimaging studies report that individuals with MDD exhibit prefrontal 'inefficiency' (e.g. hyperactivity) during difficult working memory tasks (Roiser et al 2013). Emerging evidence also suggests that dorsal and ventral prefrontal brain regions demonstrate differential responses to neurotransmitters, affecting separate cognitive and emotional processes (Arnsten and Rubia 2012).

Notwithstanding advances in psychological and pharmacological interventions, relapse and recurrence rates have been estimated to exceed 50%, with persistent inter-episodic deficits reported for both cognitive and emotional processes in MDD (Vanderhasselt 2012; Arnsten and Rubia 2012). Consequently, treatment strategies targeting depressive symptomatology alone are insufficient in the resolution of underlying cognitive and emotional processing deficits, underscoring the importance of addressing 'hot' and 'cold' cognition and implicit processes for the successful treatment of MDD (Gupta and Kar 2012; Phillips et al 2012). For example, unaffected first-degree relatives of individuals with MDD display similar alterations in the inhibition of negative emotions when compared to healthy controls (Vanderhasselt et al 2012). Moreover, individuals with remitted MDD have been reported to exhibit a persistent and enhanced susceptibility to emotional distraction by negative emotional information, which further impedes and decision making (Kerestes 2012). Hence, the presence of abnormal emotional regulation in individuals with MDD, independent of current symptomatology, severity, and other clinical characteristics, suggests that it may be a trait marker of MDD (Kanske et al 2012).

Individual differences in emotion regulation are reported to be associated with the modulation of amygdalar activity and amygdala-prefrontal cortex functional connectivity (i.e. the temporal correlation of activity between disparate, but functionally related, brain regions as part of a neural network; Whitfield-Gabrieli and Ford 2012). More specifically, greater amygdala attenuation and inverse connectivity between the amygdala and prefrontal brain regions predicted successful down-regulation of negative emotion, suggesting that individuals with greater regulatory ability are better able to engage dorsal-ventral neural circuits during tasks involving emotional regulation (Lee et al 2012). Replicated evidence suggests that increased hippocampal-amygdalar connectivity is associated with increased negative memory bias, which is mediated by the amygdala-hippocampal interaction and modulated by the emotional salience of contexts conveyed via the amygdala (Gerritsen et al 2011; Lee et al 2012). Likewise, the hippocampus may continue to cultivate negative emotional processing bias by retaining information that was conveyed as significant (e.g. negative stimuli, negative perceptions, and negative expectations), subsequently driving the amygdala to differentially respond to negative (versus positive) stimuli within the context of MDD (Gerritsen et al 2011; Lee et al 2012; Figure 4.3).

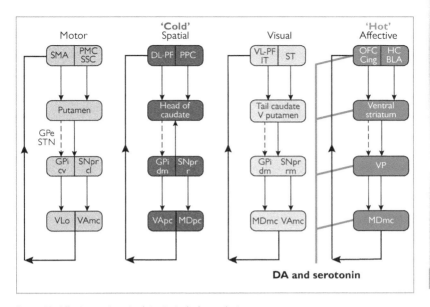

Figure 4.3 Affective corticostriatal circuits in the human brain

DA, dopamine. Republished with permission of Annual Reviews, from Alexander GE et al. Annu Rev Neurosci 1986;9:357–81; permission conveyed through Copyright Clearance Center, Inc.

4.4 Attributional style & locus of control

The internal attribution of negative events and external attribution of positive events is a causal attribution pattern commonly employed by individuals with MDD (Seidel et al 2012). In contrast, healthy individuals demonstrate a 'self-serving' bias wherein they attribute positive events to internal factors (e.g. a goal was achieved based on the individual's ability) and negative events to external causes (e.g. a goal was not achieved due to circumstance; Seidel et al 2012; Lee et al 2012). The latter causal attribution pattern often does not accurately reflect external objective evidence; however, it is posited to serve a psychologically protective function (Seidel et al 2012). For example, individuals with MDD who report blaming others rather than oneself tend to experience decreased negative emotions when compared to those who report self-blame (Green et al 2013). However, in general, depressed individuals foster associations between their performance on various tasks and negative self-related schemas that commonly lead to frank non-self-serving tendencies, posited to be a critical contributing factor to ruminative responses (Demeyer et al 2012; Seidel et al 2012).

4.5 Rumination

Rumination is frequently described as a response strategy of recurrent, self-reflective, and focused attention on global negative affect and the potential causes and/or consequences of depressive symptoms associated with the inability to inhibit and/or disengage from negative cognitions (Roiser and Sahakian 2013; Demeyer et al 2012). Recent studies examining the neural aspects of cognition in MDD suggest that negative biases are not primarily a consequence of attentional bias towards negatively valenced stimuli. A rival hypothesis is that negative biases are a consequence of the inability to disengage from negative stimuli characterized by automatic biases in attention towards negatively valenced stimuli (Foland-Ross and Gotlib 2012; Demeyer et al 2012). For example, individuals with MDD were reported to demonstrate unimpaired cognitive flexibility in the shifting of attention between different classes of neutral stimuli in a go/no-go task; these results were in contrast to the shifting of attention required between emotional categories on the affective go/no-go task, which resulted in slower response times between negatively valenced to positively valenced stimuli (Murphy et al 2012). Furthermore, abnormalities in emotional processing and the effective redirection of attention towards or away from emotionally salient information are reported to be proportional to illness severity and duration (Kerestes et al 2012; Carballedo 2011). Preliminary evidence evaluating neural correlates of negative bias suggest that disturbances in frontolimbic coupling exist during attributional decisions among individuals with MDD (Seidel et al 2012). Taken together, negative biases and depressive attributional style in MDD have been proposed to subserve processes underlying rumination (Foland-Ross and Gotlib 2012; Demeyer et al 2012).

In keeping with this view, autobiographical and personally relevant negative stimuli have been reported to generate greater interference during cognitive tasks when compared to generally negative stimuli, suggesting that a depressive attributional style further diminishes cognitive capacity (Foland-Ross and Gotlib 2012; Demeyer et al 2012). Moreover, mediational analyses suggest that rumination plays a crucial role in the prediction of future depressive symptoms among remitted individuals with MDD (Demeyer et al 2012). It is hypothesized that deficits in cognitive processes associated with emotion regulation (e.g. inhibition) are predictive of future MDEs due to difficulties in disengaging from negative thoughts, leading to depressive rumination and depressive symptom intensification (Demeyer et al 2012). Rumination utilizes cognitive resources (e.g. working memory) and interferes with effective cognitive strategies such as selection, sequencing and monitoring of internal and external incoming information, thereby negatively impacting daily functioning, and increasing risk of depression recurrence. (Joormann et al 2011; Schaefer et al 2003).

Healthy individuals have been reported to exhibit retrieval-induced forgetting mechanisms that are associated with distressing aspects of a negative event (Elliott et al 2002). Moreover, evidence for diminished cognitive resources, affecting task performance, decision making, and problem solving, in association with repetitive negative cognitions and disturbed working memory function (e.g. inhibition) among individuals with MDD, has been reported (Foland-Ross and Gotlib 2012; Demeyer et al 2012). Increasing cognitive load compromises the active removal of irrelevant negative information. Consequent abnormal bottom-up activation patterns of associative negative networks contribute to, an individual's vulnerability to, or engagement in, depressive rumination (Foland-Ross and Gotlib 2012; Phillips et al 2010).

Replicated evidence demonstrates that the inverse coupling between the frontal and limbic brain regions is associated with top-down cognitive control over reappraisals on systems responsible for inducing emotional responses; hence, aberrances in this dynamic interaction between cognitive control and emotional processing perform a crucial role in ruminative thinking (Foland-Ross and Gotlib 2012; Demeyer et al 2012). Notably, neuroimaging studies have demonstrated that cognitive reappraisal (i.e. reframing the meaning of a stimulus or event) enhances the coupling of frontal and limbic brain regions (e.g. ventro- and dorsolateral prefrontal cortex with the amygdala) to decrease emotional responses, indicating that explicit top-down cognitive processes contribute to the regulation of emotional neural systems (Foland-Ross and Gotlib 2012). Notwithstanding the bourgeoning interest in the relationship between cognitive and affective systems in MDD, there is a paucity of evidence evaluating the nonautomatic cognitive response to regulating affect (Kanske et al 2012).

Over the past decade, research has investigated whether differences in reasoning about one's experiences may lead to rumination in MDD (Kross et al 2012). In contrast to healthy individuals, who analyse negative experiences to resolve negative affect, individuals with MDD demonstrate a proclivity to ruminate and exacerbate their initial negative affective response (Kross et al 2012). The differences observed between healthy and depressed individuals following negative experiences and their analyses have been purported to result from differences in perspective-taking (Kross et al 2012). Namely, healthy individuals tend to assess negative experiences with a 'self-distanced' perspective, wherein the self that is analysing the experience is psychologically removed from the self that experienced the event using a metacognitive strategy (Kross et al 2012). Conversely, individuals with MDD tend to assess negative experiences with a 'self-immersed' perspective, wherein the self that is analysing the experience and the self that experienced the event are one and the same (Kross et al 2012). Taken together, analysing negative experiences from a self-immersed perspective represents a maladaptive cognitive strategy that predisposes the individual to emotionally arousing features of his/her experience, potentially allowing the emotional experiences to shift from a reconstructive purpose to solely recounting the experience. Using a self-distanced cognitive strategy when assessing negative experiences may mitigate some of the harmful effects associated with rumination by reducing negative thoughts over time (Kross et al 2012).

References

Arnsten AF, Rubia K. Neurobiological circuits regulating attention, cognitive control, motivation, and emotion: disruptions in neurodevelopmental psychiatric disorders. J Am Acad Child Adolesc Psychiatry 2012 Apr;51(4):356–67.

Carballedo A, Scheuerecker J, Meisenzahl E, Schoepf V, Bokde A, Moller HJ, et al. Functional connectivity of emotional processing in depression. J Affect Disord 2011 Nov;134(1–3):272–9.

Demeyer I, De Lissnyder E, Koster EH, De Raedt R. Rumination mediates the relationship between impaired cognitive control for emotional information and depressive symptoms: A prospective study in remitted depressed adults. Behav Res Ther 2012 May;50(5):292–7.

Diener C, Kuehner C, Brusniak W, Ubl B, Wessa M, Flor H. A meta-analysis of neurofunctional imaging studies of emotion and cognition in major depression. Neuroimage 2012 Jul;61(3):677–85.

Elliott R, Rubinsztein JS, Sahakian BJ, Dolan RJ. The neural basis of mood-congruent processing biases in depression. Arch Gen Psychiatry 2002 Jul;59(7):597–604.

Elliott R, Zahn R, Deakin JF, Anderson IM. Affective cognition and its disruption in mood disorders. Neuropsychopharmacology 2011 Jan;36(1):153–82.

Everaert J, Koster EH, Derakshan N. The combined cognitive bias hypothesis in depression. Clin Psychol Rev 2012 Jul;32(5):413–24.

Foland-Ross LC, Gotlib IH. Cognitive and neural aspects of information processing in major depressive disorder: an integrative perspective. Front Psychol 2012;3:489.

Gerritsen L, Rijpkema M, van Oostrom I, Buitelaar J, Franke B, Fernandez G, et al. Amygdala to hippocampal volume ratio is associated with negative memory bias in healthy subjects. Psychol Med 2011 Jul 11;1–9.

Green S, Moll J, Deakin JF, Hulleman J, Zahn R. Proneness to decreased negative emotions in major depressive disorder when blaming others rather than oneself. Psychopathology 2013;46(1):34–44.

Gupta R, Kar BR. Attention and memory biases as stable abnormalities among currently depressed and currently remitted individuals with unipolar depression. Front Psychiatry 2012;3:99.

Holmes AJ, Lee PH, Hollinshead MO, Bakst L, Roffman JL, Smoller JW, et al. Individual differences in amygdala-medial prefrontal anatomy link negative affect, impaired social functioning, and polygenic depression risk. J Neurosci 2012 Dec;32(50):18087–100.

Hu Z, Liu H, Weng X, Northoff G. Is there a valence-specific pattern in emotional conflict in major depressive disorder? An exploratory psychological study. PLoS One 2012;7(2):e31983.

Joormann J, Levens SM, Gotlib IH. Sticky thoughts: depression and rumination are associated with difficulties manipulating emotional material in working memory. Psychol Sci 2011 Aug;22(8):979–83.

Joormann J, Talbot L, Golib, IH (2007). Biased processing of emotional information in girls at risk for depression. J Abnorm Psychol 2007 Feb;116(1):135–43.

Kanske P, Heissler J, Schonfelder S, Wessa M. Neural correlates of emotion regulation deficits in remitted depression: the influence of regulation strategy, habitual regulation use, and emotional valence. Neuroimage 2012 Jul;61(3):686–93.

Kerestes R, Ladouceur CD, Meda S, Nathan PJ, Blumberg HP, Maloney K, et al. Abnormal prefrontal activity subserving attentional control of emotion in remitted depressed patients during a working memory task with emotional distracters. Psychol Med 2012 Jan;42(1):29–40.

Kross E, Gard D, Deldin P, Clifton J, Ayduk O. "Asking why" from a distance: its cognitive and emotional consequences for people with major depressive disorder. J Abnorm Psychol 2012 Aug;121(3):559–69.

Lee H, Heller AS, van Reekum CM, Nelson B, Davidson RJ. Amygdala-prefrontal coupling underlies individual differences in emotion regulation. Neuroimage 2012 Sep;62(3):1575–81.

Lu Q, Li H, Luo G, Wang Y, Tang H, Han L, et al. Impaired prefrontal-amygdala effective connectivity is responsible for the dysfunction of emotion process in major depressive disorder: a dynamic causal modeling study on MEG. Neurosci Lett 2012 Aug;523(2):125–30.

McIntyre RS, Cha DS, Soczynska JK, Woldeyohannes HO, Gallaugher LA, Kudlow P, et al. Cognitive deficitys and functional outcomes in major depressive disorder: determinants, substrates, and treatment interventions. Depress Anxiety 2013 Jun;30(6):515–27.

Murphy FC, Michael A, Sahakian BJ. Emotion modulates cognitive flexibility in patients with major depression. Psychol Med 2012 Jul;42(7):1373–82.

Phillips WJ, Hine DW, Thorsteinsson EB. Implicit cognition and depression: a meta-analysis. Clin Psychol Rev 2010 Aug;30(6):691–709.

Roiser JP, Sahakian BJ. Hot and cold cognition in depression. CNS Spectr 2013 Jun;18(3):139–49.

Schaefer A, Collette F, Philippot P, van der Linden M, Laureys S, Delfiore G, et al. Neural correlates of "hot" and "cold" emotional processing: a multilevel approach to the functional anatomy of emotion. Neuroimage 2003 Apr;18(4):938–49.

Seidel EM, Satterthwaite TD, Eickhoff SB, Schneider F, Gur RC, Wolf DH, et al. Neural correlates of depressive realism—an fMRI study on causal attribution in depression. J Affect Disord 2012 May;138(3):268–76.

Sterzer P, Hilgenfeldt T, Freudenberg P, Bermpohl F, Adli M. Access of emotional information to visual awareness in patients with major depressive disorder. Psychol Med 2011 Aug;41(8):1615–24.

Strand M, Saetrevik B, Lund A, Hammar A. The relationship between residual symptoms of depression and emotional information processing. Nord J Psychiatry 2012 Aug 21. [Epub ahead of print]

Vanderhasselt MA, De Raedt R, Dillon DG, Dutra SJ, Brooks N, Pizzagalli DA. Decreased cognitive control in response to negative information in patients with remitted depression: an event-related potential study. J Psychiatry Neurosci 2012 Jul;37(4):250–8.

Whitfield-Gabrieli S, Ford JM. Default mode network activity and connectivity in psychopathology. Annu Rev Clin Psychol 2012;8:49–76.

Chapter 5

Differentiating cognitive dysfunction from other dimensions of psychopathology in major depressive disorder

Key Points

- Cognitive dysfunction is a criterion for the diagnosis of a major depressive episode.
- Cognitive dysfunction in MDD may be moderated by other depressive symptom domains.
- The association between MDD and cognitive dysfunction persists after adjusting for other depressive domains.

5.1 Introduction

Major depressive disorder is associated with clinically significant, and not infrequent, progressive cognitive deficits (McIntyre et al 2013). The hazards posed by cognitive deficits among individuals with MDD provide the impetus for identifying clinical and neurobiological determinants of cognitive dysfunction in this population. Specifically, identifying the mediational and/or moderational effect of cognitive deficits on illness susceptibility, recurrence, chronicity, treatment outcome and functional decline is warranted. Disparate cognitive deficits have been reported in MDD across multiple cognitive domains with no pathognomonic or specific cognitive signature (McIntyre et al 2013; Vanderhasselt et al 2012; Sarapas et al 2012). For many individuals with MDD, deficits in cognitive function are an epiphenomenon of other psychopathology (e.g. fatigue, low motivation). In addition to this subgroup, many individuals with MDD exhibit 'cold' cognitive deficits as a core phenomenon of their illness. It is further observed that emotionally valenced cognitive dysfunction (i.e. 'hot' cognition) affects the majority of individuals with MDD.

5.2 Trait versus state effects of MDD on cognition

Currently, two non-mutually exclusive hypotheses have been proposed in an effort to delineate the relationship between cognitive deficits and affective dysregulation in

MDD: (1) the Trait Hypothesis posits that cognitive deficits may represent an endo-phenotype among subpopulations of depressed individuals who commonly experi-ence more severe, recurrent, and/or chronic depressive episodes with persisting inter-episodic cognitive deficits; and (2) the State Hypothesis posits that there is a causal relationship between depression symptom severity and worsening cognitive performance (Sarapas et al 2012).

The conventional assumption has been that cognitive impairments in MDD are a con-sequence or a residual feature of an acute MDE. However, some individuals with MDD have reported experiencing cognitive impairments and reduced psychosocial function-ing prior to their first MDE (McIntyre et al 2013). The presence and persistence of cog-nitive impairment in individuals with MDD prior to and following a MDE suggest that cognitive impairment can be conceptualized as a fundamental feature of MDD rather than a consequence of depressive symptoms and severity (Halvorsen et al 2012). For example, a recent study demonstrated that UHR-MDD first-degree unaffected healthy relatives of individuals with MDD displayed similar neural activations when confronted with negative information (Lisiecka et al 2012). However, differences in behavioural response were associated with sustained inhibition of negative processing character-ized by increased activation in the left inferior frontal gyrus (Lisiecka et al 2012).

The majority of studies evaluating cognitive function in MDD have included individu-als in the acute phase of depression with a paucity of studies evaluating individuals in remission and/or have taken into account the effect of current or past psychiatric/medical comorbidity (Halvorsen et al 2012; Elgamal 2010). A recent study comparing cognitive function in currently depressed (n = 37), previously depressed (n = 81), and healthy (n = 50) individuals reported that both depression groups exhibited compara-ble performance on measures of processing speed but performed significantly worse as compared to the healthy control group (Halvorsen et al 2012). Notwithstanding the ample documentation of various cognitive deficits in association with MDD, results remain mixed partly because of the heterogeneity of the disorder (see Chapter 3; McIntyre et al 2013; Elgamal et al 2010).

5.3 Other MDD domains associated with cognitive dysfunction

5.3.1 Sleep

The sleep cycle is comprised of two broad categories: (1) rapid eye movement (REM) and (2) non-rapid eye movement (NREM). The latter includes three stages: (1) N1, (2), N2, and (3) N3 (frequently described as slow-wave sleep (SWS); Figure 5.1; Walker MP and van der Helm, 2009). Abnormalities in the pattern, architecture, and duration of the sleep cycle are common among individuals with MDD, with approximately 90% reporting sleep difficulties (Reynolds and Kupfer 1987). Although a substantial body of evidence exists examining the association between MDD and general sleep disturbances (e.g. insomnia, obstructive sleep apnea, and fatigue), there is a dearth of information regarding the impact of disrupted sleep on cognition and function. It is well established that sleep is particularly important in promoting specific cognitive processes (e.g. learning and memory), commonly affected in MDD; however, the impact of sleep disturbances

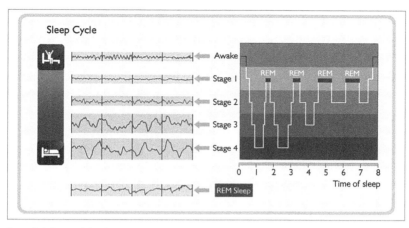

Figure 5.1 Stages of the sleep cycle
http://sleep.braintreatmentcenter.com/EEG.php

on cognition in individuals with MDD has not been well elucidated (Walker MP and van der Helm 2009). Nevertheless, there is some evidence to suggest that sleep exerts a differential effect on declarative memory (Dresler 2011).

Notwithstanding a paucity of evidence evaluating the direct effects of sleep disturbances in cognition in MDD, a recent meta-analysis investigating the validity of sleep polysomnography-based biomarkers of MDD reported that the combination of increased REM density and decreased SWS represents a stable and heritable trait; however, the utility of this endophenotype in predicting risk for MDD is not yet clear (Peterson MJ and Benca 2006). Interestingly, various antidepressants have been reported to supress REM sleep (e.g. citalopram and reboxetine) without negatively affecting cognitive performance (i.e. memory) (Edge 2010; Goder et al 2011). Furthermore, it is hypothesized that REM provides the optimal environment for 'bottom-up' processing of affectively valenced information by (1) increasing activity within limbic and paralimbic structures posited to reactivate similar brain responses to previously acquired experiences; (2) providing a coordinated neural network, free of external distractions, to integrate experienced emotional events; and (3) generating an environment limited to basal aminergic neurotransmitter concentrations (Edge 2010). Taken together, this model proposes that pathological increase in REM disproportionately amplifies and strengthens memories for negative experiences (Edge 2010).

5.3.2 Anhedonia

Anhedonia is a core diagnostic criterion for MDD and is described as the lack of interest or pleasure in response to hedonic stimuli or experiences (Liu et al 2012; Sherdell et al 2012; Dunn 2012). Available evidence suggests that anhedonia is relatively specific to depression when compared to negative affect, which is shared across various psychiatric disorders (Dunn 2012). However, currently available evidence-based treatments for MDD largely neglect anhedonia and focus on mitigating depressive symptoms rather than up-regulating and/or reinstating positive affect because of a limited understanding of neuropsychological systems that subserve disturbances in positive emotional processing (Dunn 2012). Moreover, recent studies examining the emotional response and memory performance in individuals with anhedonia have demonstrated that state and trait anhedonia are associated with attenuated experience of positive and negative information, respectively (Liu et al 2012; Chase et al 2012). Furthermore, evidence indicates that the reduced reward-seeking behaviour observed among individuals with anhedonia from deficits in anticipating pleasure rather than deficits in motivation for reward (i.e. blunted consummatory behaviour; Sherdell et al 2012).

References

Cappa SF. Subjective cognitive complaints: not to be dismissed. Eur J Neurol 2012 May;19(5):665.

Chase HW, Frank MJ, Michael A, Bullmore ET, Sahakian BJ, Robbins TW. Approach and avoidance learning in patients with major depression and healthy controls: relation to anhedonia. Psychol Med 2010 Mar;40(3):433–40.

Dunn BD. Helping depressed clients reconnect to positive emotion experience: current insights and future directions. Clin Psychol Psychother 2012 Jul;19(4):326–40.

Elgamal S, Denburg S, Marriott M, Macqueen G. Clinical factors that predict cognitive function in patients with major depression. Can J Psychiatry 2010 Oct;55(10):653–61.

Edge LC. The role of emotional brain processing during sleep in depression. J Psychiatr Ment Health Nurs 2010 Dec;17(10):857–61.

Goder R, Seeck-Hirschner M, Stingele K, Huchzermeier C, Kropp C, Palaschewski M, et al. Sleep and cognition at baseline and the eff ects of REM sleep diminution after 1 week of antidepressive treatment in patients with depression. J Sleep Res 2011 Dec;20(4):544–51.

Halvorsen M, Hoifodt RS, Myrbakk IN, Wang CE, Sundet K, Eisemann M, et al. Cognitive function in unipolar major depression: a comparison of currently depressed, previously depressed, and never depressed individuals. J Clin Exp Neuropsychol 2012;34(7):782–90.

Lisiecka DM, Carballedo A, Fagan AJ, Connolly G, Meaney J, Frodl T. Altered inhibition of negative emotions in subjects at family risk of major depressive disorder. J Psychiatr Res 2012 Feb; 46(2):181–8 .

Liu WH, Wang LZ, Zhao SH, Ning YP, Chan RC. Anhedonia and emotional word memory in patients with depression. Psychiatry Res 2012 Dec;200(2–3):361–7.

McIntyre RS, Cha DS, Soczynska JK, Woldeyohannes HO, Gallaugher LA, Kudlow P, et al. Cognitive defi cits and functional outcomes in major depressive disorder: determinants, substrates, and treatment interventions. Depress Anxiety 2013 Mar; 30(6):515–27 .

Peterson MJ, Benca RM. Sleep in mood disorders. Psychiatr Clin North Am 2006 Dec;29(4):1009–32.

Reynolds CF III, Kupfer DJ. Sleep research in aff ective illness: state of the art circa 1987. Sleep 1987 Jun;10(3):199–215.

Sarapas C, Shankman SA, Harrow M, Goldberg JF. Parsing trait and state effects of depression severity on neurocognition: evidence from a 26-year longitudinal study. J Abnorm Psychol 2012 Nov;121(4):830–7 .

Sherdell L, Waugh CE, Gotlib IH. Anticipatory pleasure predicts motivation for reward in major depression. J Abnorm Psychol 2012 Feb;121(1):51–60.

Vanderhasselt MA, De Raedt R, Dillon DG, Dutra SJ, Brooks N, Pizzagalli DA. Decreased cognitive control in response to negative information in patients with remitted depression: an eventrelated potential study. J Psychiatry Neurosci 2012 Jul;37(4):250–8.

Walker MP, van der Helm E. Overnight therapy? The role of sleep in emotional brain processing. Psychol Bull 2009 Sep;135(5):731–48.

Chapter 6

The mediational role of cognition on functional outcomes in major depressive disorder

Key Points

- MDD causes significant functional impairment.
- Cognitive function represents a critical determinant of functional outcome in MDD.
- Thirty to sixty per cent of costs attributable to MDD are due to impaired workforce performance.

6.1 Introduction

Major depressive disorder is associated with substantial disease burden and more years of life lost due to poor health when compared to all other mental, neurological, and substance use disorders (McIntyre et al 2013). Notwithstanding the burden of illness attributable to MDD, there is insufficient data that primarily aims to evaluate the impact of impaired cognition on overall psychosocial functioning, which is in contradistinction to schizophrenia and BD. Accumulating evidence suggests that deficits in cognitive function represent a critical determinant of functional outcome in MDD. For example, more than 25% of the impact of an MDE on role functioning and psychosocial outcomes in MDD has been reported to be mediated by impairments in cognition (Buist-Bouwman et al 2008). Taken together, the impact of cognitive deficits on psychosocial function and quality of life outcomes in individuals with MDD underscores the importance of identifying which individuals may be particularly vulnerable to cognitive impairments, with prevention and full functional recovery being the overarching therapeutic objectives.

6.2 Impaired workforce performance in MDD

Impairments in executive function, may be particularly relevant to functional out-comes, since components such as abstract thinking, planning, and working mem-ory are essential for optimal work performance (see Figure 3.1). Abnormalities in this domain are common in MDD and often persist during euthymic states. A sig-nificant component of the overall disability and cost associated with MDD relates to impaired work performance, underscoring the importance of cognitive func-tion in mediating and reducing productivity in individuals with MDD (Katon 2009; Figure 6.1). Altered brain activity has been reported in neuroimaging studies evaluat-ing the differences between individuals with MDD who are on long-term sick leave compared to those with MDD who are able to maintain workplace performance (Sandstrom et al 2012). More specifically, frontal hypoactivation in individuals with MDD who were on work-related long-term sick leave was significantly lower com-pared to individuals with depression (but not on long-term sick leave) and healthy controls (Sandstrom et al 2012).

Therapeutic interventions specifically targeting cognitive impairments in MDD are not currently available due to insufficient information regarding the subjectivity of these complaints. Taken together, cognitive impairments are associated with con-sistent, replicable (albeit non-specific), and clinically significant impact on the onset, course, and outcome of MDD, as well as psychosocial outcomes, notably workforce performance.

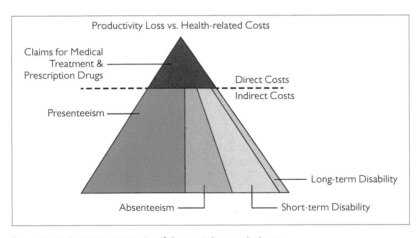

Figure 6.1 A schematic representation of direct vs indirect medical costs
Modified from Hemp Presenteeism: at work—but out of it. Harv Bus Rev 2004 Oct;82(10):49–58, 155.

References

Buist-Bouwman MA, Ormel J, de Graaf R, de Jonge P, van Sonderen E, Alonso J, et al. Mediators of the association between depression and role functioning. Acta Psychiatr Scand 2008 Dec;118(6):451–8.

Cappa SF. Subjective cognitive complaints: not to be dismissed. Eur J Neurol 2012 May;19(5):665.

Chase HW, Frank MJ, Michael A, Bullmore ET, Sahakian BJ, Robbins TW. Approach and avoidance learning in patients with major depression and healthy controls: relation to anhedonia. Psychol Med 2010 Mar;40(3):433–40.

Dunn BD. Helping depressed clients reconnect to positive emotion experience: current insights and future directions. Clin Psychol Psychother 2012 Jul;19(4):326–40.

Edge LC. The role of emotional brain processing during sleep in depression. J Psychiatr Ment Health Nurs 2010 Dec;17(10):857–61.

Goder R, Seeck-Hirschner M, Stingele K, Huchzermeier C, Kropp C, Palaschewski M, et al. Sleep and cognition at baseline and the effects of REM sleep diminution after 1 week of antidepressive treatment in patients with depression. J Sleep Res 2011 Dec;20(4):544–51.

Hemp P. Presenteeism: at work—but out of it. Harv Bus Rev 2004 Oct;82(10):49–58, 155.

Katon W. The impact of depression on workplace functioning and disability costs. Am J Manag Care 2009 Dec;15(11 Suppl):S322–27.

Liu WH, Wang LZ, Zhao SH, Ning YP, Chan RC. Anhedonia and emotional word memory in patients with depression. Psychiatry Res 2012 Dec;200(2–3):361–7.

McIntyre RS, Cha DS, Soczynska JK, Woldeyohannes HO, Gallaugher LA, Kudlow P, et al. Cognitive deficits and functional outcomes in major depressive disorder: determinants, substrates, and treatment interventions. Depress Anxiety 2013 Mar;30(6):515–27.

Peterson MJ, Benca RM. Sleep in mood disorders. Psychiatr Clin North Am 2006 Dec;29(4):1009–32.

Reynolds CF III, Kupfer DJ. Sleep research in affective illness: state of the art circa 1987. Sleep 1987 Jun;10(3):199–215.

Sandstrom A, Sall R, Peterson J, Salami A, Larsson A, Olsson T, et al. Brain activation patterns in major depressive disorder and work stress-related long-term sick leave among Swedish females. Stress 2012 Sep;15(5):503–13.

Sherdell L, Waugh CE, Gotlib IH. Anticipatory pleasure predicts motivation for reward in major depression. J Abnorm Psychol 2012 Feb;121(1):51–60.

Walker MP, van der Helm E. Overnight therapy? The role of sleep in emotional brain processing. Psychol Bull 2009 Sep;135(5):731–48.

Chapter 7

Neurobiology of cognitive dysfunction in major depressive disorder

Key Points

- Cognitive deficits in MDD are subserved by disturbances in several integrated neural, neurotransmitter, inflammatory, oxidative, and endocrine systems.
- Disturbances in 'hot' and 'cold' cognition reflect abnormal structure, function, and reciprocity in discrete and overlapping neural circuits and networks.
- Both state and trait effects pertinent to the neurobiology of cognition have been documented.

7.1 Introduction

Recent advances in neuroimaging have allowed for the study of complex neural networks that subserve cognitive and affective processes. Neural connectivity refers to networks of structural and functional connections between brain regions that are active under basal and task-related conditions. Understanding the neural connectivity between and within brain regions may elucidate the underlying pathophysiological basis of MDD and its associated cognitive deficits (Hulvershorn et al 2011). Cognitive function in MDD is also subserved by multiple interacting neurobiological systems, including neurotransmitter, inflammatory, oxidative, and endocrine.

7.1.1 Connectivity

Progress in molecular and cellular biology as well as neuroimaging have advanced the understanding of cognitive function/dysfunction and has supported the hypothesis that MDD is subserved by abnormalities in the structure, function, and chemical composition of frontosubcortical circuitry (Pizzagalli 2011). The neural basis of cognitive and emotional processes is represented by overlapping neural networks (Figure 7.1; Diener et al 2012). For example, the ventromedial prefrontal cortex has reciprocal connections to the amygdala and hippocampus, regions implicated in emotional processing and memory, as well as the dorsolateral prefrontal cortex, which has reciprocal connections to disparate brain areas (e.g. basal ganglia, cingulate cortex, parietal cortex; Wood and Grafman 2003). The ventral prefrontal cortex integrates emotional and cognitive information along with environmental stimuli, whereas the dorsolateral prefrontal cortex

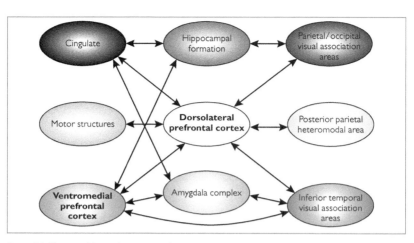

Figure 7.1 The neural basis of cognitive and emotional processes

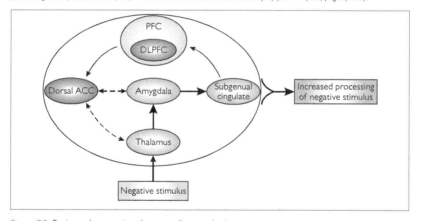

Figure 7.2 Reciprocal connections between disparate brain areas

is involved in the control and regulation of behaviour and responses to environmental stimuli (Figure 7.2; Diener et al 2012; Wood and Grafman 2003).

In individuals with MDD, several nodal structures (e.g. hippocampus, amygdala, anterior cingulate cortex) are susceptible to volumetric/functional changes as a consequence of illness duration, episode frequency, and severity (MacQueen et al 2003; Malykhin et al 2012). The most replicated volumetric abnormality in MDD is bilateral hippocampal reduction, which is a consequence of neuropil loss, decreased dendritic density, and reduced neuronal soma size (MacQueen et al 2003; Malykhin et al 2012). Hypofrontality of the prefrontal cortex and overactivity of the anterior cingulate cortex are hypothesized to be the correlate of a functional disconnection between cortical and subcortical structures (Pizzagalli 2011). These alterations may mediate deficits in measures of executive function, attention, learning and memory, as well as information processing speed.

The default mode network (DMN), is composed of several nodal structures, including the ventromedial prefrontal cortex, the posterior cingulate/retrosplenial cortex, and the bilateral inferior parietal lobe and may be relevant to cognitive function in MDD. (Minzenberg et al 2011). The DMN is active during the 'resting state' rather than during the execution of a cognitive, emotional, and/or motor task (Mars et al 2012). The DMN has been implicated in pathological disease states of which cognitive deficits are a defining feature (e.g. Alzheimer's disease (AD); Kenna 2012). For example, abnormalities in insulin resistance are accompanied by disturbances in the DMN in individuals at risk for AD. Moreover, modafinil, an agent with established procognitive effects via the enhancement of catecholamine neurotransmission, augments deactivation in the major nodal structures of the DMN (Minzenberg et al 2011).

7.3 Monoamines

For over six decades, the prevailing hypothesis has been that disturbances in monoamines are central to the pathophysiology of MDD. The monoaminergic system is also salient to normal brain function, including affective and cognitive processes. The extensive network of serotoninergic, noradrenergic, and dopaminergic neuronal projections from the raphe nucleus, locus coeruleus, and ventral tegmentum/substantia nigra, to disparate cortical areas allows for the modulation of a wide range of cognitive functions (Chamberlain and Robbins 2013).

Serotonin, although renowned for its role in affect, also modulates mechanisms that subserve cognitive function. Varying serotonin receptor subtypes and isoforms are widely expressed in brain regions involved in disparate cognitive processes (e.g. prefrontal cortex, hippocampus, and septum). Evidence from tryptophan (TRP; the amino acid precursor of serotonin) depletion studies implicates serotonin in several cognitive functions, including 'cold' and 'hot' cognitive processes. Across studies, the most consistent and robust effect of TRP depletion is impairment in the consolidation of episodic memories, notably those requiring verbal learning (Buhot et al 2000; Olivier 2009; Cowen and Sherwood 2013). Smaller effects on the encoding aspect of verbal information have also been described following TRP depletion, whereas the domains of executive function (planning, decision making, and response inhibition) and attention (e.g. sustained, selective, and divided) as well as spatial, semantic, and working memory remain largely unaffected (Mendelsohn et al 2009). Depletion of TRP has also been shown to modify emotional processing (e.g. reduce positive attentional bias) as well as increase activity in the ventral striatum, hippocampal complex, insula, and ventrolateral prefrontal cortex in response to emotionally valenced stimuli. Distinct functional variants of the serotonin transporter may also differentially affect cognitive function. For example, the short genotype of the serotonin transporter, a less active variant, has been associated with impaired verbal recall in healthy volunteers following TRP depletion, whereas individuals with the long genotype of the serotonin transporter did not exhibit these decrements. The short allele of the serotonin transporter promoter polymorphism is also associated with greater depression severity and suicidality in response to stressful life events (see Sahakian and Morein-Zamir 2013 for review).

Consistent with these observations, results from a functional magnetic resonance imaging (fMRI) study demonstrate that TRP depletion in healthy male volunteers is associated with reduced activity in the right hippocampus, a region normally activated during encoding, whereas regions implicated in information retrieval along with other cognitive functions (frontal, parietal, temporal, cingulate, striatal, and cerebellar regions) were unaffected (Cowen and Sherwood 2013). In recovered individuals with MDD, low-dose TRP depletion has been associated with deficits in immediate recall, autobiographical memory, and negative cognitive bias, without adversely affecting mood (Cowen and Sherwood 2013; Haddad et al 2009). In healthy individuals with a family history of depression, TRP depletion has been associated with impaired autobiographical memory (i.e. specificity), which is consistent with outcomes in individuals with declared or diagnosed MDD (see Cowen and Sherwood 2013 for a review).

The noradrenergic system plays a significant role in the stress response via activation of the sympathetic nervous system and 'fight or flight' reaction. In addition to enhanced arousal and alertness, the noradrenergic system participates in other cognitive functions, including working memory, memory consolidation, attention, and executive function (Chamberlain and Robbins 2013; Wang et al 2013). Evidence also indicates that the noradrenergic system exerts effects on cognition via an inverted-U–shaped pattern, with both hyper- and hyponoradrenergic activation leading to cognitive impairment, implying that superior cognitive performance likely occurs at an optimal level of noradrenergic activity (Chamberlain and Robbins 2013). For example, it is well documented that individuals under highly stressful conditions exhibit decreased cognitive performance (Wang et al 2013; Hermans et al 2011), while exposure to moderate stress levels has been shown to augment cognitive performance. Higher cerebral spinal fluid (CSF) norepinephrine levels have also been associated with poor performance on tests of attention and executive function (Figure 7.3; Wang et al 2013).

The synthesis of norepinephrine is intrinsically linked to dopamine (DA; i.e. DA, once hydroxylated, gives rise to norepinephrine; Chamberlain and Robbins 2013). Dopamine is salient to cognition, motivated behaviour, and reward response (Cole et al 2013). The effect of DA on cognitive function follows a linear as well as an inverted-U–shaped pattern. The inverted-U–shape effect is particularly evident with respect to the modulatory role of the DA receptor D_1 over working memory within the prefrontal cortex, with both agonist and antagonist treatment leading to impairment (Floresco 2013). Conversely, excess stimulation of DA receptors (i.e. D_1 and D_2) does not appear to exert detrimental effects on set-shifting (Floresco 2013). Dopamine manipulation in healthy individuals also exerts both linear and inverted-U–shaped effects on functional connectivity in a basal ganglia resting state network (i.e. lateral frontoparietal and medial frontal neocortical areas, respectively). The non-linearity might reflect the involvement of higher-order cognitive, emotional, or motivational processes. However, the optimum level of dopaminergic activity may differ between individuals, type of cognitive processes, and brain regions (Cole et al 2013). In individuals with MDD, acute phenylalanine and tyrosine depletion (APTD), a dietary intervention that selectively lowers DA synthesis, was associated with worse performance on a decision-making task (Figure 7.4; Cole et al 2013).

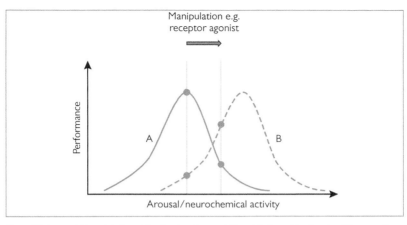

Figure 7.3 Inverted-U model of cognitive function. Inverted-U model of cognitive function. The example shown considers two cognitive functions, (A) and (B), which exhibit different optimal set points, in terms of arousal/neurochemical activity. An increase in neurochemical activity (e.g. via administration of a receptor agonist) shifts cognitive function (A) beyond its optimal point, leading to behavioural impairment. In contrast, the manipulation shifts cognitive function (B) towards its optimal point, leading to behavioural improvement.

Adapted from Chamberlain SR, Robbins TW. Noradrenergic modulation of cognition: Therapeutic implications. J Psychopharmacol 2013 March 21.

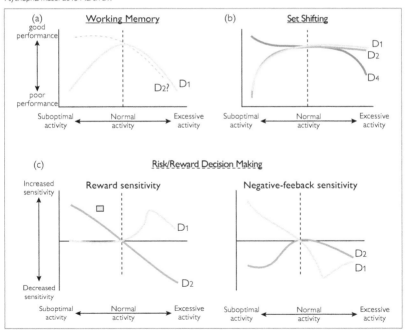

Figure 7.4 Optimum level of dopaminergic activity may differ between individuals

Adapted from Floresco SB. Prefrontal dopamine and behavioral flexibility: shifting from an "inverted-U" toward a family of functions. Front Neurosci 2013;7:62. Copyright © 2013 Floresco.

Cognitive domain	Acetyl-choline	Dopa-mine	GABA	Gluta-mate	Hista-mine	Noradrena-line	Serotonin
Attention	X	X		X	X	X	
Executive function		X	X	X			X
Memory	X	X	X	X	X	X	X
Psychomotor speed				X	X		

Table 7.1 Neurotransmitters involved in cognitive processes

Monoamines may influence cognition directly or indirectly via interactions with other pertinent neurotransmitter systems [e.g. cholinergic, glutamatergic, gamma-aminobutyric acid (GABA)ergic]. Serotonin may influence cognitive function indirectly by acting on other neurotransmitter systems that are more directly involved in cognition (e.g. glumatergic, cholinergic; Buhot et al 2000; Cowen and Sherwood 2013). For example, in the prefrontal cortex, the serotonin receptor 5-hydroxytryptamine $(5-HT)_3$ is primarily found on GABAergic interneurons and contributes to the regulation of cholinergic, dopaminergic, and glutamatergic activity (Cowen and Sherwood 2013). Preclinical evidence indicates that selective antagonism of $5-HT_7$, expressed on glutamatergic neurons in regions pertinent to cognitive function (e.g. the suprachiasmatic nucleus of the hypothalamus, hippocampus, cortex, thalamus, and raphe nuclei), prevents decrements in working memory induced by dizocilpine (an antagonist of the N-methyl-D-aspartate (NMDA) glutamate receptor; Cowen and Sherwood 2013).

In summary, monoamines exert a widespread influence in the CNS and play a role in disparate cognitive processes. The type of cognitive function, however, may be more specific or more highly influenced by specific neurotransmitters. For example, serotonin has been shown to exert more control over reversal learning, whereas norepinephrine and DA are more pertinent to executive function (e.g. set-shifting; Table 7.1; Robbins and Roberts 2007).

7.4 Glutamate

Glutamate is an excitatory neurotransmitter that is critical for normal and abnormal cognitive processes (e.g. regulating neuroplasticity, dendritic branching, learning and memory; Mathews and Henter 2012). For example, long-term potentiation (LTP), the neurobiological substrate of learning and memory, is a glutamate-dependent process. Moreover, the administration of phencyclidine (PCP), an NMDA antagonist, is reported to decrease cognitive performance in the novel object recognition paradigm. Glutamate excitotoxicity has been proposed as a model of disease pathogenesis in subsets of individuals with MDD (Mathews and Henter 2012). A derivative of this observation is the proposition that glutamatergic signalling disturbances may contribute to abnormalities in neural systems subserving cognition in MDD. Several mechanistically dissimilar agents, capable of mitigating depressive symptoms in individuals with mood disorders, exhibit direct and/or indirect effects on glutamatergic signalling.

The metabotropic glutamate (mGlu) receptor has recently been implicated in memory and synaptic plasticity (Mukherjee and Manahan-Vaughan 2013). Pharmacological modulation of mGlu1 or mGlu2 has been shown to improve cognitive function in preclinical models (Hikichi et al 2013; Goeldner et al 2013). Increased mGlu2 receptor has also been reported in post-mortem cortical tissues of individuals with MDD (Goeldner et al 2013). It has also been observed that higher prefrontal glutamate/glutamine and aminobutyric acid concentrations are associated with impairments in memory, executive function, and psychomotor speed, as well as with mild depression in individuals with diabetes mellitus type I (Lyoo et al 2009).

7.5 GABA

Gamma-aminobutyric acid is the principal inhibitory neurotransmitter in the CNS. Although it has not been a principal focus of mechanistic studies in MDD, recent preclinical evidence suggests that metabotropic $GABA_B$ receptors may be involved in MDD and/or its cognitive deficits. Preclinical evidence suggests that $GABA_B$ receptor antagonists exert antidepressant-like effects and improve cognitive function (e.g. spatial working memory). Cognition enhancement with $GABA_B$ antagonists may occur via increased neural activity in the hippocampus as well as increased brain-derived neurotrophic factor (BDNF) levels in the hippocampus and disparate cortical regions. There is a paucity of clinical studies in cohorts of individuals with MDD (Cryan and Slattery 2010).

7.6 Acetylcholine

The cholinergic system is composed of neurons that use the neurotransmitter acetylcholine. Acetylcholine acts on both the ionotropic nicotinic acetylcholine receptors (nAChRs) as well as metabotropic muscarinic receptors. The ionotropic receptors belong to a supergene family that includes $GABA_A$, $GABA_C$, and the $5\text{-}HT_3$ receptors, while the metabotropic belong to the multigene family that includes monoaminergic receptors (Graef et al 2011).

The role of the central cholinergic system on attention and learning is well established. The muscarinic receptor subtypes have been implicated in memory as well as attentional processes, whereas subtypes of AChR (i.e. nicotinic $\alpha7$, $\alpha4\beta2$) appear to function in working memory and attention. Consistent with these observations, nAChRs are widely distributed throughout the CNS, including brain regions subserving cognitive function (e.g. the hippocampus and frontal cortex; Warburton 1992). Moreover, alterations in the central signalling of acetylcholine are hypothesized to be a canonical neurochemical abnormality in AD (Jones et al 2006).

Preclinical studies indicate that the administration of nicotinic ligands improves measures of attention and working memory performance (Levin and Rezvani 2002). These observations are hypothesized to be related to an increase in ventral tegmental dopaminergic neuron burst activity (Knott 1999). Moreover, cognitive deficits are well documented in healthy individuals as well as those with psychiatric disorders receiving anticholinergic treatments.

7.7 Inflammation

Disturbances in immuno-inflammatory networks, which are largely mediated by inflammatory cytokines, are well documented in individuals with MDD. For example, abnormalities in pro- and anti-inflammatory cytokines (e.g. Interleukin (IL)-1β, IL-6, tumour necrosis factor (TNFα)) have frequently been reported in MDD during affective and euthymic states (Leonard et al 2012). It has also been hypothesized that abnormalities in immuno-inflammatory networks may mediate the relationship between cognitive deficits and MDD and may also explain the increased risk for dementia in this population (Leonard 2007). It is increasingly recognized that pro-inflammatory cytokines and immune cells play a role in physiological and pathophysiological cognitive processes via an inverted-U-shaped function (McAfoose and Baune 2009).

An extensive body of preclinical evidence supports the roles of IL-1β, IL-6, and TNF-α in cognitive processes, notably hippocampal-dependent learning and memory (e.g. spatial, contextual fear conditioning) at the levels of synaptic plasticity, neurogenesis, and neuromodulation (McAfoose and Baune 2009). Impairments in hippocampal-dependent learning and memory are reported in the absence of, and elevation in, pro-inflammatory cytokines (McAfoose and Baune 2009). This foregoing observation comports with the modulatory role of IL-1β in synaptic transmission, with inhibition of LTP occurring at pathophysiological levels and LTP maintenance at physiologic levels (McAfoose and Baune 2009). Conversely, administration of an IL-1 receptor antagonist has been shown to reverse stress-induced inhibition of hippocampal-dependent conditioning and improve avoidance memory at low doses, as well as impair memory in other studies (Table 7.2; Khairova 2009).

Interleukin-6 may also regulate synaptic plasticity, with LTP suppression described with exogenous IL-6 administration. This pro-inflammatory cytokine also plays a significant role in the regulation of neurogenesis, with overexpression being associated with compromised neurogenesis in the hippocampal dentate gyrus. However, its role in healthy cognitive function remains inconclusive, as both improved and impaired cognitive performance are reported in transgenic mice that do not express IL-6 (Table 7.2; McAfoose and Baune 2009). In women with recurrent MDD, higher levels of the pro-inflammatory cytokine IL-6 have been significantly associated with reduced performance in verbal learning and memory (i.e. immediate and delayed verbal recall; Grassi-Oliveira 2011). Similar results were also observed in elderly individuals with MDD, with reduced encoding and recall, but not executive function and attention/processing, being associated with higher levels of IL-6. This relationship, however, was also observed in the elderly control group, with no significant difference between the two groups (Elderkin-Thompson 2012). Others have reported a similar effect for IL-6 as well as IL-1β and TNF-α in animal models, the aging population, and other neurologic and medical conditions (e.g. AD, heart failure; McAfoose and Baune 2009; Wilson et al 2002; Athilingam 2013).

Tumour necrosis factor is implicated in physiological, neuroprotective, and neurodegenerative processes with the effect being largely dependent on the type of TNF receptor being activated (i.e. TNF-R1 leading to caspase activation and apoptosis whereas the converse effect is reported for TNF-R2). Tumour necrosis factor has been identified to be salient for normal learning and memory function (McAfoose and Baune

Table 7.2 Regulation of synaptic plasticity and behavioural correlates by pro-inflammatory cytokines

Cytokine	Effect on synaptic plasticity	Behavioural correlates
Physiological levels of TNF-α	upregulation of AMPA receptor trafficking; increased synaptic strength (Beattie et al 2002; Stellwagen et al 2005)	?
High pathological levels of TNF-α	inhibition of LTP (Butler et al 2004; Coogan et al 1999; Cunningham et al 1996; Tancredi et al 1992)	'depressive-like' behaviour, impaired learning and memory in animal models (Aloe et al 1999; Dantzer, 2001; Golan et al 2004)depressive symptoms, anxiety and memory impairments in mood disorders (Dantzer et al 2008; Raison et al 2006)
Physiological levels of IL-1	maintenance of short-term plasticity and LTP (Avital et al 2003; Goshen et al 2007, 2008; Yirmiya et al 2002)	improved hippocampal-dependent memory (Avital et al 2003; Brennan et al 2003; Song et al 2003)
High pathological levels of IL-1 or IL-18	impaired LTP (Coogan et al 1999; Curran and O'Connor 2001; Goshen et al 2007)	'depressive-like' behaviour and impaired hippocampal-dependent memory in animal models (Dantzer 2001; Dantzer et al 2008; Gibertini et al 1995)
Increased levels of IL-6	decreased glutamate release (D'Arcangelo et al 2000); decreased expression of LTP (Tancredi et al 2000)	'depressive-like' behaviour, impaired learning and memory in animal models (Balschun et al 2004; Bluthe et al 1999; Heyser et al 1997)marked cognitive disturbances and depression symptoms in MDD (Capuron et al 2001a; Raison et al 2006)
Increased levels of INF-α and INF-γ	decreased dendritic AMPA receptor clustering (Vikman et al 2001); inhibition of glutamate-mediated excitatory post-synaptic potentials and LTP (Mendoza-Fernandez et al 2000)	anxiety and learning deficits in animal models (Fahey et al 2008; Myint et al 2007)depressive symptoms and cognitive defects in MDD (Gabbay et al 2008; Raison et al 2006)

Abbreviations: AMPA, α-amino-3-hydroxy-5-methyl-4-isoxazoleproprionic acid

Source: Adapted with permission from Khairova RA, Machado-Vieira R, Du J, Manji HK. A potential role for pro-inflammatory cytokines in regulating synaptic plasticity in major depressive disorder. Int J Neuropsychopharmacol 2009 May;12(4):561–78. © Cambridge University Press

2009); however, cognitive impairment has also been demonstrated in transgenic mice over-expressing TNF. Tumour necrosis factor-associated LTP inhibition in the dentate gyrus has also been reported (McAfoose and Baune 2009), although not in all studies (Stellwagen and Malenka 2006). Improved spatial memory has been reported in TNF-α knockout mice, and impaired spatial memory has been shown in TNF-α–overexpressing mice. Moreover, intracerebroventricular administration of TNF-α antibodies has been shown to have antidepressant-like effects in animal models (Table 7.2; Khairova 2009).

Immune cells (e.g. T cells, microglia) and anti-inflammatory cytokines may also exert an inverted-U-shaped influence on cognitive function (Rook et al 2012; Ziv et al 2006;

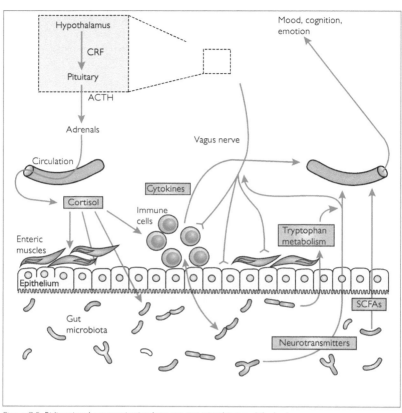

Figure 7.5 Bidirectional communication between gut microbiota and the brain

Reproduced with permission from Cryan JF, Dinan TG. Mind-altering microorganisms: the impact of the gut microbiota on brain and behaviour. Nat Rev Neurosci 2012 October;13(10):701–12.

Kipnis et al 2004). For example, IL-4-producing T cells accumulate in the meninges of rodents following performance of a cognitive task and lead to subsequent increases in IL-10 and decreases in TNF-α. Cognitive deficits are also evident in IL-4 knockout mice or following T cell depletion; deficits that can be remedied with T cell restoration (Ziv et al 2006; Kipnis et al 2004). Microglia, the primary immune cells of the CNS and a source of inflammatory cytokines, contribute to both normal and abnormal learning and memory. In the healthy state, hippocampal neurogenesis induced by an enriched environment has been associated with the recruitment of T cells and the activation of microglia, an effect that was markedly impaired in immune-deficient animals (Ziv et al 2006).

Several physical and psychosocial stressors have been hypothesized to induce exaggerated and/or persistent neuroinflammatory reactions leading to decrements in cognitive function (Barrientos et al 2010). Exposure to microbes has also been proposed to play a role in cognitive pathophysiology. Infectious burden (i.e. composite of exposure to *Chlamydia pneumoniae*, *Helicobacter pylori*, *cytomegalovirus*, and *herpes simplex virus* 1 and 2) was associated with decreased cognitive performance in a large multi-ethnic cohort study (Katan et al 2013). Healthy older individuals are more likely to suffer memory impairments following stressful life events (e.g. bacterial infection, surgery, severe psychosocial stressor) than younger adults (Barrientos et al 2010). Early-life infection has been associated with cognitive impairment in young adulthood in the presence of an inflammatory challenge. For example, neonatally infected rats also exhibit memory impairments at 16 months of age (Bilbo 2010).

An important observation regarding the pathoetiological contribution of inflammation to MDD is its non-specificity, insofar as disturbances in individual inflammatory cytokines, notably IL-1β, IL-6, and TNFα, are not unique to MDD but appear to be involved in many chronic conditions, including psychiatric, neurological, and medical. Future research will need to parse out unique neurobiological signatures capable of distinguishing between disorders as well as identify whether any specific signature is unique to individuals with MDD who manifest persisting cognitive deficits.

7.8 Oxidative stress

Oxidative stress is a consequence of the accumulation of reactive oxygen and nitrosative species, which are produced in response to intracellular functional and metabolic changes that result from exposure to damaging or unhealthy conditions (e.g. disease, excess caloric intake or unhealthy nutrition, physical inactivity; Naviaux 2012). Although reactive oxygen and nitrosative species have evolved to exert protective functions (e.g. activating apoptotic pathways to prevent damage to neighbouring cells), excessive and/or uncontrolled oxidative accumulation can have long-term consequences on the structure and function of cellular proteins, lipids, and nucleic acids; the brain being particularly vulnerable due to its high oxygen utilization. Oxidative stress has been hypothesized to be a causative factor in the onset and progression of several neurodegenerative diseases (e.g. Parkinson's disease, AD). Others, however, argue that it is a consequence, and not the cause, of the disease (Naviaux 2012).

Oxidative stress may mediate/moderate the relationship between MDD and cognitive impairment. Available evidence indicates that markers of oxidative stress are associated with impairments in short- and long-term declarative memory, verbal fluency, and working memory (Talarowska 2012). Although antioxidant treatment has been reported to be beneficial in some preclinical and clinical studies in either healthy aging or neurodegenerative conditions (enhancing cognition or slowing progression), compelling evidence of antioxidants providing robust improvement in cognitive performance is limited (Head 2009).

7.9 Cortisol

It has been frequently reported that abnormalities in hypothalamic-pituitary-adrenal (HPA) axis function may cause or exacerbate both cognitive impairment and depressive symptoms (McQuade and Young 2000; Rush 1996). Patients with Cushing's syndrome, characterized by chronic elevation of endogenous cortisol levels, have consistently been reported to manifest cognitive impairment and a high incidence of depression, which resolves with the correction of hypercortisolemia (Forget et al 2002). In healthy volunteers, both acute and subchronic administration of the synthetic steroid hydrocortisone has been reported to result in reversible impairments in cognitive function (Newcomer et al 1999).

Abnormalities in the HPA axis are frequently reported in individuals with MDD (Rush et al 1996), with up to 50% of individuals not being effective suppressors of dexamethasone (synthetic glucocorticoid family) on the dexamethasone suppression test. Peripheral cortisol levels have also been reported to be correlated with depressive symptoms and cognitive deficits (Reus and Wolkowitz 2001). It has been hypothesized that proximal and distal stressors (e.g. childhood maltreatment) may contribute to HPA axis dysregulation and consequent cognitive dysfunction in MDD. However, due to the hormetic effect of stress on cognition, it is not surprising that both positive and negative effects of stress or HPA axis activity have been reported in MDD (Krogh et al 2012; Zobel et al 2004; Goldberg et al 2013).

In keeping with this observation, the antiglucocorticoid mifepristone (RU-486) was evaluated for its procognitive and antidepressant properties in individuals with BD (Young et al 2004). At high doses, the progesterone antagonist mifepristone is an antagonist of the glucocorticoid receptor (GR) subtype of corticosteroid receptors. Preliminary reports indicate that mifepristone and the novel GR antagonist ORG-34517 have antidepressant effects in both psychotic and nonpsychotic MDD, notably in subjects with high rates of hypercortisolemia (Young et al 2004).

7.10 Microbiota

Following the germ theory of disease proposed by Louis Pasteur in the nineteenth century, hypotheses first linking gut microbiota to mental health, notably how microbes in the colon contribute to fatigue, melancholia, and neuroses, continued to emerge in to the early twentieth century. In 1930, Stokes and Pillsbury reviewed putative gastrointestinal mechanisms (including alterations to intestinal microbiota) that could explain the co-occurrence of emotional disorders and inflammatory skin conditions. These theories were largely abandoned until the twenty-first century (Smith 2012; Bested et al 2013).

Contemporary evidence indicates that the gut microbiota make up a diverse and dynamic ecosystem, with more microorganisms residing within the human gastrointestinal tract than the total number of cells or genes found in the human body (Cryan and Dinan 2012; Bercik 2011). Several studies indicate that the link between diet and heart disease may be mediated by toxic metabolites producted by gut microbita (Tang et al 2013). A bidirectional communication between the gut and the CNS, via neuronal, humoral, and immunological pathways, has been documented (Cryan and Dinan 2012; Bercik 2011). Disturbances in the gut-brain axis may have pathophysiologic consequences resulting in behavioural, affective, and cognitive changes (Cryan and Dinan 2012; for mechanisms of gut-CNS communications, see Figure 7.6).

Individuals with MDD exhibit high rates of metabolic and gastrointestinal co-morbidity (e.g. obesity, irritable bowel syndrome), and vice versa. Recent evidence indicates that stressful experiences are associated with altered composition of the intestinal microbiota in animals, with *bifidobacteria* being particularly vulnerable to the effects of emotional stress in humans (Desbonnet 2010). It has also been demonstrated that increasing the concentration of probiotics in the gut (i.e. *Bifidobacterium and Lactobacillus*) mitigates symptoms of anxiety and regulates mood in individuals with chronic fatigue syndrome and irritable bowel syndrome. A recent study that utilized the maternal separation animal model of depression reported that the probiotic *Bifidobacterium infantis* normalized the immune response, improved performance on the forced swim test, and restored basal levels of norepinephrine in the brainstem (Desbonnet 2010).

Evidence linking microbiota to cognition is in its infancy. Nevertheless, evidence from animal models suggests that, in the presence of stress, the absence of microbes in the gastrointestinal tract is associated with alterations in systems pertinent to cognition and MDD (i.e. reduced expression of BDNF in the cortex and hippocampus as well as an exaggerated HPA axis response; Rook et al 2012; Desbonnet 2010). Moreover, mice exposed to stress and intestinal *Citrobacter rodentium* infection exhibit memory impairment, which can be prevented with probiotic treatment (Gareau et al 2011); similar effects on learning and memory were demonstrated by Li et al with diet-induced alteration in microbiota (Li et al 2009). Germ-free mice also exhibit non-spatial and working memory deficits (Cryan and Dinan 2012), and restoration of the gut microbiome is associated with increased plasma serotonin (Rook et al 2012).

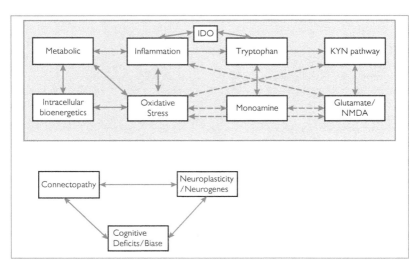

Figure 7.6 Interacting systems involved in cognitive deficits in MDD

7.11 Insulin

The localization of insulin receptors in the hippocampus and mediotemporal cortex suggests that cellular insulin and glucose homeostasis are relevant to normal memory function (Craft and Watson 2004). For example, LTP, the neurobiological model of learning, is modulated by insulin (Skeberdis et al 2001; Kopf and Baratti 1999; Figlewicz 1993) . Insulin is also synthesized by hippocampal neurons, promotes axonal growth in fetal brain cells, induces NMDA receptor potentiation, and increases NMDA receptor density at synapses (Skeberdis et al 2001; Kopf and Baratti 1999; Figlewicz 1993). It has been reported that insulin deficiency in diabetic rats triggers a retraction of dendrites and reduces NMDA transmission in hippocampal neurons, resulting in atrophy of the hippocampus and diminished memory performance. These data suggest that CNS insulin may protect and support the development of neuronal connectivity in hippocampal brain regions, resulting in improved memory functioning (Skeberdis et al 2001; Kopf and Baratti 1999; Figlewicz 1993).

Insulin resistance and reduced insulin effectiveness may exert a negative influence on memory. For example, impairments in operant learning and classical conditioning have been demonstrated in an animal model of type I diabetes. Similarly, deficits in psychomotor performance and memory (verbal/visual), along with reduced volumes of hippocampal and amygdalar structures, have been reported in euglycemic patients with type II diabetes mellitus (Craft and Watson 2004; Greenwood and Winocur 2001; Richardson 1990). Insulin dysregulation has also been implicated in the pathophysiology of several neurodegenerative conditions (e.g. AD, Parkinson's disease). The infusion of insulin has been shown to lower the plasma concentration of amyloid precursor protein in healthy volunteers and in persons with AD. Recent evidence indicates that intranasal administration of insulin for 8 weeks to euthymic individuals with BD improves certain executive functions (i.e. cognitive set-shifting; McIntyre et al 2012).

7.12 Incretins

The incretin glucagon-like peptide-1 (GLP-1), a gastrointestinal hormone, has been documented to cross the blood-brain barrier and exert functions in the CNS. In humans, GLP-1 receptors have been found in disparate brain regions (cerebral cortex, hypothalamus, hippocampus, thalamus, caudate-putamen, globus pallidum). In addition to its roles in regulating gastric emptying, control of food intake, and glucose homeostasis via modulatory effects on insulin secretion and glucose production, GLP-1 mediates the hypothalamic stress response and autonomic regulation of blood pressure and heart rate. Recently, GLP-1 and GLP-1 analogues have been reported to have neurotrophic and neuroprotective effects. Improvement of cognitive function, including learning and memory, following GLP-1 treatment has recently been described (McIntyre et al 2013).

7.13 Psychiatric and medical co-morbidity

Major depressive disorder is associated with high rates of psychiatric and medical co-morbidity. Age-related medical conditions (e.g. diabetes mellitus type II, cardiovascular disease) in MDD often manifest at an earlier age and consequently contribute

Table 7.3 Cognitive performance on the RBANS vs co-morbidity						
Co-morbidity†	Immediate memory (Mean ± SE)	Visuospatial (Mean ± SE)	Language (Mean ± SE)	Attention (Mean ± SE)	Delayed memory (Mean ± SE)	Total score (Mean ± SE)
None	84.9 ± 3.8	100.4 ± 3.9	106.6 ± 3.1	96.6 ± 3.9	96.3 ± 3.4	96.1 ± 3.5
Medical only	86.5 ± 5.2	95.1 ± 5.3	105.7 ± 4.2	88.4 ± 5.4	97.2 ± 4.7	92.1 ± 4.7
Psychiatric only	85.01 ± 2.8	90.5 ± 2.9	97.1 ± 2.3	89.0 ± 2.9	90.4 ± 2.6	86.5 ± 2.6
Psychiatric plus medical	85.4 ± 3.4	88.5 ± 3.5	102.7 ± 2.8	91.7 ± 3.5	91.5 ± 3.1	87.4 ± 3.1

†Mutually exclusive categories.

Abbreviations: RBANS, Repeatable Battery for the Assessment of Neuropsychological Status, SE, standard error.

Source: Baune BT, McAfoose J, Leach G, Quirk F, Mitchell D. Impact of psychiatric and medical comorbidity on cognitive function in depression. Psychiatry Clin Neurosci 2009 June;63(3):392–400.

to premature mortality. It could be hypothesized that co-morbidities in MDD could exert a compounding or synergistic effect on cognitive dysfunction. Baune et al (2009) reported that only 20.8% of currently depressed individuals with MDD or BD participants did not meet criteria for another co-occurring condition. In individuals with co-morbidity, 67.7% were psychiatric and 39.6% were medical. The presence of psychiatric co-morbidity was associated with decreased cognitive performance in the visuospatial/constructional and language domains as compared to individuals without any co-morbidities. Moreover, increasing number of psychiatric co-morbidities was associated with worse decrements in cognitive function. Interestingly, the presence of medical co-morbidity in the absence of a psychiatric co-morbidity did not exert a negative influence on cognitive performance. However, an additive effect of both medical and psychiatric co-occurring syndromes in this population was identified, particularly with respect to visuospatial/constructional cognitive abilities (Table 7.3; Baune et al 2009).

Taken together, cognitive deficits in MDD are subserved by heterogeneous pathoetiological substrates characterized by disturbances in neuroinflammation, neuroendocrinology, and monoamine signalling. It is proposed that the cognitive deficits observed are a phenotypic manifestation of abnormalities in the structure, function, and chemical composition of neural and glial cells within discrete brain circuits and distributed networks. The heterogeneous substrates subserving cognition indicate several possible mechanisms that may cause and maintain cognitive impairment and further indicate that mechanistically dissimilar interventions targeting multiple effector systems may be capable of providing procognitive effects in individuals with MDD (see Figure 7.5 for an integrative model of neurobiological systems involved in cognitive deficits in MDD).

References

Athilingam P, Moynihan J, Chen L, D'Aoust R, Groer M, Kip K. Elevated levels of interleukin 6 and C-reactive protein associated with cognitive impairment in heart failure. Congest Heart Fail 2013 Mar;19(2):92–8.

Barrientos RM, Frank MG, Watkins LR, Maier SF. Memory impairments in healthy aging: Role of aging-induced microglial sensitization. Aging Dis 2010 Jan;1(3):212–31.

Baune BT, McAfoose J, Leach G, Quirk F, Mitchell D. Impact of psychiatric and medical comorbidity on cognitive function in depression. Psychiatry Clin Neurosci 2009 Jun;63(3):392–400.

Bercik P. The microbiota-gut-brain axis: learning from intestinal bacteria? Gut 2011 Mar;60(3):288–9.

Bested AC, Logan AC, Selhub EM. Intestinal microbiota, probiotics and mental health: from Metchnikoff to modern advances: Part I—autointoxication revisited. Gut Pathog 2013 Mar;5(1):5.

Bilbo SD. Early-life infection is a vulnerability factor for aging-related glial alterations and cognitive decline. Neurobiol Learn Mem 2010 Jul;94(1):57–64.

Buhot MC, Martin S, Segu L. Role of serotonin in memory impairment. Ann Med 2000 Apr;32(3):210–21.

Chamberlain SR, Robbins TW. Noradrenergic modulation of cognition: Therapeutic implications. J Psychopharmacol 2013 Mar 21. [Epub ahead of print]

Cole DM, Beckmann CF, Oei NY, Both S, van Gerven JM, Rombouts SA. Differential and distributed effects of dopamine neuromodulations on resting-state network connectivity. Neuroimage 2013 Apr;78C:59–67.

Cowen P, Sherwood AC. The role of serotonin in cognitive function: evidence from recent studies and implications for understanding depression. J Psychopharmacol 2013 Jul;27(7): 575–83.

Craft S, Watson GS. Insulin and neurodegenerative disease: shared and specific mechanisms. Lancet Neurol 2004 Mar;3(3):169–78.

Cryan JF, Dinan TG. Mind-altering microorganisms: the impact of the gut microbiota on brain and behaviour. Nat Rev Neurosci 2012 Oct;13(10):701–12.

Cryan JF, Slattery DA. GABAB receptors and depression. Current status. Adv Pharmacol 2010;58:427–51.

Desbonnet L, Garrett L, Clarke G, Kiely B, Cryan JF, Dinan TG. Effects of the probiotic Bifidobacterium infantis in the maternal separation model of depression. Neuroscience 2010 Nov;170(4):1179–88.

Diener C, Kuehner C, Brusniak W, Ubl B, Wessa M, Flor H. A meta-analysis of neurofunctional imaging studies of emotion and cognition in major depression. Neuroimage 2012 Jul;61(3):677–85.

Disner SG, Beevers CG, Haigh EA, Beck AT. Neural mechanisms of the cognitive model of depression. Nat Rev Neurosci 2011 Aug;12(8):467–77.

Elderkin-Thompson V, Irwin MR, Hellemann G, Kumar A. Interleukin-6 and memory functions of encoding and recall in healthy and depressed elderly adults. Am J Geriatr Psychiatry 2012 Sep;20(9):753–63.

Figlewicz DP, Szot P, Israel PA, Payne C, Dorsa DM. Insulin reduces norepinephrine transporter mRNA in vivo in rat locus coeruleus. Brain Res 1993 Jan;602(1):161–4.

Floresco SB. Prefrontal dopamine and behavioral flexibility: shifting from an "inverted-U" toward a family of functions. Front Neurosci 2013;7:62.

Forget H, Lacroix A, Cohen H. Persistent cognitive impairment following surgical treatment of Cushing's syndrome. Psychoneuroendocrinology 2002 Apr;27(3):367–83.

Gareau MG, Wine E, Rodrigues DM, Cho JH, Whary MT, Philpott DJ, Macqueen G, Sherman PM. Bacterial infection causes stress-induced memory dysfunction in mice. Gut 2011 Mar;60(3):307–17.

Goeldner C, Ballard TM, Knoflach F, Wichmann J, Gatti S, Umbricht D. Cognitive impairment in major depression and the mGlu2 receptor as a therapeutic target. Neuropharmacology 2013 Jan;64:337–46.

Goldberg X, Fatjo-Vilas M, Alemany S, Nenadic I, Gasto C, Fananas L. Gene-environment interaction on cognition: A twin study of childhood maltreatment and COMT variability. J Psychiatr Res 2013 Jul;47(7):989–94.

Graef S, Schonknecht P, Sabri O, Hegerl U. Cholinergic receptor subtypes and their role in cognition, emotion, and vigilance control: an overview of preclinical and clinical findings. Psychopharmacology (Berl) 2011 May;215(2):205–29.

Grassi-Oliveira R, Bauer ME, Pezzi JC, Teixeira AL, Brietzke E. Interleukin-6 and verbal memory in recurrent major depressive disorder. Neuro Endocrinol Lett 2011;32(4):540–4.

Greenwood CE, Winocur G. Glucose treatment reduces memory deficits in young adult rats fed high-fat diets. Neurobiol Learn Mem 2001 Mar;75(2):179–89.

Haddad AD, Williams JM, McTavish SF, Harmer CJ. Low-dose tryptophan depletion in recovered depressed women induces impairments in autobiographical memory specificity. Psychopharmacology (Berl) 2009 Dec;207(3):499–508.

Head E. Oxidative damage and cognitive dysfunction: antioxidant treatments to promote healthy brain aging. Neurochem Res 2009 Apr;34(4):670–8.

Hermans EJ, van Marle HJ, Ossewaarde L, Henckens MJ, Qin S, van Kesteren MT, Schoots VC, Cousijn H, Rijpkema M, Oostenveld R, Fernandez G. Stress-related noradrenergic activity prompts large-scale neural network reconfiguration. Science 2011 Nov;334(6059):1151–3.

Hikichi H, Kaku A, Karasawa JI, Chaki S. Stimulation of Metabotropic Glutamate (mGlu) 2 Receptor and Blockade of mGlu1 Receptor Improve Social Memory Impairment Elicited by MK-801 in Rats. J Pharmacol Sci 2013 May;122(1):10–16.

Hulvershorn LA, Cullen K, Anand A. Toward dysfunctional connectivity: a review of neuroimaging findings in pediatric major depressive disorder. Brain Imaging Behav 2011 Dec;5(4):307–28.

Jones IW, Westmacott A, Chan E, Jones RW, Dineley K, O'Neill MJ, Wonnacott S. Alpha7 nicotinic acetylcholine receptor expression in Alzheimer's disease: receptor densities in brain regions of the APP(SWE) mouse model and in human peripheral blood lymphocytes. J Mol Neurosci 2006;30(1–2):83–4.

Katan M, Moon YP, Paik MC, Sacco RL, Wright CB, Elkind MS. Infectious burden and cognitive function: the Northern Manhattan Study. Neurology 2013 Mar;80(13):1209–15.

Kenna H, Hoeft F, Kelley R, Wroolie T, Demuth B, Reiss A, Rasgon N. Fasting plasma insulin and the default mode network in women at risk for Alzheimer's disease. Neurobiol Aging 2012 Mar;34(3):641–9.

Khairova RA, Machado-Vieira R, Du J, Manji HK. A potential role for pro-inflammatory cytokines in regulating synaptic plasticity in major depressive disorder. Int J Neuropsychopharmacol 2009 May;12(4):561–78.

Kipnis J, Cohen H, Cardon M, Ziv Y, Schwartz M. T cell deficiency leads to cognitive dysfunction: implications for therapeutic vaccination for schizophrenia and other psychiatric conditions. Proc Natl Acad Sci U S A 2004 May;101(21):8180–5.

Knott V, Bosman M, Mahoney C, Ilivitsky V, Quirt K. Transdermal nicotine: single dose effects on mood, EEG, performance, and event-related potentials. Pharmacol Biochem Behav 1999 June;63(2):253–61.

Kopf SR, Baratti CM. Effects of posttraining administration of insulin on retention of a habituation response in mice: participation of a central cholinergic mechanism. Neurobiol Learn Mem 1999 Jan;71(1):50–61.

Krogh J, Videbech P, Renvillard SG, Garde AH, Jorgensen MB, Nordentoft M. Cognition and HPA axis reactivity in mildly to moderately depressed outpatients: a case-control study. Nord J Psychiatry 2012 Dec;66(6):414–21.

Leonard B, Maes M. Mechanistic explanations how cell-mediated immune activation, inflammation and oxidative and nitrosative stress pathways and their sequels and concomitants play a role in the pathophysiology of unipolar depression. Neurosci Biobehav Rev 2012 Feb;36(2):764–85.

Leonard BE. Inflammation, depression and dementia: are they connected? Neurochem Res 2007 Oct;32(10):1749–56.

Levin ED, Rezvani AH. Nicotinic treatment for cognitive dysfunction. Curr Drug Targets CNS Neurol Disord 2002 Aug;1(4):423–31.

Li W, Dowd SE, Scurlock B, Acosta-Martinez V, Lyte M. Memory and learning behavior in mice is temporally associated with diet-induced alterations in gut bacteria. Physiol Behav 2009 Mar 23;96(4–5):557–67.

Lyoo IK, Yoon SJ, Musen G, Simonson DC, Weinger K, Bolo N, Ryan CM, Kim JE, Renshaw PF, Jacobson AM. Altered prefrontal glutamate-glutamine-gamma-aminobutyric acid levels and relation to low cognitive performance and depressive symptoms in type 1 diabetes mellitus. Arch Gen Psychiatry 2009 Aug;66(8):878–87.

MacQueen GM, Campbell S, McEwen BS, Macdonald K, Amano S, Joffe RT, Nahmias C, Young LT. Course of illness, hippocampal function, and hippocampal volume in major depression. Proc Natl Acad Sci U S A 2003 Feb;100(3):1387–92.

Malykhin NV, Carter R, Hegadoren KM, Seres P, Coupland NJ. Fronto-limbic volumetric changes in major depressive disorder. J Affect Disord 2012 Feb;136(3):1104–13.

Mars RB, Neubert FX, Noonan MP, Sallet J, Toni I, Rushworth MF. On the relationship between the "default mode network" and the "social brain". Front Hum Neurosci 2012;6:189.

Mathews DC, Henter ID, Zarate CA. Targeting the glutamatergic system to treat major depressive disorder: rationale and progress to date. Drugs 2012 Jul;72(10):1313–33.

McAfoose J, Baune BT. Evidence for a cytokine model of cognitive function. Neurosci Biobehav Rev 2009 Mar;33(3):355–66.

McIntyre RS, Powell AM, Kaidanovich-Beilin O, Soczynska JK, Alsuwaidan M, Woldeyohannes HO, Kim AS, Gallaugher LA. The neuroprotective effects of GLP-1: Possible treatments for cognitive deficits in individuals with mood disorders. Behav Brain Res 2013 Jan;237:164–71.

McIntyre RS, Soczynska JK, Woldeyohannes HO, Miranda A, Vaccarino A, Macqueen G, Lewis GF, Kennedy SH. A randomized, double-blind, controlled trial evaluating the effect of intranasal insulin on neurocognitive function in euthymic patients with bipolar disorder. Bipolar Disord 2012 Nov;14(7):697–706.

McQuade R, Young AH. Future therapeutic targets in mood disorders: the glucocorticoid receptor. Br J Psychiatry 2000 Nov;177:390–5.

Mendelsohn D, Riedel WJ, Sambeth A. Effects of acute tryptophan depletion on memory, attention and executive functions: a systematic review. Neurosci Biobehav Rev 2009 Jun;33(6):926–52.

Minzenberg MJ, Yoon JH, Carter CS. Modafinil modulation of the default mode network. Psychopharmacology (Berl) 2011 May;215(1):23–31.

Mukherjee S, Manahan-Vaughan D. Role of metabotropic glutamate receptors in persistent forms of hippocampal plasticity and learning. Neuropharmacology 2013 Mar;66:65–81.

Naviaux RK. Oxidative shielding or oxidative stress? J Pharmacol Exp Ther 2012 Sep;342(3):608–18.

Newcomer JW, Selke G, Melson AK, Hershey T, Craft S, Richards K, Alderson AL. Decreased memory performance in healthy humans induced by stress-level cortisol treatment. Arch Gen Psychiatry 1999 Jun;56(6):527–33.

Olivier JD, Jans LA, Blokland A, Broers NJ, Homberg JR, Ellenbroek BA, Cools AR. Serotonin transporter deficiency in rats contributes to impaired object memory. Genes Brain Behav 2009 Nov;8(8):829–34.

Pizzagalli DA. Frontocingulate dysfunction in depression: toward biomarkers of treatment response. Neuropsychopharmacology 2011 Jan;36(1):183–206.

Reus VI, Wolkowitz OM. Antiglucocorticoid drugs in the treatment of depression. Expert Opin Investig Drugs 2001 Oct;10(10):1789–96.

Richardson JT. Cognitive function in diabetes mellitus. Neurosci Biobehav Rev 1990 Winter;14(4):385–8.

Robbins T, Roberts A. Differential Regulation of Fronto-Executive Function by the Monoamines and Acetylcholine. Cerebral Cortex 2007 Jun;17(Supplement 1):i151–i160.

Rook GA, Raison CL, Lowry CA. Can we vaccinate against depression? Drug Discov Today 2012 May;17(9–10):451–8.

Rush AJ, Giles DE, Schlesser MA, Orsulak PJ, Parker CR, Jr., Weissenburger JE, Crowley GT, Khatami M, Vasavada N. The dexamethasone suppression test in patients with mood disorders. J Clin Psychiatry 1996 Oct;57(10):470–84.

Sahakian B, Morein-Zamir S. Depression and resilience: insights from cognitive, neuroimaging, and psychopharmacological studies. In: Delgado MR, Phelps EA, Robbins T, editors. Decision Making, Affect, and Learning. Oxford University Press: 2013. p. 327–53.

Skeberdis VA, Lan J, Zheng X, Zukin RS, Bennett MV. Insulin promotes rapid delivery of N-methyl-D- aspartate receptors to the cell surface by exocytosis. Proc Natl Acad Sci U S A 2001 March 13;98(6):3561–6.

Smith KA. Louis pasteur, the father of immunology? Front Immunol 2012;3:68.

Stellwagen D, Malenka RC. Synaptic scaling mediated by glial TNF-alpha. Nature 2006;440(7087):1054–9.

Talarowska M, Galecki P, Maes M, Bobinska K, Kowalczyk E. Total antioxidant status correlates with cognitive impairment in patients with recurrent depressive disorder. Neurochem Res 2012 Aug;37(8):1761–7.

Tang WH, Wang Z, Levison BS, Koeth RA, Britt EB, Fu X, Wu Y, Hazen SL. Intestinal microbial metabolism of phosphatidylcholine and cardiovascular risk. N Engl J Med 2013 Apr;368(17):1575–84.

Wang LY, Murphy RR, Hanscom B, Li G, Millard SP, Petrie EC, Galasko DR, Sikkema C, Raskind MA, Wilkinson CW, Peskind ER. Cerebrospinal fluid norepinephrine and cognition in subjects across the adult age span. Neurobiol Aging 2013 April 30. [Epub ahead of print]

Warburton DM. Nicotine as a cognitive enhancer. Prog Neuropsychopharmacol Biol Psychiatry 1992 Mar;16(2):181–91.

Wilson CJ, Finch CE, Cohen HJ. Cytokines and cognition—the case for a head-to-toe inflammatory paradigm. J Am Geriatr Soc 2002 Dec;50(12):2041–56.

Wood JN, Grafman J. Human prefrontal cortex: processing and representational perspectives. Nat Rev Neurosci 2003 Feb;4(2):139–47.

Young AH, Gallagher P, Watson S, Del Estal D, Owen BM, Nicol FI. Improvements in neurocognitive function and mood following adjunctive treatment with mifepristone (RU-486) in bipolar disorder. Neuropsychopharmacology 2004 Aug;29(8):1538–45.

Ziv Y, Ron N, Butovsky O, Landa G, Sudai E, Greenberg N, Cohen H, Kipnis J, Schwartz M. Immune cells contribute to the maintenance of neurogenesis and spatial learning abilities in adulthood. Nat Neurosci 2006 Feb;9(2):268–75.

Zobel AW, Schulze-Rauschenbach S, von Widdern OC, Metten M, Freymann N, Grasmader K, Pfeiffer U, Schnell S, Wagner M, Maier W. Improvement of working but not declarative memory is correlated with HPA normalization during antidepressant treatment. J Psychiatr Res 2004 Jul;38(4):377–83.

Chapter 8

Prevention and resilience of cognitive deficits in major depressive disorder

Key Points

- Cognitive reserve mediates the relationship between MDD pathology and cognitive performance.
- Psychosocial factors, premorbid intelligence, nutrition and physical activity are some examples of proposed indicators of cognitive reserve.

8.1 Introduction

Cognitive performance in MDD is highly variable between individuals, with some individuals exhibiting average or above average performance given similar illness characteristics. This suggests the existence of protective factors against cognitive decline. A hypothetical construct that is believed to mediate the relationship between pathology and cognitive performance is referred to as cognitive reserve (Figure 8.1).

8.2 Cognitive reserve

High cognitive reserve is associated with lower cognitive decline in the presence of increasing pathology (Figure 8.2). Cognitive reserve and neuroplasticity are interrelated constructs, based on the hypothesis that changes in cognitive function are related to continuous brain adaptations (e.g. synaptic connections, neurogenesis) that occur in response to disparate environmental demands (Fu and Zuo 2011). These adaptations can be either positive or negative, resulting in morphological changes that either maintain, protect or impair cognitive function (Vance et al 2010). Various proposed proxies of cognitive reserve have been described (Figure 8.3). In preclinical studies, it has been demonstrated that environmental enrichment protects animals from the effects of chronic stress on cognitive function and hippocampal integrity (Hutchison 2012—Hutchinson KM, McLaughlin KJ, Wright RL, Bryce Ortiz J, Anouti DP, Mika A, Diamond DM, Conrad CD. Neurobiol Learn Mem. 2012 Feb;97(2):250–60). Psychosocial factors have been shown to impact plasticity and may account for some of the interindividual variability in cognitive function in MDD (Jones et al 2011). Higher psychosocial functioning has also been suggested to delay or ameliorate

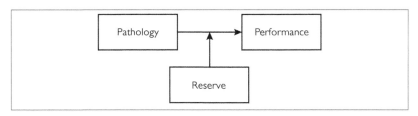

Figure 8.1 Cognitive reserve influences the relationship between pathology and performance

Jones RN, Manly J, Glymour MM, Rentz DM, Jefferson AL, Stern Y. Conceptual and measurement challenges in research on cognitive reserve. J Int Neuropsychol Soc 2011 July;17(4):593–601. Reproduced with permission of Cambridge University Press.

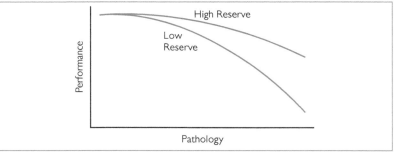

Figure 8.2 The relationship between pathology and performance. The relationship between pathology and performance is attenuated in high relative to low reserve.

Jones RN, Manly J, Glymour MM, Rentz DM, Jefferson AL, Stern Y. Conceptual and measurement challenges in research on cognitive reserve. J Int Neuropsychol Soc 2011 July;17(4):593–601. Reproduced with permission of Cambridge University Press.

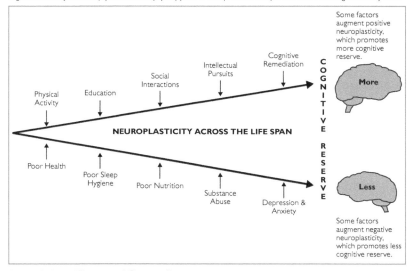

Figure 8.3 Neural Plasticity and Cognitive Reserve

Adapted from Vance et al. Journal of Psychosocial Nursing and Mental Health Services. 48; 4, 23–30. Copyright © 2013, Slack Inc. Reproduced with permission of SLACK Incorporated.

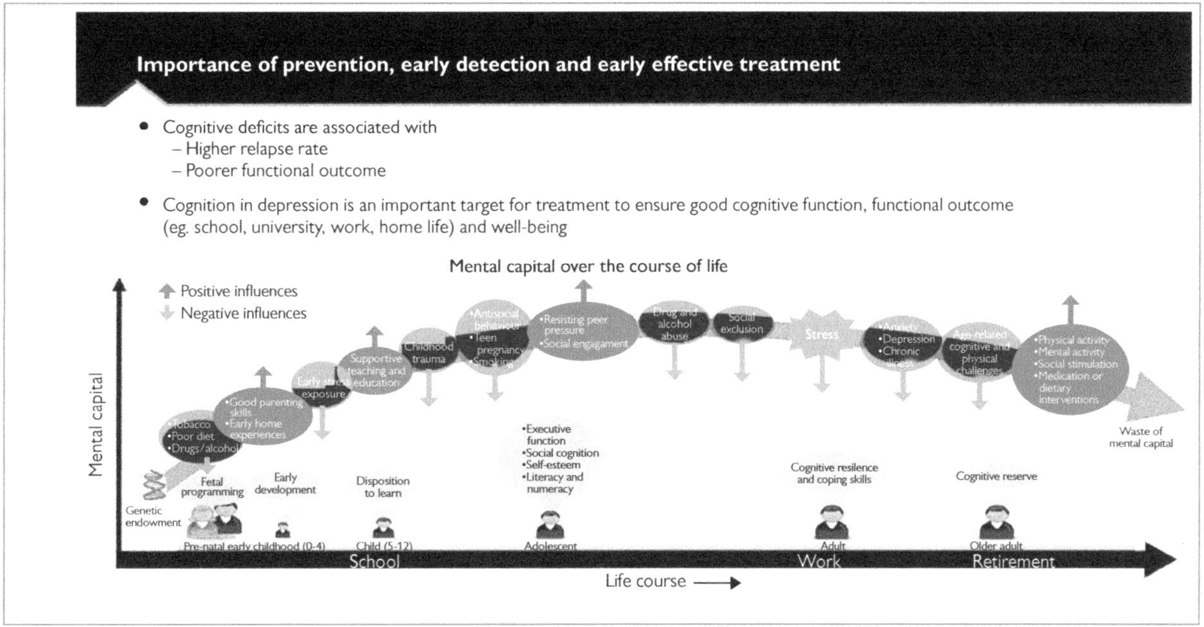

Figure 8.4 The importance of prevention, early detection and treatment for cognitive dysfunction in MDD

Reprinted by permission from Macmillan Publishers Ltd: Nature (Beddington et al, The mental wealth of nations. Vol 455, 23 October 2008), Copyright (2008).

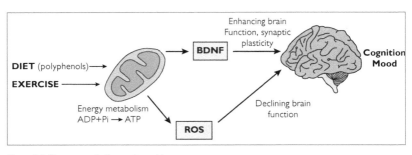

Figure 8.5 Energy metabolism and cognition

Reproduced with permission from Gomez-Pinilla, Fernando; Nguyen, Trang T J, 'Natural mood foods: The actions of polyphenols against psychiatric and cognitive disorders', Nutritional Neuroscience, Volume 15, Issue no.3, 2012, pp. 127–133(7), Figure 1. © W.S. Maney & Son Ltd 2012.

cognitive impairment associated with neurodegeneration in the non-MDD population. Education may promote synaptic growth and foster development of new cognitive strategies. It has also been reported that highly skilled occupational status is associated with a 50% reduction in the risk of dementia (Jones et al 2011; Valenzuela et al 2006) (Figure 8.4). Moreover, a healthy diet (e.g. Mediterranean) and regular physical activity regulates energy metabolism, reduces inflammation, reactive oxygen species, and increases neurotrophic factors, thereby enhancing synaptic plasticity and cognitive function (Figure 8.5) (Gomez-Pinilla and Nguyen 2012; Verghese et al 2009). Future research will need to elucidate which factors render individuals with MDD more resilient to cognitive decline, aiding in the development of more effective prevention and remediation strategies.

References

Fu M, Zuo Y. Experience-dependent structural plasticity in the cortex. Trends Neurosci 2011 Apr;34(4):177–87.

Gomez-Pinilla F, Nguyen TT. Natural mood foods: the actions of polyphenols against psychiatric and cognitive disorders. Nutr Neurosci 2012 May;15(3):127–33.

Jones RN, Manly J, Glymour MM, Rentz DM, Jefferson AL, Stern Y. Conceptual and measurement challenges in research on cognitive reserve. J Int Neuropsychol Soc 2011 Jul;17(4):593–601.

Valenzuela MJ, Sachdev P. Brain reserve and dementia: a systematic review. Psychol Med 2006 Apr;36(4):441–54.

Vance DE, Roberson AJ, McGuinness TM, Fazeli PL. How neuroplasticity and cognitive reserve protect cognitive functioning. J Psychosoc Nurs Ment Health Serv 2010 Apr;48(4):23–30.

Verghese J, Cuiling W, Katz MJ, Sanders A, Lipton RB. Leisure activities and risk of vascular cognitive impairment in older adults. J Geriatr Psychiatry Neurol 2009 Jun;22(2):110–8.

Chapter 9

The effect of treatment on cognitive deficits in major depressive disorder

Key Points

- Antidepressant pharmacotherapy exerts procognitive effects in major depressive disorder.
- As of yet, there is no with no compelling evidence that any class of antidepressants is more superior over another.
- Several lines of evidence indicate that some neurostimulatory approaches are associated with less cognitive impairment and may even exert procognitive effects in some individuals with MDD.

9.1 Introduction

Conventionally, the efficacy of treatment in MDD has been premised on the intervention's ability to reduce global depression severity over and above expectancy factors. Until recently, interventional research in MDD has not primarily aimed to mitigate cognitive deficits in individuals with MDD. The determination that cognitive deficits cause and maintain functional impairment in MDD underscores the relevance of identifying treatments that can mitigate and/or prevent cognitive deficits.

The hypothesized heterogeneous pathways subserving cognitive function in MDD provide the rationale for mechanistically dissimilar therapeutic interventions including, but not limited to, psychopharmacological, behavioural, neuromodulatory, and nutraceutical. It is well established that manually based psychotherapies (e.g. cognitive behavioural therapy) are effective in correcting abnormal emotional-cognitive processing in MDD. Notwithstanding, there is insufficient evidence to indicate that manually based psychotherapies have a direct effect on 'cold' cognitive functioning while some preliminary evidence suggests that psychotherapy (e.g., Cognitive Behavioural Therapy) may be beneficial for aspects of 'hot' cognition (e.g., emotional processing) (Baker R et al 2012).

Table 9.1 Mean change from baseline in depressive symptoms and cognition after 8 weeks of vortioxetine

Efficiency variables	LOCF, ANCOVA		MMRM	
	Lu AA21004	DUL	Lu AA21004	DUL
HRSD-24[a]	−3.32**	−5.48***	−3.82***	−6.12***
MADRS	−4.29***	−6.83***	−4.74***	−7.59***
HAM-A	−2.35**	−3.54***	−2.69***	−3.89***
CGI-S	−0.60***	−1.02***	−0.65***	−1.11***
CGI-I[b]	−0.56***	−0.84***	−0.62***	−0.92***
DSST (#correct)	2.79*	0.77	−	−
RAVLT acquisition	1.14*	1.41**	−	−
RAVLT delayed recall	0.47*	0.64**	−	−

Abbreviations: ANCOVA, analysis of covariance; CGI-I, Clinical Global Impression-Improvement; CGI-S, Clinical Global Impression-Severity; HAM-A, Hamilton Rating Scale for Anxiety; HRSD-24, Hamilton Rating Scale for Depression (24 items); LOCF, last observation carried forward; MADRS, Montgomery-Åsberg Depression Rating Scale; MMRM, mixed model for repeated measures.

[a] Primary efficacy variable: HAM-D24, full analysis set (FAS), LOCF, ANCOVA; other efficiency assessments are outside the statistical testing hierarchy and nominal P-values are shown.

[b] Treatment difference from placebo in mean CGI-I score at week 8. *$P < 0.05$; **$P < 0.01$; ***$P < 0.001$ vs placebo.

9.2 Psychopharmacology

9.2.1 Antidepressants

All commercially available and approved conventional antidepressants augment central indolamine and/or catecholamine neurotransmission (Muller and Schwarz 2007; Xu et al 2011). The monoaminergic effects of antidepressant therapy provides the basis for hypothesizing that, in addition to ameliorative effects on depressed mood and vegetative function, they could also offer a procognitive effect in individuals with MDD (Furtado et al 2012; Wagner et al 2012; McLennan and Mathias 2010). However, in the context of no compelling evidence of procognitive effects in healthy populations, a plausible hypothesis is that antidepressants likely rehabilitate cognitive function in the presence of psychopathology.

Each of the conventional classes of antidepressants (e.g. SSRIs, SNRIs, DA modulators (bupropion), norepinephrine reuptake inhibitors (reboxetine) and multimodal agents) have all been demonstrated to improve cognitive performance across heterogeneous samples of adults with MDD (Table 9.2; Herrera-Guzman 2010; Herrera-Guzman 2008; Ferguson et al 2003). A clinical impression has been that antidepressants that principally target catecholaminergic transmission and/or are multimodal in their putative mechanism of action may be more likely to offer a benefit across multiple cognitive domains (Herrera-Guzman 2010; Herrera-Guzman 2008; Ferguson et al 2003). Notwithstanding this observation, there are insufficient data to indicate that any specific category or any class of antidepressants is more likely to be procognitive.

Table 9.2 The effect of antidepressants on cognition in MDDa						
Author	**Sample**	**Age (mean, SD)**	**Method**	**Treatment & Dosing**	**Cognitive Measures**	**Results**
Culang-Reinlieb et al. (2012)	MDD: N = 63

Sertraline (n=33)

Nortriptyline (n=30) | Total Sample 64.19, 8.47

Sertraline 64.85, 8.83

Nortriptyline 63.47, 8.15 | 12 week, double-blind, randomized clinical trial comparing the effects of nortriptyline to sertraline on cognitive function

Randomization was preceded by a 1 week, single-blind placebo lead-in to ensure placebo did not reduce the HRSD score by ≥25% | *Sertraline* Week 1 = 50 mg Weeks 2–5 = 100 mg (if remission not observed: HRSD < 10) Week 9 = 200 mg (if no evidence of response)

Nortriptyline Dose calculated at 1 mg/kg Days 1–3: 1/3 of calculated dose mg/kg Days 3–6: 2/3 of calculated dose mg/kg Day 7: Full calculated dose at 1 mg/kg (Note: doses adjusted for plasma levels within 80–120 ng/mL) | MMSE, Purdue Pegboard, CPT, TMT A, TMT B, SRT, SCWT, Buschke Selective Reminding Test | *Sertraline* Significant improvement in verbal learning and memory on the Buschke Selective Reminding Test observed

Nortriptyline No significant change observed on any cognitive measure |
| Hinkelmann et al. (2012) | MDD: N = 52

Controls: N =50 | MDD: 35, 11.5

Controls: 35, 11.6 | 3 week, double-blind, randomized, controlled clinical trial

5 day lead-in washout period for subjects treated with antidepressants | Following antidepressant-free baseline testing, all subjects received escitalopram 10 mg (increased to 15 mg or 20 mg if required during following weeks) | RAVLT, Digit Span Test, ROCF, Taylor complex figure test, Letter Cancellation Test (d2), TMT A, TMT B | Significant improvement in verbal and non-verbal memory when compared to healthy controls |

(Continued)

Table 9.2 (Continued)

Author	Sample	Age (mean, SD)	Method	Treatment & Dosing	Cognitive Measures	Results
Herrera-Guzmán et al. (2010)	MDD: N = 73 *Escitalopram* (n=36) *Duloxetine* (n=37) Controls: N = 37	*Escitalopram* 32.91, 8.73 *Duloxetine* 33.21, 8.61 Controls 33.05, 8.04	24 week, open-label clinical trial evaluating the effects of escitalopram and duloxetine on cognitive function	Escitalopram 10 mg/day Duloxetine 60 mg/day Treatments were not titrated	CANTAB: Rapid Visual Information Processing, Match to Sample Visual Search, Stroop Test, Intra-Extra-Dimensional Set Shift, Stockings of Cambridge	Significant improvement in attention and executive function in both the escitalopram and duloxetine groups
Herrera-Guzmán et al. (2009)	MDD: N = 73 *Escitalopram* (n=36) *Duloxetine* (n=37) Controls: N = 37	*Escitalopram* 32.91, 8.73 *Duloxetine* 33.21, 8.61 Controls 33.05, 8.04	24 week, open-label clinical trial evaluating the effects of escitalopram and duloxetine on cognitive function	Escitalopram 10 mg/day Duloxetine 60 mg/day Treatments were not titrated	WAIS III vocabulary subtest, WAIS III digit span, RAVLT, CANTAB: SSP, SWM, PRM, PAL, DMS, SRM, RTI, Stroop Test	Significant improvement in visual and verbal memory in both escitalopram and duloxetine groups
Culang et al. (2009)	MDD: N = 174 *Citalopram* (n=84) *Placebo* (n=90)	Total Sample 79.57, 4.36 *Citalopram* 79.82, 3.97 *Placebo* 79.33, 4.69	8 week, double-blind, placebo-controlled trial evaluating the effect of citalopram on cognitive function 1 week single-blind placebo lead-in	Citalopram 20 mg/day End of Week 4: Patients with an HRSD score >10 had the dose increased to 40mg/day Placebo	MMSE, WAIS III digit symbol subtest, CRT, JOLO, SCWT, Buschke Selective Reminding Test	Significant improvement in the citalopram group on the Buschke Selective Reminding Test when adjusting for baseline MMSE and Digit Symbol

Fales et al. (2009)	MDD: N = 23	MDD 36.4, 9.4	fMRI of patients with MDD before and after antidepressant treatment	Escitalopram 10mg/day immediately following first fMRI scan; subsequent doses determined by clinical response	The emotion-interference task	No significant improvement in response time
	Controls: N = 18	Controls 33.4, 8.2		Other antidepressants prescribed due to patient preference: Sertraline 150 mg (n=3) Sertraline 100 mg (n=1) Paroxetine 20 mg (n=2)		Subjects demonstrated a significant increase in recruitment of dorsolateral prefrontal cortex following antidepressant treatment, suggesting this brain region's involvement in emotional interference tasks
Wise et al. (2007)	MDD: N = 311 Duloxetine (n=207) Placebo (n=104)	MDD without comorbidity 71.3, 5.3 MDD with comorbidity 73.4, 5.7	Double-blind, placebo-controlled study evaluating the impact of medical comorbidity on the efficacy and tolerability of duloxetine. 1 week double-blind placebo phase prior to randomization to duloxetine or placebo	Duloxetine 60 mg/day followed by a 1-week, double-blind, discontinuation phase wherein duloxetine was tapered to 30mg/day for 4 days	Verbal Learning and Recall Test, DSST, Two-Digit Cancellation Test, LNS	Significant improvement in the composite cognitive score among duloxetine-treated subjects when compared to placebo. No significant difference occurred in composite cognitive score when comparing subjects with vs. without a medical comorbidity

(Continued)

Table 9.2 (*Continued*)

Author	Sample	Age (mean, SD)	Method	Treatment & Dosing	Cognitive Measures	Results
Gualtieri et al. (2007)	MDD: N = 81 Controls: N = 27	MDD: 18–65 Bupropion 44.00 Venlafaxine 46.11 SSRIs 43.81	Naturalistic, cross-sectional study evaluating cognitive differences among individuals using antidepressants	Bupropion (n=27) Venlafaxine (n=27) SSRIs (n=27) Dosing not available	CNS Vital Signs: Verbal memory, visual memory, finger tapping, symbol-digit coding, the stroop test, the shifting attention test, continuous performance test	No significant difference in cognitive performance when comparing the bupropion group to controls Individuals taking SSRIs scored significantly lower on tests of psychomotor speed, cognitive flexibility, and reaction time when compared to controls The venlafaxine group score worse than controls on measures of reaction time
Raskin et al. (2007)	MDD: N = 311 *Duloxetine* (n=207) *Placebo* (n=104)	*Duloxetine* 72.6, 5.7 *Placebo* 73.3, 5.7	8-week, double-blind, placebo-controlled clinical trial evaluating the efficacy of duloxetine on cognition	Duloxetine 60 mg/ day followed by a 1-week, double-blind, discontinuation phase wherein duloxetine was tapered to 30mg/day for 4 days, followed by replacement with placebo	Composite cognitive score based on: the Verbal Learning and Recall Test, DSST, Two-Digit Cancellation Test, LNS	Significant improvement was observed in the composite cognitive score (particularly verbal learning and memory) in the duloxetine group when compared to placebo

Study	Sample	Age	Study design	Dosing	Measures	Results
Gualtieri et al. (2006)	MDD: N = 69 MDD drug-free (n=38) MDD Rx (n=31) Controls: N = 69	MDD drug-free 38.11, 9.95 MDD Rx 43.55, 10.68 Controls 41.30, 11.40	Naturalistic, cross-sectional study evaluating of cognitive deficits among individuals with MDD	No dosing available. MDD Rx subjects were those who responded to antidepressant monotherapy	CNS Vital Signs: verbal memory, visual memory, finger tapping test, DSST, the stroop test, the shifting attention test, the continuous performance test	Post hoc analysis demonstrated significant differences between the MDD drug-free and MDD Rx groups in domains of complex attention and vigilance
Constant et al. (2005)	MDD: N = 20 Controls: N = 26	MDD: 21–74 47.65 Controls: 21–69 48.85	7-week, open-label study evaluating the effects of sertraline on attention and executive function	Sertraline 50 mg (n=1, sertraline 75 mg; n=1, sertraline 75 mg after 3 weeks of treatment)	Phasic Alertness Task, Classic Stroop, The Supraliminal and Subliminal Emotional Stroop, Autoquestionnaires	Sertraline demonstrated improvement on psychomotor slowing and executive function
Trick et al. (2004)	MDD: N = 88 Venlafaxine (n=45) Dosulepin (n=43)	Venlafaxine 71.5, 7.3 Dosulepin 71.0, 5.7	26-week, prospective, double-blind, randomized, active comparator controlled study comparing the cognitive and psychomotor effects and efficacy of venlafaxine and Dosulepin 1 week, post-study follow-up period	Venlafaxine 37.5 mg twice per day OR Dosulepin 25 mg in the morning followed by 50 mg in the evening	Critical Flicker Fusion Threshold, Short-term Memory – Kim's Game, Serial subtraction of numbers, Cognitive Failures Questionnaire	No significant difference was observed between the two treatment groups in short-term memory or on the Cognitive Failures Questionnaire Results suggest venlafaxine at doses of 37.5 mg b.i.d. in this population do not disrupt cognitive function or psychomotor performance

(Continued)

Author	Sample	Age (mean, SD)	Method	Treatment & Dosing	Cognitive Measures	Results
Ferguson et al. (2003)	MDD: N = 74	MDD: 18–65	8-week, randomized, double-blind, placebo- and active-treatment-controlled, fixed/flexible dose comparisons of reboxetine, paroxetine, and placebo	Reboxetine 8–10 mg/day Paroxetine 20–40 mg/day Placebo	Cognitive Drug Research test battery: Immediate Word Presentation, RTI, Digit Vigilance Task, CRT, Numeric Working Memory, Word Recognition, Critical Flicker Fusion Threshold	Significant improvements were reported for reboxetine in sustaining attention and in speed of cognitive functioning at day 56 compared with baseline No significant changes were observed among subjects receiving either paroxetine or placebo

Table 9.2 (Continued)

^aStudies listed in reverse chronological order.

Herrera-Guzman et al (2010; 2009) reported on the cognitive effects of the SSRI escitalopram and the SNRI duloxetine in adults (aged 20–50 years) with MDD. The authors observed that both treatments improved working memory, attention, and executive function as well as psychomotor processing speed (Herrera-Guzman 2010; Herrera-Guzman 2009). It was found that duloxetine, however, was superior to escitalopram on episodic and working memory, whereas no significant differences between the antidepressants were observed on measures of attention and executive function (Herrera-Guzman 2010; Herrera-Guzman 2009).

Raskin et al (2007) sought to determine the effect of duloxetine vs placebo on several measures of cognitive performance. They reported that, when compared to placebo, 8 weeks of treatment with duloxetine significantly improved the global cognitive composite score in cognitively intact individuals with MDD ($\geq$65 years old; Raskin et al 2007). The improvement in the duloxetine-treated group was largely accounted for by an improvement in measures of verbal learning and recall, with trending between-group differences noted on measures of attention, psychomotor processing speed (i.e. Digit Symbol Substitution Test (DSST)), visual attention, and executive function (i.e. Two Digit Cancellation Test), and working memory/executive function (i.e. Letter-Number Sequencing (LNS) Test; Figure 9.1; Raskin et al 2007). Path analyses showed that improvements on the cognitive composite score was a 90.9% direct effect and a 90.1% indirect effect through improvement in a Geriatric Depression Scale Total Score (Figure 9.1; Raskin et al 2007).

Vortioxetine (Lu AA21004) is a novel multimodal antidepressant that is hypothesized to mitigate depressive symptom severity via its effects on the serotonin transporter and serotonergic receptors (Figure 9.2). More specifically, vortioxetine exhibits $5\text{-}HT_3$ and $5HT_7$ receptor antagonism, $5\text{-}HT_{1A}$ receptor agonism, $5\text{-}HT_{1B}$ receptor partial agonism, and inhibition of the 5-HT transporter. In addition to its effect on indolamine signalling, vortioxetine also increases central neurotransmission of norepinephrine, dopamine, acetylcholine, and histamine (Bang-Andersen et al 2011). Vortioxetine has demonstrated antidepressant efficacy in adult and elderly individuals with MDD populations as well as in adults with generalized anxiety disorder (Figure 9.2; Katona et al 2012; Alvarez et al 2012).

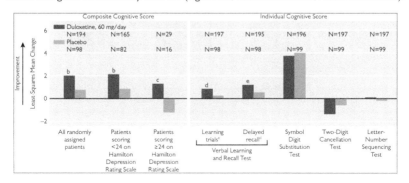

Figure 9.1 Cognitive score changes among depressed patients randomly assigned to DUL or placebo

Efficacy of Duloxetine on Cognition, Depression, and Pain in Elderly Patients With Major Depressive Disorder: An 8-Week, Double-Blind, Placebo-Controlled Trial Joel Raskin, M.D., F.R.C.P.C.; Curtis G. Wiltse, Ph.D.; Alan Siegal, M.D., M.B.A.; Javaid Sheikh, M.D.; Jimmy Xu, Ph.D.; James J. Dinkel; Benjamin T. Rotz, R.Ph.; Richard C. Mohs, Ph.D. Am J Psychiatry 2007;164:900-909. Reprinted with permission from The American Journal of Psychiatry, (Copyright ©2007). American Psychiatric Association.

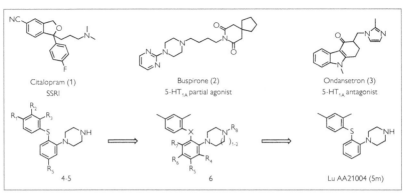

Figure 9.2 Structures of citalopram (1), buspirone (2), ondansetron (3), target compounds (4–6) and vortioxetine (5m)

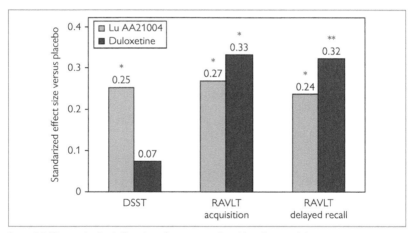

Figure 9.3 The standardized effect sizes of vortioxetine 5 mg/day relative to duloxetine

Author	Sample	Age (mean, SD)	Method	Treatment	Cognitive Measures	Results
Electroconvulsive Therapy						
Brakemeier et al. (2011)	MDD: N = 90 Bilateral (n = 18) Right unilateral (n = 18)	MDD Bilateral 50.6, 18.8 Right unilateral 54.4, 16.0	Prospective, randomized, double-masked trial comparing the effects of pulse width and electrode placement on cognitive outcomes	ECT conditions differed in electrode placement (bilateral vs. right unilateral) and pulse width (0.3 vs. 1.5 ms)	Cognitive Failures Questionnaire, Squire Memory Complaint Questionnaire, Global Self-Evaluation – Memory)	Cognitive Failures Questionnaire and Squire Memory Complaint Questionnaire scores improved post-ECT when compared to pre-ECT Post-ECT Global Self-Evaluation – Memory scores improved when treated with ultrabrief right unilateral ECT Global Self-Evaluation – Memory post right unilateral ECT were poorer when compared to those who received bilateral ECT
Roepke et al. (2011)	MDD: N = 40 40-Hz Group (n = 20) 100-Hz Group (n = 20)	40-Hz Group 54.3, 15.5 100-Hz Group 57.9, 14.3	Stimulus pulse-frequency-dependent efficacy study on cognitive outcomes	Either 40-Hz or 100-Hz stimulation with equal initial stimulus doses in 9 sessions of suprathreshold right unilateral, ultrabrief-pulse ECT	Verbal Learning Memory Test (VLMT): immediate, delayed, recognition, WMS-R, Regensburger Wortflüssigkeits Test	No significant difference between frequency groups (40-Hz vs. 100-Hz) on measures of verbal or working memory

(Continued)

Table 9.3 (Continued)

Author	Sample	Age (mean, SD)	Method	Treatment	Cognitive Measures	Results
Warnell et al. (2011)	MDD: N=20	MDD 57.8, 8.2	2-week, randomized clinical study evaluating the effects of 0.5 ms ECT on cognition	Stimuli were 0.5 ms pulse width and 900 milliamperes (mA) current, bitemporal electrode placement, frequency ranged from 10–70 Hz (mean 31.5Hz), mean charge was 192.5 millicoulombs (mC) at 900 mA Administered across 6 sessions, 3 times a week	MMSE	No significant difference in MMSE scores pre- to post-ECT
Sackeim et al. (1993)	MDD: N=96 **Right Unilateral** Low dose (n=23) High dose (n=23) **Bilateral** Low dose (n=23) High dose (n=27)	**Right Unilateral** Low dose 57, 16 High dose 56, 15 **Bilateral** Low dose 60, 13 High dose 53, 15	Randomized, double-blind study evaluating the effects of stimulus intensity and electrode placement on cognition following ECT 2 month follow up. Treatment responders were followed for 1 year to assess relapse rates	ECT was administered 3 times per week Low dose The electrical intensity that resulted in a generalized seizure in the 1st session was the one administered at each subsequent treatment session High dose At the 2nd and subsequent treatment sessions each subject received stimulation of an electrical intensity that in units of charge was 2.5 times the threshold identified in the 1st treatment session Frequency of brief pulses ranged from 20-Hz to 140-Hz	MMSE, recognition memory of paired words and faces, the Buschke Selective Reminding Test, Autobiographical Memory Interview, Squire Subjective Memory Questionnaire	No difference in cognitive effects were reported between the two treatment groups two months following treatment Unilateral treatment associated with fewer severe deficits in cognition following treatment

Transcranial Magnetic Stimulation

Isserles et al. (2011)	MDD: N = 46 No cognitive-emotional reactivation group (n = 20) Positive cognitive-emotional reactivation group (n = 14) Negative cognitive-emotional reactivation group (n = 11)	No cognitive-emotional reactivation group 45.40, 13.18 Positive cognitive-emotional reactivation group 45.93, 12.98 Negative cognitive-emotional reactivation group 41.75, 12.70	4-week, randomized study using deep transcranial magnetic stimulation	4 weeks of daily 20-Hz stimulation sessions plus an additional 4 weekly sessions as a brief maintenance phase Two groups received either positive or negative cognitive-emotional reactivation during stimulation sessions	Mindstreams (computerized system designed to assess cognitive performance)	Negative cognitive-emotional reactivation can disrupt the antidepressant effects of deep transcranial magnetic stimulation
Fitzgerald et al. (2009)	MDD: N = 27 High frequency (left-sided) (n = 15) Low frequency (right-sided) (n = 11)	High frequency (left-sided) 42.12, 9.32 Low frequency (right-sided) 46.54, 11.43	3-week, two-arm, randomized, double-blind study of high frequency left-sided rTMS and low frequency right-sided rTMS	High frequency 30 stimulation trains of 5s in duration at 100% of the resting motor threshold Low frequency Stimulation applied in 4 trains of 180s duration (30s inter-train interval) at 110% of the resting motor threshold 3 weeks of treatment with a possible 1 week extension. Non-responders were offered the opportunity to cross over to the alternate treatment	Brief Visuospatial Memory Test, HVLT-R, COWAT, WAIS-R	Significant improvements were observed on measures of immediate verbal memory and verbal fluency. Improvements on these cognitive measures were not dependent on the type of rTMS received

(Continued)

Table 9.3 (Continued)

Author	Sample	Age (mean, SD)	Method	Treatment	Cognitive Measures	Results
Levkovitz et al. (2009)	MDD: N=65 H1-120% (n=24) H2-120% (n=22) H1L-120% (n=11) H1L-110% (n=8)	MDD: 18–65	4-week, randomized, double-blind study evaluating cognitive effects of deep transcranial magnetic stimulation	Three groups were treated with 20-Hz trains at a stimulation intensity of 120% motor threshold or a lower intensity of 110% motor threshold The three groups differed in stimulation laterality (H1-coil [H1-120%]: greater stimulation over the left prefrontal cortex; H2-coil [H2-120%]: induces bilateral stimulation; and H1L-coil [H1L-120%/ H1L-110%]: induces stimulation exclusively over the left prefrontal cortex) Each TMS session consisted of 42-second trains with an inter-train interval of 20s	CANTAB	Individuals receiving deep left-lateralized treatments demonstrated normalization of significantly impaired cognitive domains at baseline
Vander-hasselt et al. (2009)	MDD: N=16	MDD: 22–61	Double-blind, placebo-controlled, crossover, within subjects design	20 minutes of 10-Hz or placebo (sham) rTMS over the left dorsolateral prefrontal cortex	Visual Analogue Mood Scales, Decision Time, Movement Time	Following a high frequency vs. placebo rTMS over the left dorsolateral prefrontal cortex, beneficial effects were observed for task-switching performance
Mosimann et al. (2004)	MDD: N=24	MDD 62, 12	2-week, randomized, double-blind, placebo-controlled study of rTMS	10 daily rTMS sessions (20-Hz, 2s trains, 28 intertrain intervals, 100% of motor threshold) Sham Coil tilted to 90°	MMSE, measurement for verbal memory with a verbal learning task, the stroop test, TMT A, TMT B, word fluency test	No differences were observed on the MMSE from baseline to endpoint

Padberg et al. (1999)	MDD: N=18	MDD: 51.2, 16.1	1-week, randomized, double-blind, sham controlled study of rTMS in pharmacotherapy-refractory MDD	Fast rTMS 10-Hz Slow rTMS 0.3-Hz Sham	Verbal learning task for explicit verbal learning and the retrieval of verbal information from short-term memory	Verbal memory performance improved significantly following fast (10-Hz) rTMS treatment Significant improvement in learning performance following fast rTMS, but not slow rTMS was reported.

Deep Brain Stimulation

Grubert et al. (2011)	MDD: N=10	MDD: 32–65	1 year, prospective, longitudinal study for the neuropsychological safety of DBS in MDD	Stimulation applied with permanent pulse-train stimulation from 2 to 4 volts in steps of 1 volt Pulse width (90 μs), frequency (130-Hz), and electrode settings (contacts 1 and 2 negative against case) were kept constant for the first 4 weeks Parameters ranged from 1.5–10.0 volts, 100–150 Hz, and 60–210 μs	MMSE, d2 Aufmerksamkeits-Belastungstest, Verbal Learning and Memory Test, the Rey Visual Design Learning test, the HAWIE lexis tests, WMS-R, TMT A, TMT B, the Five-Point Test, the stroop test, SCWT, the Hooper Visual Organization Test	Significant improvements in domains of attention, learning and memory, executive function, and visual perception were reported
McNeely et al. (2008)	MDD: N=6	MDD 46, 8	1 year, prospective, longitudinal study	Contacts were established 1 per hemisphere. Stimulation parameters were monopolar 3.0 to 4.5 volts, 130-Hz, and 60 μs pulse widths	Object Alternation Test, Iowa Gambling Task	Cognitive domains exhibiting impairments at baseline improved over follow-up and these changes did not correlated with improvements in mood

(Continued)

Table 9.3 (*Continued*)

Author	Sample	Age (mean, SD)	Method	Treatment	Cognitive Measures	Results
High Fr3quency Magnetic Seizure Therapy						
Fitzgerald et al. (2013)	MDD: N=13	MDD 46.77, 14.82	Open label, cross-sectional study evaluating the effects of high-frequency magnetic seizure therapy on cognition in MDD	18 sessions Single stimulation trains at 100% machine output applied at 100-Hz Stimulation for treatment was provided using a train 4s longer than the individual ST	WTAR, Autobiographical Memory Interview – Short Form, RAVLT, Brief Visuospatial Memory Test, Prose Passages, WAIS-R, Rey Complex Figure Test, Digit Span, Digit Symbol Coding, TMT, SCWT, COWAT	Non-significant improvements on most cognitive measures were observed after controlling for multiple comparisons

ªStudies listed in reverse chronological order.

In addition to antidepressant properties, vortioxetine has demonstrated cognitive-enhancing properties in animal and human studies (Mork et al 2012). For example, vortioxetine showed statistically significant improvement on acquisition and delayed recall (as measured by The Rey Auditory Verbal Learning Test (RAVLT)), as well as attention and processing speed (as measured by DSST) in a group of elderly (n = 453) non-demented individuals with MDD (subjects receiving duloxetine exhibited significant improvement only on the RAVLT but not DSST) (Table 9.3; Katona et al 2012) (see Table 9.1 and Figure 9.3). A path analysis concluded that 83% of the improvement on DSST with vortioxetine was a direct effect (duloxetine 26%), whereas the direct effect of vortioxetine on acquisition and delayed recall was 71% and 72%, respectively (for duloxetine, 65% and 66%, respectively; Katona et al 2012). The beneficial effects of vortioxetine in this study on cognitive performance were a secondary outcome, providing an empirical basis for evaluating the effect of this agent on cognition and, consequently, psychosocial function as a primary outcome in younger adults with MDD (Katona et al 2012).

Several lines of evidence indicate that some antidepressants may exert an adverse effect on cognitive performance, notably agents with prominent sedative/somnologenic effects, as well as agents with prominent *in vivo* anticholinergic properties. A not infrequent observation in individuals receiving maintenance treatment with SSRI therapy is subjective complaints of cognitive dulling, apathy, and reduced processing speed. An observation has been made that the beneficial effects of SSRIs on cognition may in part be mediated by a history of stressful life events, insofar as individuals reporting stressful life events are more likely to demonstrate improvement in cognitive performance when receiving tricyclic antidepressants (TCAs). It has been hypothesized that the reciprocal interaction between central serotonergic and catecholaminergic systems may mediate this phenomenon. Table 9.2 summarizes antidepressant treatments with preliminary evidence for procognitive effects in individuals with MDD. None of the treatments, however, have been subjected to sufficient and rigorous evaluation in replicated, large, double-blind, placebo-controlled trials.

9.3 Aerobic Exercise

Aerobic exercise exerts favourable effects on multiple interacting physiological systems implicated in MDD pathophysiology. Moreover, preclinical models indicate that aerobic exercise promotes neurogenesis, synaptic plasticity, and neuronal cell survival. Manualized aerobic exercise interventions have been evaluated across mixed psychiatric populations of variable age. For example, aerobic exercise has been documented to mitigate depressive symptoms in SSRI non-responders with MDD and to improve cognitive performance in middle-aged adults with schizophrenia. However, there is a paucity of studies, that have primarily aimed to document the procognitive effects of aerobic exercise in adults with MDD. A compelling rationale for evaluating the procognitive effects of aerobic exercise in the non-geriatric MDD population, could be made based on observations of improved cognitive performance in geriatric individuals with MDD and its documented antidepressant properties in younger adults with MDD, (Knöchel et al 2012).

Table 9.4 Neuromodulatory therapeutic approaches	
Neuromodulatory Approach	**Description**
Electroconvulsive Therapy (ECT)	• Brief electrical pulse administered to induce a generalized seizure • Typically followed by maintenance pharmacology
Repetitive Transcranial Magnetic Stimulation (rTMS)	• An electrical current is passed through a metal coil placed on the patient's scalp; the current generates a magnetic field that passes through the scalp and skull to the brain to induce neuronal depolarization • Less invasive when compared to other neuromodulatory approaches
Deep Brain Stimulation (DBS)	• Drilling of burr holes using stereotactic guidance to implant electrodes into simulation sites within the brain; electrodes are connected to a pacemaker • The stimulator remains active 24h per day and can be adjusted using a telemetric wand
Magnetic Seizure Therapy (MST)	• A brain stimulation method using transcranial magnetic stimulation to induce therapeutic seizures using convulsive parameters in the same setting used for electroconvulsive therapy (ECT) • Preliminary evidence suggests that magnetic seizure therapy may be superior to ECT on measures of attention, retrograde amnesia, and category fluency

9.4 Neuromodulation

Neuromodulatory therapeutic approaches (e.g. electroconvulsive therapy (ECT), repetitive transcranial magnetic stimulation (rTMS), deep brain stimulation (DBS), magnetic seizure therapy (MST), and vagus nerve stimulation (VNS)) are frequently employed in patients with severe depression who, despite multiple antidepressant/augmentation trials (of sufficient duration and intensity) in combination with psychotherapy, fail to achieve therapeutic response (Table 9.3; for a brief description of conventional neuromodulatory therapies see Table 9.4).

The detrimental effects of ECT on cognitive function (e.g. retrograde amnesia) are well established and are among the principal reasons cited for poor patient acceptability and adherence. The hazardous effects of ECT on neurocognition along with the established beneficial effects of other neurostimulatory treatments provides the impetus for determining whether a beneficial effect on cognition could be obtained. Preliminary results support transcranial direct current stimulation (tDCS) for working memory.

Several studies have evaluated the effect of rTMS, MST, tDCS, VNS, and DBS across several cognitive domains. Taken together, non-ECT neurostimulatory approaches are associated with less cognitive impairment than ECT. There are some lines of evidence to suggest that, in subpopulations of individuals with MDD, the foregoing neurostimulatory approaches may benefit cognition. For example, McNeely et al. (2008) have reported improvements in verbal and visual memory, motor speed, and verbal learning in individuals receiving up to 12 months of DBS for treatment-resistant depression.

Table 9.5 Effects of alternative, behavioural, and nutraceutical treatments on cognitive function[a]

Author	Sample	Age (mean, SD)	Method	Treatment	Cognitive Measures	Results
Antypa et al. (2012)	History of at least one MDE: N = 71 Omega-3 (n=36) Placebo (n=35)	Omega-3 25.8, 11.8 Placebo 23.5, 6.0	4-week, randomized, double-blind study evaluating the effects of omega-3 on mood and emotional infor-mation processing in recovered depressed individuals	Daily: Fish oil (2.3 g of n-3 PUFA (including 1.74 g eicosapentae-noic acid [EPA] + 0.25 g DHA) Placebo (olive oil)	Affective Go/No-Go task, Go/No-Go task, Facial Expression Recognition task, Decision-Making (Gambling) task	No significant effects were observed for memory, attention, or cognitive reactivity
Levkovitz et al. (2012)	MDD: N = 46 Adjunct SAMe (n = 27) Adjunct Placebo (n = 19)	Adjunct SAMe 54.3, 13.5 Adjunct Placebo 50.5, 9.7	Secondary Analysis of a 6-week, double-blind, randomized clinical trial of adjunctive S-adenosylmethionine (SAMe) for MDD	400 mg SAMe pill b.i.d. Placebo (dummy pills) b.i.d.	General Hospital Cognitive and Physical Functioning Questionnaire	Significant improve-ment was reported for information recall
Lavretsky et al. (2011)	MDD: N = 112 Escitalopram + Tai Chi (n = 36) Escitalopram + Health Education (n = 37)	MDD: Escitalopram + Tai Chi 69.1, 7.0 Escitalopram + Health Education 72.0, 7.4	10-week, rand-omized, single-blind trial evaluating the complementary use of Tai Chi to aug-ment escitalopram treatment	Initial dose of escitalopram 10 mg/day (responders able to opt for increasing dosage to 20 mg/day following 4 weeks of treatment) Randomized to either Tai Chi or Health Education for 2 hours per week	Brief neuropsy-chological assess-ment battery developed and administered by one of the authors of the study	Significant improvement reported for individuals in the escitalopram + Tai Chi group for memory

Author	Sample	Age (mean, SD)	Method	Treatment	Cognitive Measures	Results
Hvas et al. (2004)	MDD: N = 140 *Vitamin B-12* (n=70) *Placebo* (n=70) (Individuals with increased plasma methylmalonic acid (0.40–2.00µmol/L) not previously treated with vitamin B-12 were included for this study)	*Vitamin B-12* 75 *Placebo* 74	Randomized, placebo-controlled study evaluating the effect of vitamin B-12 treatment on cognitive function in depression	Participants received injections of either: 1 mg Cyanocobalamin OR 1 mL Isotonic sodium chloride Weekly for 4 weeks	Cambridge Cognitive Examination (CAMCOG), MMSE, 12-words learning test	No significant effects were observed between the treatment and placebo group on any cognitive measure
Michalon et al. (1997)	MDD: N = 30 (Seasonal Affective Disorder [SAD]) Controls: N = 29	*MDD* *White light* 38.5 Range: 20–54 *Red light* 39.1 Range: 20–56 *Controls* 36.7 Range: 22–59	2-week cross-sectional study evaluating the effects of light therapy on neuropsychological function and mood in seasonal affective disorder	Light therapy was administered between 06:00 and 08:00 daily during 2 consecutive weeks between October and February	ROCF, Taylor Complex Figure Test, Facial Recognition, Cognitive Failures Questionnaire	No differences were observed between red and white light on any neuropsychological measures Significant improvement was reported for performance on visual memory tests following both light therapies (although scores remained below control group levels)

Table 9.5 (*Continued*)

Cognitive Remediation					
Siegle et al. (2007)	MDD: N = 31 *Cognitive Control Training (CCT)* (n=10) *Treatment-as-usual* (n=10)	MDD: 18–55	2-week, single-blind, adjunctive intervention study using "cognitive control training" for severe MDD	*Cognitive Control Training:* Two components geared towards activating the prefrontal cortex: (1) exercising prefrontal function in the context of likely automatic ruminative cognitions; (2) prefrontal control and use of working memory in the presence of frustration (Note: Full intervention protocol and computer-based stimuli available from authors upon request) *Treatment as usual:* Intensive out-patient day-treatment program (IOP) for 3h, 3 days per week and meet with the program psychiatrist once a week for clinical management	Wells's (2000) Attention Training, the Paced Auditory Serial Addition task (PASAT)	Significant improvement on the non-adaptive PASAT task was observed pre- to post-treatment among subjects who had received adjunctive CCT when compared to healthy controls Six subjects with MDD had completed fMRI assessment pre- and post-CCT. Individuals with MDD demonstrated disruptions in amygdala activity on emotional tasks and the DLPFC during cognitive tasks (i.e. digit sorting). Following CCT treatment, subjects demonstrated increased amygdala response to positive words, but decreased to negative and neutral words. Moreover, following CCT treatment the left DLPFC responses decreased during easier conditions and increased in more difficult conditions.

(Continued)

Table 9.5 (Continued)

Author	Sample	Age (mean, SD)	Method	Treatment	Cognitive Measures	Results
Elgamal et al. (2007)	MDD: N = 24 *MDD: receiving cognitive retraining package* (n = 12) *MDD: matched sample* (n = 12) Controls: N = 22	*MDD: receiving cognitive retraining package* 50.26, 6.41 *MDD: matched sample* 47.42, 6.78 Controls 49.1, 9.0	10-week course of cognitive remediation in patients with long-term MDD cross-sectionally evaluating four targeted cognitive domains	The treatment group completed the training sessions independently, using the PSSCogReHab cognitive remediation software program 2 times per week over a 10 week period	CVLT, Ruff's 2&7 Selective Attention Test, WAIS-R: Digit Span Forwards and Backwards, and Similarities subtests, TMT A, TMT B, COWAT	Subjects receiving cognitive training improved cognition in areas relevant to attention, verbal learn and memory, psychomotor speed, and executive function. Improvements across the foregoing cognitive domains exceeded those of the matched MDD sample that did not receive cognitive training
Naismith et al. (2008)	MDD: N = 44 *Immediate treatment* (n = 24) *Wait-list control* (n = 20)	Total Sample 64.8, 8.5	10-week, wait-list control study evaluating the effects of combined psychoeducation and cognitive training in late-life depression	The intervention was conducted in a group format comprising of less than 10 subjects *Psychoeducation:* 10 one hour sessions per week covering: the brain and neuropsychological assessment, keeping the brain healthy, sleep, diet & exercise, attention & processing speed, learning & memory, executive functions and general cognitive strategies, depression in older adults and pharmacological treatments, non-pharmacological treatments for depression in older adults, anxiety	RAVLT, TMT A, TMT B, D-KEFS, Rey Complex Figure Test, WMS-R	Significant improvements in learning and memory were observed in older people with a history of MDD and cognitive decline without dementia by participating in a combined psychoeducation and cognitive therapy program

			Cognitive Therapy: sessions lasted 1 hour, conducted once per week over a 10-week period using Medalia's Neuropsychological Educational Approach to Remediation (NEAR)		Significant improvement reported for memory encoding and retention
Naismith et al. (2010)	MDD: N = 16 *Cognitive Training* (n=8) *Waitlist* (n=8)	Total Sample 33.5, 9.9 *Cognitive Training* 33.6, 9.2 *Waitlist* 33.4, 11.2	10-week, non-randomized, waitlist control study evaluating cognitive training in affective disorders for improving memory	The intervention was Medalia and Revheim's (1999) Neuropsychological Educational Approach to Remediation (NEAR) Training utilizes commercially available computer games tailored to an individual's strengths and weaknesses, but delivered in a group-setting	Standardized battery of neuropsychological tasks based on extant evidence with alternate forms available for counter-balanced administration
Choi et al. (2011)	MDD: N = 8 (subjects received right unilateral ECT prior to cognitive remediation therapy)	MDD 49.35, 8.43	Open pilot cohort study to evaluate the treatment feasibility of cognitive remediation in individuals with MDD who have undergone ECT	Memory Training for ECT (Mem-ECT): Consists of 7 training sessions, each lasting approximately 20–30 min, incorporating paper-and-pencil as well as computer-based training tasks Mem-ECT is designed to provide education about possible memory difficulties associated with ECT and then teach and practice memory strategies Training modules targeting sustained and selective attention as well as working memory were added since deficits in these cognitive domains are frequently observed/reported in severe depression	NART, modified MMSE, Goldberg Remote Memory Questionnaire, Continuous Performance Test-IP d prime, Buschke Selective Reminding Test, Autobiographical Memory Interview – Short Form	This pilot study demonstrated that cognitive training is well tolerated by severely depressed individuals recovering from right unilateral ECT; however, due to small sample size the efficacy of this cognitive training program has yet to be elucidated

(Continued)

Author	Sample	Age (mean, SD)	Method	Treatment	Cognitive Measures	Results
Meusel (2012)	Total Sample: N = 58 MDD: n = 28 BD: n = 12 Controls: n = 18	MDD & BD 49.9, 7.9 Controls 44.5, 12.3	10-week, non-randomized, controlled study examining the effectiveness of computer-assisted cognitive remediation in patients with mood disorders	Computer-Assisted Cognitive Remediation: The intervention consisted of 20 computer tasks from PSSCogRehab12 Training was hierarchically organized, requiring successful completion of one module prior to the start of a new module Task difficulty was modified according to the subjects' individual abilities Training took place during one hour sessions, 3 times a week for 10 weeks.	DSST, COWAT, TMT A, TMT B, Digit Span Forward and Backward, Ruff's 2 & 7 Selective Attention Test, HVLT	Subjects improved on measures of working memory and delayed recall following treatment. Positive correlations between improved cognitive functioning and psychosocial functioning were reported Depressive symptom severity and engagement in treatment were reported to be predictive of improvement
Bowie et al. (2013)	MDD: N = 33 Cognitive Remediation (n=17) Waitlist (n=16)	Cognitive Remediation 49.2, 11.8 Waitlist 42.2, 13.4	10-week, randomized, waitlist control study evaluating cognitive remediation in treatment resistant depression	Cognitive remediation was conducted for 90 min. during one session each week over a 10 week period	The Symbol Coding Task, the Continuous Performance Test-Identical Pairs Version, COWAT, TMT A, TMT B, HVLT, the Letter Number Sequencing Test, SCWT	A significant time by treatment interaction was reported for attention/processing speed and verbal memory Changes in function were not significant, although improved cognition predicted improvements in overall function

Table 9.5 (Continued)

ᵃStudies listed in reverse chronological order.

Table 9.6. Alternative therapeutic strategies to address cognitive symptoms

Therapeutic approach	Influence on emotional symptoms*	Influence on cognitive impairment*	Psychiatric disorders targeted
Currently available pharmacotheraphy	+ ────────→	−/0/+	Schizophrenia, depression, bipolar disorder, anxiety disorders
Deep-brain sitmulation or electroconvulsive therapy	+ ────────→	0/−	Major depression
Repetitive transcranial magnetic stimulation	0/+ ────────→	+/0	Mainly depression (autism, schizophrenia)
Cognitive behavioural theraphy	+ ────────→	0	Mainly depression (anxiety disorders)
Cognitive remediation theraphy	0/+ ────────→	+	Mainly schizophrenia (depression)
Improved drugs (alone and in combination with above strategies)	+ ⇄	+	Dependent on mechanism of action

*The '+' symbol corresponds to improvement; the '−' symbol corresponds to worsening'; and '0' corresponds to no marked change.

Reprinted by permission from Macmillan Publishers Ltd: Nat Rev Drug Discov (Millan ML, et. al. Nat Rev Drug Discov. 2012 Feb 1;11(2):141–68), Copyright (2012)

Moreover, improvements in cognitive measures (e.g. verbal working memory) have been reported in individuals with medication-resistant MDD receiving rTMS. (For review of alternative psychopharmacological, behavioural, and nutraceutical treatments on cognitive function, see Tables 9.5 and 9.6).

9.5 Cognitive Remediation

In keeping with the observation that some individuals with MDD remain resilient to cognitive decline, increasing efforts have been made towards the development of non-pharmacological approaches that may render the individual more resistant to insults that negatively influence brain and cognitive function. It has been hypothesized that increasing cognitive reserve may induce morphological changes that support healthy cognitive functioning. In keeping with this view, it could be hypothesized that similar behavioural/cognitive strategies may not only prevent but remedy cognitive function via positive effects on neuroplasiticity. Cognitive remediation is an approach initially developed for individuals with Traumatic Brain Injury and has been shown to facilitate development of compensatory skills and strategies . Rigorous clinical trials utilizing cognitive remediation in MDD are lacking. Nevertheless, several small scale studies have documented improvements is several domains of cognitive function (See Table 9.5). Recent evidence suggests that improvement in cognitive function following cognitive remediation is also positively associated with psychosocial functioning. Future research will need to elucidate which cognitive remediation strategies are most effective for improving cognitive function, and most importantly which interventions are most pertinent to functional recovery in MDD.

References

Alvarez E, Perez V, Dragheim M, Loft H, Artigas F. A double-blind, randomized, placebo-controlled, active reference study of Lu AA21004 in patients with major depressive disorder. Int J Neuropsychopharmacol 2012 Jun;15(5):589–600.

Baker R, Owens M, Thomas S, Whittlesea A, Abbey G, Gower P, Tosunlar L, Corrigan E, Thomas PW. Does CBT facilitate emotional processing? Behav Cogn Psychother. 2012 Jan;40(1):19–37.

Bang-Andersen B, Ruhland T, Jorgensen M, Smith G, Frederiksen K, Jensen KG, et al. Discovery of 1-[2-(2,4-dimethylphenylsulfanyl)phenyl]piperazine (Lu AA21004): a novel multimodal compound for the treatment of major depressive disorder. J Med Chem 2011 May;54(9):3206–21.

Bowie CR, Gupta M, Holshausen K, Jokic R, Best M, Milev R. Cognitive remediation for treatment-resistant depression: effects on cognition and functioning and the role of online homework. J Nerv Ment Dis. 2013 Aug;201(8):680–5.

Bowie CR, Gupta M, Holshausen K. Cognitive remediation therapy for mood disorders: rationale, early evidence, and future directions. Can J Psychiatry. 2013 Jun;58(6):319–25.

Choi J, Lisanby SH, Medalia A, Prudic J. A conceptual introduction to cognitive remediation for memory deficits associated with right unilateral electroconvulsive therapy. J ECT. 2011 Dec;27(4):286–91.

Elgamal S, McKinnon MC, Ramakrishnan K, Joffe RT, MacQueen G. Successful computer-assisted cognitive remediation therapy in patients with unipolar depression: a proof of principle study. Psychol Med. 2007 Sep;37(9):1229–38.

Ferguson JM, Wesnes KA, Schwartz GE. Reboxetine versus paroxetine versus placebo: effects on cognitive functioning in depressed patients. Int Clin Psychopharmacol 2003 Jan;18(1):9–14.

Furtado CP, Hoy KE, Maller JJ, Savage G, Daskalakis ZJ, Fitzgerald PB. An investigation of medial temporal lobe changes and cognition following antidepressant response: A prospective rTMS study. Brain Stimul May;6(3):346–54.

Greg J. Siegle, Frank Ghinassi, Michael E. Thase. Neurobehavioral Therapies in the 21st Century: Summary of an Emerging Field and an Extended Example of Cognitive Control Training for Depression. Cognitive Therapy and Research. April 2007, Volume 31, Issue 2, pp 235–262

Herrera-Guzman I, Gudayol-Ferre E, Herrera-Guzman D, Guardia-Olmos J, Hinojosa-Calvo E, Herrera-Abarca JE. Effects of selective serotonin reuptake and dual serotonergic-noradrenergic reuptake treatments on memory and mental processing speed in patients with major depressive disorder. J Psychiatr Res 2009 Jun;43(9):855–63.

Herrera-Guzman I, Gudayol-Ferre E, Lira-Mandujano J, Herrera-Abarca J, Herrera-Guzman D, Montoya-Perez K, et al. Cognitive predictors of treatment response to bupropion and cognitive effects of bupropion in patients with major depressive disorder. Psychiatry Res 2008 Jul;160(1):72–82.

Herrera-Guzman I, Herrera-Abarca JE, Gudayol-Ferre E, Herrera-Guzman D, Gomez-Carbajal L, Pena-Olvira M, et al. Effects of selective serotonin reuptake and dual serotonergic-noradrenergic reuptake treatments on attention and executive functions in patients with major depressive disorder. Psychiatry Res 2010 May;177(3):323–9.

Katona C, Hansen T, Olsen CK. A randomized, double-blind, placebo-controlled, duloxetine-referenced, fixed-dose study comparing the efficacy and safety of Lu AA21004 in elderly patients with major depressive disorder. Int Clin Psychopharmacol 2012 Jul;27(4):215–23.

Knöchel C, Oertel-Knöchel V, O'Dwyer L, Prvulovic D, Alves G, Kollmann B, Hampel H. Cognitive and behavioural effects of physical exercise in psychiatric patients. Prog Neurobiol. 2012 Jan;96(1):46–68.

McLennan SN, Mathias JL. The depression-executive dysfunction (DED) syndrome and response to antidepressants: a meta-analytic review. Int J Geriatr Psychiatry 2010 Oct;25(10):933–44.

Mork A, Pehrson A, Brennum LT, Nielsen SM, Zhong H, Lassen AB, et al. Pharmacological effects of Lu AA21004: a novel multimodal compound for the treatment of major depressive disorder. J Pharmacol Exp Ther 2012 Mar;340(3):666–75.

Meusel LAC. Cognitive Remediation in Patients with Mood Disorders: Behavioural and Neural Correlates. 4-1-2012. McMaster University. Open Access Dissertations and Theses.

Muller N, Schwarz MJ. [Immunological aspects of depressive disorders]. Nervenarzt 2007 Nov;78(11):1261–73.

Naismith SL, Redoblado-Hodge MA, Lewis SJ, Scott EM, Hickie IB. Cognitive training in affective disorders improves memory: a preliminary study using the NEAR approach. J Affect Disord. 2010 Mar;121(3):258–62.

Raskin J, Wiltse CG, Siegal A, Sheikh J, Xu J, Dinkel JJ, et al. Efficacy of duloxetine on cognition, depression, and pain in elderly patients with major depressive disorder: an 8-week, double-blind, placebo-controlled trial. Am J Psychiatry 2007 Jun;164(6):900–9.

Wagner S, Doering B, Helmreich I, Lieb K, Tadic A. A meta-analysis of executive dysfunctions in unipolar major depressive disorder without psychotic symptoms and their changes during antidepressant treatment. Acta Psychiatr Scand 2012 Apr;125(4):281–92.

Xu Z, Zhang Z, Shi Y, Pu M, Yuan Y, Zhang X, et al. Influence and interaction of genetic polymorphisms in catecholamine neurotransmitter systems and early life stress on antidepressant drug response. J Affect Disord 2011 Sep;133(1–2):165–73.

Index